Heinz-Gerd Joosten, Alfred Golloch, Jörg Flock, Susan Killewald
Atomic Emission Spectrometry

## Also of interest

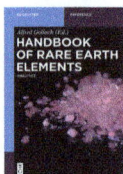

*Handbook of Rare Earth Elements*
*Analytics*
Golloch (Ed.), 2017
ISBN 978-3-11-036523-8, e-ISBN 978-3-11-036508-5

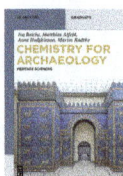

*Chemistry for Archaeology*
*Heritage Sciences*
Reiche, Alfeld, Radtke, Hodgkinson, 2019
ISBN 978-3-11-044214-4, e-ISBN 978-3-11-044216-8

*Chemical Analysis in Cultural Heritage*
Sabbatini, van der Werf (Eds.), 2019
ISBN 978-3-11-045641-7, e-ISBN 978-3-11-045753-7

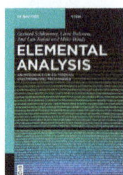

*Elemental Analysis*
*An Introduction to Modern Spectrometric Techniques*
Schlemmer, Balcaen, Todolí, 2019
ISBN 978-3-11-050107-0, e-ISBN 978-3-11-050108-7

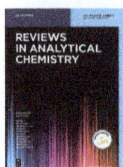

*Reviews in Analytical Chemistry*
Schlechter, Israel (Editor-in-Chief)
e-ISSN 2191-0189

Heinz-Gerd Joosten, Alfred Golloch,
Jörg Flock, Susan Killewald

# Atomic Emission Spectrometry

AES – Spark, Arc, Laser Excitation

**DE GRUYTER**

**Authors**
**Dr. Heinz-Gerd Joosten**
Nissingstr. 16
47559 Kranenburg

**Prof. Dr. Alfred Golloch**
Schönauer Bach 21
52072 Aachen
Germany

**Dr. Jörg Flock**
Hengstenbergstr. 10a
58239 Schwerte
Germany

**Susan Killewald**
Heerstr. 11
47533 Kleve
Germany

ISBN 978-3-11-052768-1
e-ISBN (PDF) 978-3-11-052969-2
e-ISBN (EPUB) 978-3-11-052775-9

**Library of Congress Control Number: 2020931604**

**Bibliographic information published by the Deutsche Nationalbibliothek**
The Deutsche Nationalbibliothek lists this publication in the Deutsche Nationalbibliografie;
detailed bibliographic data are available on the Internet at http://dnb.dnb.de.

© 2020 Walter de Gruyter GmbH, Berlin/Boston
Cover image: Comstock / Stockbyte / Getty Images
Typesetting: Integra Software Services Pvt. Ltd.
Printing and binding: CPI books GmbH, Leck

www.degruyter.com

# Preface

Almost 120 years ago, Heinrich Kayser postulated in his handbook of spectroscopy: "So, I come to the conclusion that quantitative spectroscopic analysis has proven to be unfeasible." Twenty years later, evidence to the contrary had already been attained, but atomic spectrometry with spark and arc excitation was first established in the laboratories of the metal producing and processing industries in the 1950s. With its high speed of analysis, this new analytical technique made possible a boost in innovation in metallurgical production that continues to the present. Today, thousands of spectrometers are in operation worldwide especially for the production and processing of metallic materials. While numerous publications and books were published during the initial stages of the development of this new analytical technique, it has been somewhat neglected in the specialist literature of the last decade despite its economic importance. This is regrettable since rapid technical developments in the areas of electronics, information technology and optics have steadily improved the performance of spectrometers in recent years.

And so, the authors' idea arose to provide users of atomic emission spectrometry using arc, spark or laser excitation with a current, comprehensive presentation of the technology. Practical application of this method was to be in focus in addition to the theoretical background.

On one hand, the concept for this book is to provide beginners and students with an introduction to the theory of the method. On the other hand, it is intended to support the practical user in understanding the design and function of the instrument in order to ensure trouble-free operation. The chapter on the history of atomic emission spectrometry is intended to give the reader an impression as to how much time was required to advance the method until it has become today's high-performance technology. This was significantly influenced by the theories with which the origin of the spectra could be explained. The theoretical fundamentals of spectral formation are only explained to the extent that the principles of the processes can be understood and the parameters whose variation have an influence on the formation of the spectra, and thus on the quality of the analytical results can be evaluated. The main focus of this book is formed by Chapters 3 and 7, which are devoted to the design and operating principle of spectrometers. The detailed description of stationary and mobile spectrometers and their operation reflects the current state of technology. Exact knowledge of the hardware assists in optimizing testing methods, thus improving the precision and accuracy of the measurement results. Therefore, for application of the method, the most important techniques for the sample taking and sample preparation of metallic samples are described. By explaining the analytical indicators, an attempt has been made to provide the user with criteria to assess applications possible for the methods and instruments.

https://doi.org/10.1515/9783110529692-201

This book would not have been conceivable without the many stimulating discussions and projects with numerous colleagues, the results of which have become a part of this work. Ms. Karin Dietze, Dr. Jörg Niederstrasser, Dr. David Poerschke, PD Dr. Rainer Joosten, Mr. Markus Neienhuis and Mr. Gerd Fischer stand out particularly here. Colleagues and various instrument manufacturers have also made available many images on a wide range of topics. Special thanks are due to all who have provided support.

Finally, the authors thank the team at the publisher De Gruyter, especially Lena Stoll and Dr. Ria Sengbusch, for the technical support and unproblematic realization of this project.

# Contents

# 1 Introduction

## 1.1 Definitions

Observation of the interaction of electromagnetic radiation and matter is summarized under the term *spectroscopy*. In this context, matter consists of atoms, ions, molecules or combinations of such particles, such as metals or liquids. Many of the interactions involve transitions between defined energy states, which are associated with the emission of radiation. Such interactions occur only in the gas phase in free atoms or ions, which led to the terminology atomic emission spectroscopy (AES). The emitted line spectra deliver information about the chemical elements whose atoms or ions are present in the gas phase. Spectroscopy is thus suitable as an analyst's tool. *Spectrometry* is a narrowing of the term *spectroscopy*; it is used when electromagnetic radiation is quantitatively measured.

In addition to the observable spectroscopic effects, there are interactions between radiation and matter, which are not associated with energy transitions. These interactions include refraction, diffraction, reflection and scattering, which change the direction, phase or polarization of radiation [1] (Figure 1.1).

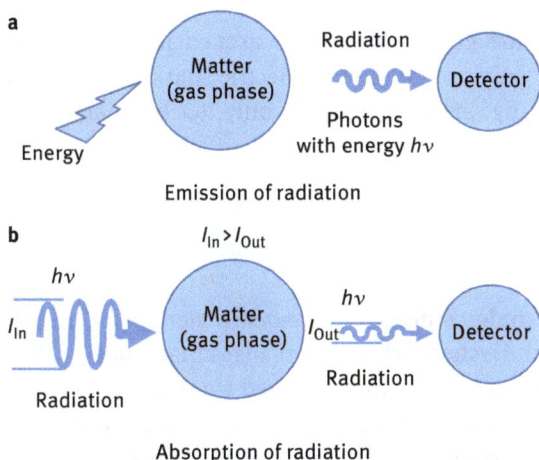

Figure 1.1: Interaction between radiation and matter.

The energy of emitted electromagnetic radiation can be distributed over a large range of frequencies (see Table 1.1). The choice of spectroscopic method determines the frequencies to be used.

The line spectra ranging from 115 nm (vacuum UV) over the visible range (390–770 nm) to approximately 1,000 nm (near infrared) are evaluated in AES, which is the subject of this book.

https://doi.org/10.1515/9783110529692-001

Table 1.1: The electromagnetic spectrum.

| Wavelength range | λ | Transition |
|---|---|---|
| Gamma radiation | <0.005 nm | Nucleus |
| X-rays | 0.005–100 nm | Inner electrons |
| Vacuum UV | 100–180 nm | Valence electrons |
| Near UV | 180–390 nm | Valence electrons |
| Visible region | 390–770 nm | Valence electrons |
| Near infrared | 770–2,500 nm | Valence electrons and molecular vibration |
| Medium infrared | 2.5–50 μm | Molecular vibration |
| Far infrared | 50–1.000 μm | Molecular rotation |
| Microwave range | 1–1,000 mm | Molecular rotation, electron and nuclear spin |
| Radio wave range | >1,000 mm | Nuclear spin |

The atoms and ions present in the gas phase can be excited in a variety of ways. Within the scope of this book, only excitation with electric spark and arc is to be discussed in detail. Excitation by laser sources is only to be marginally addressed.

### 1.1.1 Spectrometry for material analysis for research and production

The importance of AES is frequently underestimated as a modern method for instrumental analysis. This is difficult to understand when one realizes that 3,000–4,000 mobile and stationary spectrometers are bought and delivered every year. It is estimated that about 50,000 instruments are currently (as of 2017) in use around the world. These figures demonstrate the economic importance of arc/spark AES as an analytical tool.

The manufacture and processing of modern materials depends on competent analytics. The production process for metallic raw materials cannot be controlled without quick monitoring of the melt's material composition. The precise production of high-quality alloys with defined proportions of alloying elements requires rapid and accurate analysis.

The processing of materials into products leads to good results when continuous quality control is ensured with the help of AES. Efficient machines and components for the aerospace and automotive industries can only be manufactured under these conditions.

Other important application areas for arc/spark AES include the following:
- Metal identification can be performed with a 100% testing of the composition of the material components.
- Recycled material can be examined to determine which materials are present. The sorted materials can be reused separately.
- Basic research on new metallic materials involves a great deal of analytical effort. Experiments for alloy composition can be supported by fast and highly sensitive arc/spark spectrometry.

AES with arc and spark offers many advantages, above all, for material analysis and especially for quality and process control, making it the method of choice [2]:
- A direct analysis of solid samples is possible without needing to dissolve them.
- Practically all relevant elements can be simultaneously determined.
- The measurement proceeds automatically and delivers the result in just a few seconds.
- Detection sensitivity and reproducibility are good.
- A wide concentration range, from traces to 100%, can be covered.

## 1.2 About the history of atomic emission spectrometry

### 1.2.1 First observations of atomic emission spectra

The first steps of AES, also frequently referred to as "spectral analysis," began a long time ago [3].

Newton's observation of the solar spectrum in 1672 was groundbreaking. Representation of this spectrum was performed by blocking a portion of the sunlight using a round hole and then splitting the light with a prism. The spectrum became visible on a white surface behind the prism. However, the spectrum obtained in this way offered only insufficient resolution. Important details became visible only through the use of a narrow entrance slit by Wollaston in 1802. Thin black lines could be seen within the continuous color gradient. The sunlight delivered light throughout the visible range of light, whereby the radiation is missing for various narrowly limited spectral regions. At first, there was no explanation for this phenomenon. Even Fraunhofer [4], who later examined these mysterious black lines, could only measure their positions.

It was only in 1859 after the discoveries from Bunsen and Kirchhoff that it was recognized that the lines could be assigned to chemical elements. Previously, beginning in 1820, the formation of emission spectra in flames was investigated by several researchers, who placed different chemical compounds into the flames [5–8]. At that time, it was already established that this method could be used to detect the smallest of sample quantities.

At that point, electrical sparks were already being used, in addition to flames, to produce spectra [9–12]. The spectra from metals as well as compounds were analyzed.

### 1.2.2 Fundamental work by Bunsen and Kirchhoff dealing with atomic emission spectrometry

In 1859, after a large amount of preliminary work as to the formation of spectra, Bunsen and Kirchhoff explained the basics of spectral analysis in a publication [13]. They discovered that the bright lines in the spectrum of a glowing gas are caused exclusively by its chemical components. They also determined that the chemical binding has no influence on the spectrum. Bunsen and Kirchhoff described the spectra of the alkaline and alkaline earth elements and demonstrated the detection sensitivity of spectral analysis. Using spectral analysis, they discovered the new elements cesium and rubidium. The ultraviolet and infrared ranges of the spectra were found through intense occupation with spectral analysis, and methods for photographing spectra in these ranges [14–17] were developed. Bunsen and Kirchhoff's results explained the formation of the Fraunhofer lines and could be used to determine the composition of the sun and its atmosphere. The number of investigations dealing with absorption spectral analysis also increased greatly. At the turn of the century, chemical spectral analysis with flames, electric arcs and sparks had become established methods for chemical analysis. There had been preparatory work dealing with the quantitative determination of elements by Hartley, among others. A breakthrough to an accepted method could not, however, be achieved. For about 30 years, a procedure whose potential was unquestionable was available, but it was not possible to take advantage of it because the theoretical foundation was missing. This unsatisfactory state lasted until the release of a revolutionary publication from Gerlach and Schweitzer [18].

### 1.2.3 Development of atomic emission spectrometry for quantitative analysis

As previously indicated, this time period was influenced by Gerlach and Schweitzer's experiments, which were compiled in a book about the fundamentals of chemical spectral analysis [18]. The terms "analytical sensitivity" and "detectability" of a spectral line were covered in detail in Chapter 4 of the book. Building on the findings of this dialogue, Chapter 5 discusses the possibility of absolute intensity analysis and several important methods for quantitative analysis, such as the method of the comparison of spectra and methods using homologous line pairs.

These studies were supplemented by the experiments of other researchers, who investigated perfecting quantitative analysis using photometric intensity measurements. Recording spectra, according to Scheibe [19], proved to be a significant improvement in photometric detection.

Developments in the then fledgling spark technology also supported spectrometry in the 1930s. Feussner's spark generator [33–35], based on the spark gap transmitter used to transmit news, was applied. The users of spectroscopy could resort to proven, reliable technology and no longer had to worry about electrotechnical details.

After Gerlach and Schweitzer's results and publications led to useful methods for the quantitative evaluation of emission spectra, it is necessary to follow the development of AES from different points of view. From that time on, the potential of AES was utilized much better.

Because the method had become more widespread, it was improved in many ways:
- Measurement of the spectral lines and procedures for quantitative evaluation
- Division of AES into several technical categories based on the different excitation sources
- Technical developments on the instruments within these categories

Recording of the spectra, development of the photographic plates and density measurements using densitometer was a time-consuming process that made quick analyses impossible. Photoelectric radiation recording led to a decisive improvement in the measurement of line intensity. The vacuum photocell had already been invented and researched in 1890 by Julius Elster and Hans Geitel [20–23]. In 1905, Albert Einstein presented a theory that explained the photoelectric effect, for which he was awarded the Nobel Prize in 1921. In 1929, Koller and Campbell developed the S-1 photocathode material that is still being used today; significantly increasing the radiation sensitivity of vacuum photocells [25]. According to Ohls [24], Lundegardh had already used photocells to record spectra in 1930.

Investigations by Thanheiser and Heyes, who were also occupied with inclusion analysis, reported on attempts to directly measure line intensities in 1939 [26].

In the beginning, the technology of this evaluation method was still in need of improvement. Decisive progress was achieved using photomultiplier tubes (PMTs) instead of simple photocells. In the 1930s, intense research on the principle of secondary electron multiplication was already being conducted [27]. Tube technology was a major focus for technical development in this decade, in which radio was made accessible to an extensive range of the population in general. Advanced tube technology led to serial production of PMTs suitable for routine work [28, 29].

Utilization of photomultiplier tubes led to the development of the first "direct reading instruments." Commercially successful multiple element spectrometers (quantometer), which enabled automatic analysis, arose from these humble beginnings. Starting in the 1970s, attempts were made to supplement or replace the large and expensive phototubes, which allowed the measurement of only a few spectral lines per optical system, with semiconductor chips. These sensors also became available for spectrometry only after development for the mass market. The low-noise transmission free-of-loss of the measured signal to the output amplifiers became possible with so-called CCD (charge coupled devices) chips. The technique was perfected to make image capture systems, scanners and barcode readers based on semiconductors possible. In addition to a reduction in production costs, the use of CCDs in spectrometry enabled capturing of a continuous spectral range. The main advantage of the PMT technology, namely rapid signal detection, could be combined with that of photographic capture, the recording of complete spectra.

Beginning in the mid-1960s, development of semiconductor technology made integrated semiconductor-based operational amplifiers available. Introduction of the µA 709 operational amplifier by the Fairchild Semiconductor company in 1965 was a major milestone in this field. It was now possible to automate the mathematical steps from line intensities to concentrations using analog computing circuits, further accelerating the evaluation and eliminating the risk of calculation errors by the user. Easier instrument operation was an additional advantage. For the first time, operation of the systems no longer required highly specialized experts.

However, the possibilities of analog computing are limited. Every function requires a hardware circuit that can only be changed by rewiring. For this reason, minicomputers that provided greater flexibility were soon used. While such systems were powerful and could be quickly modified with changes to the software, they were also extremely expensive with prices in the five-digit US dollar range.

A further development on the mass markets provided a remedy. In 1971, Intel introduced the 4004 chip, which is considered to be the first fully integrated microprocessor. It became possible to equip spectrometer systems with microprocessor-based computers, and that at a cost that was less than a tenth of the price of a minicomputer.

Spectrometers equipped with microcomputers were small and efficient, could be adjusted to customer needs with software changes and had acceptable production costs. All these factors, together with the increased quality requirements for metallic materials, led to expansion of the market from the quantities discussed above to several thousand instruments per year.

Through optimization of the existing excitation units and development of new systems, various methods of AES emerged.

Excitation sources are as follows:
- Flames
- Electrical arc
- Electrical spark
- Glow discharge
- Inductively coupled and microwave plasmas
- Laser

As this book deals mainly with AES with spark and arc excitation, the development of these methods is more fully described in the following.

### 1.2.4  Development of atomic emission spectrometry with excitation using electrical arc

The first use of arc excitation goes back to the nineteenth century. In the period after 1970, the arc discharge was advanced in various forms as an excitation source, becoming more commercialized [2]. Electric arcs were used especially for trace analysis because highly sensitive analytical procedures could be realized with them. However, if they burn freely, they are not a stable excitation source because the conditions on the electrodes and in the plasma can vary greatly, as explained in more detail in Section 3.2.1.1.

In addition to free-burning direct current arcs, stabilized direct current and alternating current arcs were developed:
- Free-burning direct current arcs are preferred for the analysis of electrically non-conducting samples. During routine operation, the arc burns between two carbon electrodes. One electrode is used as a carrier electrode, i.e., the non-conducting analyte is filled into the electrode and vaporized during operation of the arc, thus landing in the plasma, and is excited to the transmission of a spectrum. The free-burning direct current arc is also used to sort metals and for the screening analyses of some low-alloy materials.
- Stabilized direct current arcs are more suited to the analysis of liquids. The liquid can be introduced as an aerosol or as drops on a carrier.
- Alternating current arcs have the advantage that the electrodes become less heated due to the interruptions in the arc.

Although the various types of electric arc provided solutions to many analytical problems, today no laboratory spectrometers with arc excitation are being manufactured in large numbers. One reason for this is the competition from systems with inductively coupled plasma (ICP) as the excitation source, which came onto the market in the 1980s.

However, the free-burning direct current arc is still used in mobile spectrometers for sorting, analysis and material identification. Such systems are discussed in Chapter 7.

A comprehensive compilation dealing with arc excitation can be found by Ohls [24].

### 1.2.5 Atomic emission spectrometry with spark excitation

The potential use of the electric spark as an excitation source for spectral analysis was recognized very early after application of the flame. It has already been mentioned that the electric spark was used as a source of excitation at a very early stage. According to Görlich, spark spectra were observed by Emil Du Bois-Reymond in 1859 [30]. In 1901, Charles C. Schenk published a paper that described the electric spark and also depicted spectra [31]. In 1969, Walters presented and commented on the most important facts known at that time for the understanding of the physics and chemistry of spark discharges under normal pressure in a comprehensive article [32].

He covered the following phenomena:
- Spark channeling
- Sample vaporization
- Processes on the electrodes
- Formation of excited states

The impact that technical developments for the mass markets had on the construction of spectrometers has already been explained above. In recent decades, further improvements have been incorporated into the systems:
- Optimization of the spark generators
- Flushing of the spark stand with argon
- Evacuation of the optic or flushing with an inert gas
- Development of complex algorithms for measurement value recording and processing

A wide range of instruments is available today as the result of technical developments to the spark spectrometer. The flagship precision instruments used for process control especially in modern steelworks are the most powerful spark instruments. They must not only deliver precise results but also be robust and reliable. Compact instruments used in foundries and metal processing plants make up the middle segment of the market. So-called handheld instruments are small and mobile. This type of arc/spark spectrometer is frequently used for material identification and sorting.

# Bibliography

[1]    Ingle JR, James D, Crouch SR. Spectrochemical Analysis, Englewood Cliffs, New Jersey 07632, Prentice Hall Inc, 1988.

[2]    Laqua K. Emissionsspektroskopie in Ullmanns Encyklopädie der technischen Chemie, 4. edition, Vol. 5, edited by H. Kelker Analysen-und Meßverfahren, Deerfield Beach, Florida, Basel, Verlag Chemie Weinheim, 1980, S, 441–450.

[3]    Formanek J. Die Qualitative Spektralanalyse Anorganischer und Organischer Körper, 2nd enhanced edition, Berlin, Verlag von Rudolf Mückenberger, 1905.

[4]    Fraunhofer J, Denk d.k. Akad. der Wissensch. zu München 1814, München, 1815.

[5]    Herschel JFW. J. Trans. Soc. Edinb. 1823, 9, Pogg. Ann. 1829, 16.

[6]    Talbot WHF. Brewster's J. Sci. 1825, 5, 77.

[7]    Miller WA. Brit. Assoc. Rep, 1845.

[8]    Swan Transact. Roy. Soc. Edinburgh 1857, 21, 411.

[9]    Wheatstone CH. Phil. Mag. 1835, 7, 299.

[10]    Foucault L. L'Institut. 1849, 44.

[11]    Masson A. Ann. Chim. Phys. 1851, 31, 295.

[12]    Angstrom AJ. Pogg. Ann. 1855, 94, 141.

[13]    Kirchhoff G, Bunsen R. Pogg. Ann. 1860, 110, 161–189.

[14]    Roscoe HE, Clifton R B. Proc. Lit. and Phil. Soc, 1862, Manchester.

[15]    Liweing, Dewar. Phil. Trans. of the Roy. Soc, 174(187), Proc. Roy. Soc. 34, 119.

[16]    Kayser H, Runge C. Berl. Akad. Wissensch, 1892.

[17]    Rydberg JR. Compt. Rend. 1890, 110, 394.

[18]    Gerlach W, Schweitzer E. Die chemische Emissionsspektralanalyse, Bd. 1, Leipzig, Verlag Leopold Voss, 1930.

[19]    Scheibe G, Neuhäuser A. Z., Angew. Chem.. 1928, 41, 1218.

[20]    Elster J, Geitel H. Über die Verwendung des Natriumamalgams zu lichtelectrischen Versuchen. Annalen der Physik und Chemie. 1890, NF 41, 161–165.

[21]    Elster J, Geitel H. Notiz über eine neue Form der Apparate zur Demonstration der lichtelectrischen Entladung durch Tageslicht. Annalen der Physik und Chemie. 1891, NF 42, 564–567.

[22]    Elster J, Geitel H. Lichtelectrische Versuche. Annalen der Physik und Chemie. 1892, NF 46, 281–291.

[23]    Elster J, Geitel H. Lichtelectrische Versuche. Annalen der Physik und Chemie. 1894, NF 52, 433–454.

[24]    Ohls K. Analytische Chemie-Entwicklung und Zukunft, Weinheim, Wiley-VCH Verlag GmbH & Co. KGaA, 2010.

[25]    Koller LR. Photoelectric emission from thin films of caesium. Phys. Rev. 1930, 36, 1639.

[26]    Thanheiser G, Heyes J. Mitt. Kaiser-Wilhelm-Inst. Eisenforsch. 1939, 21, 327.

[27]    Weiss G, Peter O. Anlaufstromgesteuerter Vervielfacher als übersteiles Verstärkerrohr. Zeitschrift für technische Physik 1938, 11, 444–451.

[28]    Hasler MF, Dietert HW. J. opt. Soc. America 1944, 34, 751.

[29]    Sauderson JL, Caldecourt VL, Peterson EW. J. opt. Soc. America 1945, 35, 681.

[30]    Kelker H (editor) Laqua K, Ullmanns Enzyklopädie der technischen Chemie 4, updated and enhanced edition, vol.5, Analysenverfahren und Meßverfahren, 441–500. Verlag Chemie, Weinheim-Deerfield Beach, Florida, Basel 1980.

[31]    Görlich P. Einhundert Jahre Wissenschaftliche Spektralanalyse. Berlin, Akademie Verlag, 1960.

[32]    Schenk CC. Astrophys. J. 1901, 19, 116.

[33]    Walters JP. Appl. Spectrosc. 1969, 23(4), 317–331.

# 2 Atomic emission spectrometry: fundamentals

The following facts are fundamental to atomic emission spectrometry (AES):
- Material can be vaporized, atomized and ionized when sufficient energy is applied using, for example, an electrical arc or spark.
- The energy input causes the atoms and ions to emit radiation.
- The radiation is not regularly distributed over the entire spectral range but occurs only in a finite number of narrow wavelength ranges. If a spectral apparatus breaks the spectrum down into a band of radiation in a way that the shortest wavelengths appear on the left edge and the longest wavelengths on the right edge, then the previously mentioned wavelength intervals appear within this spectrum as vertical lines with different positions and intensities.
- Atoms and ions from every element generate spectra that are characteristic in respect to position for the given element.
- The elemental content can be determined from the radiation intensity.

This chapter roughly outlines how these line spectra, which are so fundamental for the method of AES, occur. The theoretical foundations outlined in Chapter 2 serve the general understanding of the key relationships. Spectral lines for arc/spark spectrometry are empirically examined for their suitability. The purely practice-oriented reader can omit this chapter.

## 2.1 Researching the hydrogen spectrum in the nineteenth century

The emission spectra for hydrogen were measured in the nineteenth century in the course of the development of AES. The discovery of series of lines in the spectrum led to equation-like relationships between the wavelengths of these signals.

In 1885, Balmer described a series that followed the equation:

$$\lambda = A \frac{n^2}{n^2 - 4} \tag{2.1}$$

Here, $A$ is at first an empirically determined constant length of 364.56 nm, $n$ is an integer $\geq 3$ and $\lambda$ is the wavelength in nanometers.

In molecular spectroscopy, the notation of wave numbers as oscillations per centimeter is more common. However, wave numbers $\bar{\nu}$ are expressed exclusively in $m^{-1}$ in this chapter to avoid confusion. The wave number $\bar{\nu}$ is then the reciprocal value of the wavelength $\lambda$ noted in m. The wavelength (mostly noted in nm) is commonly used to refer to spectral lines in spark spectrometry.

https://doi.org/10.1515/9783110529692-002

Therefore, $\bar{v} = \frac{1}{\lambda}$ $[\text{m}^{-1}]$.

The Swedish physicist Johannes Rydberg revised Balmer's equation by using the constant $R$ in place of the constant $A$, whereby $R = \frac{4}{A}$. The value of the constant $R$ is 10973731.5685 $\text{m}^{-1}$. Because of its discoverer, R is called the Rydberg constant.

The following wave numbers (in $\text{m}^{-1}$) are obtained for the Balmer series:

$$\bar{v} = R\left(\frac{1}{2^2} - \frac{1}{n^2}\right) \qquad (2.2)$$

This revision itself was no advance, but Rydberg's generalization as shown in eq. (2.3) was:

$$\bar{v} = R\left(\frac{1}{m^2} - \frac{1}{n^2}\right) \qquad (2.3)$$

Here, $m$ represents a natural number that must always be smaller than $n$. In eq. (2.3), $n$ may have the value of 2. When Rydberg introduced eq. (2.3) in 1888, only lines in the visible spectral range could be observed. Equation (2.3) makes it possible to predict where in the spectrum additional hydrogen lines should appear. The hydrogen lines discovered in the ultraviolet spectral range by Theodore Lyman in 1906 and the near infrared lines recorded by Friedrich Paschen in 1908 were found at the positions where hydrogen lines should be found according to eq. (2.3). The same was true for the series of infrared lines later discovered by Brackett and Pfund.

## 2.2 Hydrogen spectrum and Bohr's atomic model

Equation (2.3) correctly predicts where hydrogen lines should be within the spectrum. However, it does not in the least answer the question as to why spectral lines are found at exactly these positions. An answer to this question cannot be found within the framework of classical physics.

Bohr's atomic model was the first to provide an explanation. Bohr based his model on older atomic models, such as that of Rutherford. In this model, electrons orbit the nucleus like planets the sun. The weak point with such models is the fact that according to classical electrodynamics, energy is continually released when a charge is in circular motion. According to the classical interpretation, electrons should spiral toward the nucleus and finally land on it. This is not the case.

Bohr recognized that other laws must be in force at the atomic level and proposed three fundamental assumptions (postulates):
- Electrons circle the nucleus in fixed orbits. Only orbits for which the magnitude of the angular momentum of the electrons is a multiple of $\frac{h}{2\pi}$ are allowed, whereby $h$ *is* Planck's constant ($6.626 \times 10^{-34}$ Js) discovered by Max Planck in 1899. Thus, allowed angular momentums have the form $n\frac{h}{2\pi}$ with integers $n \geq 1$.

It is important to remember that, other than in classical mechanics, not every orbit diameter is allowed.
- If an electron is located in such an orbit, no energy is emitted.
- Energy is only emitted when an electron in an allowed orbit with a higher energy E jumps to another allowed orbit with a lower energy $E'$. In this case, electromagnetic radiation with a frequency $E - E' = h\nu$ is released. The reverse applies for the absorption of radiation: To jump from a low energy level $E'$ to a higher level $E$, exactly the energy $E - E' = h\nu$ must be absorbed.

The emitted "radiation packets" are called photons. Their frequency $\nu$ can converted to wavelengths with the unit meter using the relationship $\lambda = \frac{c}{\nu}$, whereby $c$ is the speed of light (299,792,458 m s$^{-1}$). The described Bohr model provides exactly the same wavelengths that are calculated using Equation 2.3. A preliminary explanation for the hydrogen atom's line spectrum was established.

## 2.3 Schrödinger equation and quantum numbers

Bohr's atomic model is able to predict the structure of the spectra from hydrogen and other elements for which a single electron is located in the outer shell. This is the case for elements in the first group of the periodic table of elements. The model fails for elements that have more than one electron in the outer shell. In addition, the fine structure of spectra that, for example, exhibit closely spaced doublets and triplets instead of single spectral lines cannot be explained using the Bohr model.

However, quantum mechanics provides an explanation for this. The Schrödinger equation, derived by Erwin Schrödinger in 1926, which describes the quantum mechanical state of a system, forms its core. Shortly thereafter, Paul Dirac's modifications made it possible to take the electron spin into consideration.

The time-independent version of the Schrödinger equation is as follows:

$$H\psi(r) = E\psi(r) \qquad (2.4)$$

This equation looks very simple, but this is deceptive; everything that constitutes quantum mechanics is in it.

The terms have the following meanings:
- $r$ is a point in space.
- The wave function $\psi$ of a particle, in this case an electron, describes the quantum mechanical state of the particle when it is a solution for eq. (2.4). $\psi$ enables calculation of the probability that the electron can be found in a given spatial volume.
- $H$ is the so-called Hamilton operator. Here, an operator is understood to be a function that has functions as arguments and whose function values are

again functions. The Hamilton operator has wavefunctions as arguments and results in the same wave function multiplied by a scalar quantity E, the energy of the state described by the wave function. The left-hand side of eq. (2.4), $H\psi(r)$, designates the total energy of the quantum mechanical system, in this case, the electrons.
- Not all values possible on the left-hand side are allowed. This is typical for the quantum world. The restriction to discrete orbits around the atomic nucleus was already described during the discussion of Bohr's atomic model. The right-hand side of the Schrödinger equation, $E\psi(r)$, determines which energies are allowed. Mathematically speaking, $E$ is an eigenenergy of the operator $H$ with respect to the function $E\psi(r)$. It is important to note that, unlike in the macroscopic world, only certain discrete states are possible.

In the 1920s, a set of four quantum numbers was introduced to completely describe the allowed energetic states of an electron and, thus, the solutions to the Schrödinger equation:
- The main quantum number $n$ is a natural number >0. The larger $n$ becomes, the further the electron probably is away from the nucleus and therefore, the lower the binding energy to the nucleus.
- The orbital angular momentum quantum number $l$ indicates the angular momentum state of the electron. States between 0 and $n - 1$ are allowed.
- The magnetic quantum number $m$ is an integer between $-l$ and $+l$.
- The spin quantum number $s$ always has an amount of 1/2, its projection $s_z$ on a reference axis z can have the values of +1/2 or −1/2. This will be shown in Section 2.5. It simplifies the explanations when $s_z$ is also considered to be a quantum number.

In 1925, Pauli discovered that no two electrons in an atom ever have the same set of quantum numbers. The value ranges for the quantum numbers and the Pauli exclusion principle limit the possible energy values for the electron, thus, fulfilling the same function as the left side of eq. (2.4).

## 2.4 Quantum mechanically explained rough spectrum for hydrogen

There are two deciding factors for the energetic state of an electron:
1. The forces of attraction between electrons and the atomic nucleus
2. The interactions between the electrons

Because hydrogen has only one electron, the second factor does not apply. As a result, the quantum number $m$ has no and the quantum number $l$ only a very small influence on the energetic state of the electron. However, its energy varies greatly for differing main quantum numbers $n$.

Equation (2.5) describes the allowed energy values (in units of J) for different $n$ according to the Schrödinger equation.

$$E_n = - \frac{m^* e^4}{8^* h^{2*} \varepsilon_0^2 {}^* n^2} \tag{2.5}$$

Here, $\varepsilon_0$ designates the strength of the electrical field, $m$ the mass of the electrons, $e$ the elemental charge, $c$ the speed of light and $h$ Planck's constant. Equation (2.6) converts these energies into wave numbers:

$$\bar{v}_n = - \frac{m^* e^4}{8^* h^{3*} c^* \varepsilon_0^2 {}^* n^2} \tag{2.6}$$

This assumes the following relationships:

- The photon energy $E_{Photon}$ is the product of Planck's quantum of action and the oscillation frequency of the photon; that is, $E_{Photon} = hv$.
- The wave number is the quotient of oscillation frequency of the photon and the speed of light; that is, $\bar{v} = \frac{v}{c}$.

The physical constants from eq. (2.6) can be combined into the Rydberg constant discussed in Section 2.1. This results in:

$$\bar{v}_n = - \frac{R}{n^2} \tag{2.7}$$

If eq. (2.5) is applied to two energy states $n'$ and $n$, in which an electron is located before and after excitation ($n' < n$, meaning that energy is absorbed) and the difference in energies $\Delta E$ between the two states is determined, then the following is true:

$$\Delta E = \left( - \frac{R^* c^* h}{n^2} \right) - \left( - \frac{R^* c^* h}{n'^2} \right)$$

$$\Delta E = \left( \frac{R^* c^* h}{n'^2} \right) - \left( \frac{R^* c^* h}{n^2} \right)$$

$$\Delta E = R^* c^* h^* \left( \frac{1}{n'^2} - \frac{1}{n^2} \right)$$

When this is converted to a wave number (by dividing by $h$ and inserting $c = \lambda^* v$), the following is obtained:

$$\bar{v} = R\left(\frac{1}{n'^2} - \frac{1}{n^2}\right) \tag{2.8}$$

This is exactly the Rydberg formula (eq. (2.3)), which was introduced in Section 2.1.

It was substantiated by a theoretical foundation with the principles of quantum mechanics.

A photon with the wavenumber $\bar{v}$ is absorbed to raise the electron from the state $n'$ to the state $n$. Conversely, when changing from $n$ to $n'$, a photon with the wavenumber $\bar{v}$ is emitted.

The first five wavenumbers of the Lyman series are calculated by inserting the number 1 for n' and the numbers from 2 to 6 for n.

$$\bar{v}_{\text{Lyman}} = \frac{3}{4}R, \frac{8}{9}R, \frac{15}{16}R, R\frac{24}{25}, R\frac{35}{36}, \ldots \tag{2.9}$$

If the value of 2 is used for $n'$ in the formula, one obtains the Balmer series. Other values for $n'$ lead to other series of spectral lines for hydrogen. Table 2.1 gives an overview.

Table 2.1: Spectral line series of hydrogen.

| Name of series | $n'$ | n | Spectral range |
|---|---|---|---|
| Lyman series | 1 | 2, 3, 4, ... | 121 nm–91 nm |
| Balmer series | 2 | 3, 4, 5, ... | 656 nm–365 nm |
| Paschen series | 3 | 4, 5, 6, ... | 1,875 nm–820 nm |
| Brackett-series | 4 | 5, 6, 7, ... | 4,050 nm–1,460 nm |
| Pfund series | 5 | 6, 7, 8, ... | 7,457 nm–2,280 nm |

A part of the Balmer series of lines is shown in Figure 2.1 as it would be seen on the focal curve of a spectrometer optic. Figure 2.2 sketches a part of the transitions of the Lyman, Balmer, Brackett and Pfund series.

The other series named in Table 2.1 are the result of transitions between states of different n and an n' that is specific for the series (see Figure 2.2).

The energy associated with the Rydberg constant $E = hcR$ (with Planck's quantum of action $h$ and speed of light $c$) is exactly the amount of energy that is required to detach the electron from the hydrogen nucleus. The hydrogen ion $H^+$ is positively charged after removal of the electron. This is why the energy $E = hcR$ is known as the ionization energy of hydrogen. It is 13.6057 eV. The unit eV (electron volt) is practical to indicate the often small energies found on an atomic level. 1 eV is the energy that an electron absorbs when passing through a field of one volt. An eV is equivalent to $1.60218 \times 10^{-19}$ Joule.

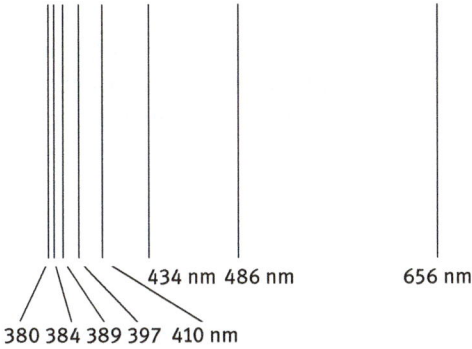

Figure 2.1: Balmer series on the focal curve of a spectrometer.

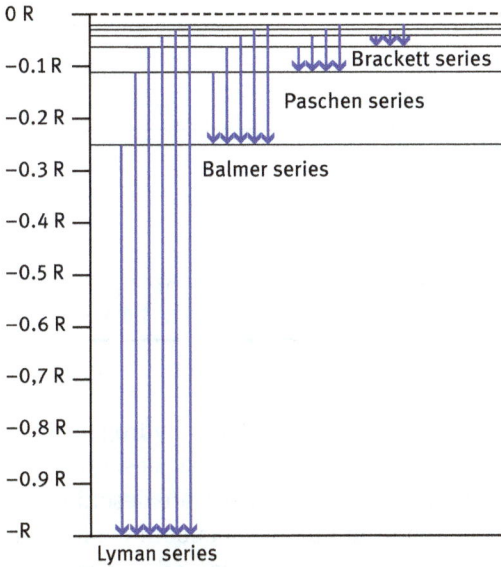

Figure 2.2: Energy transitions in Lyman, Balmer, Paschen and Brackett series.

Table 2.2 shows the quantum states for the hydrogen atom up to $n = 4$. These quantum states are obtained when the four quantum numbers are combined according to the rules discussed at the end of Section 2.3. The quantum numbers $n$, $l$ and $m$ and the corresponding values are shown in blue. The possible combinations for the fourth quantum number, the spin, are indicated by arrows pointing up or down. For the orbital angular-momentum quantum numbers, the letters s, p, d, f, g, h, ... (continued alphabetically) are frequently used instead of the numbers 0, 1, 2,

3, 4, 5, ... . These designations are shown in black in Table 2.2. For the hydrogen atom, the states in each table row have approximately the same energy level. Minimal deviations are due to the coupling of the orbital angular momentum with the spin to form a total angular momentum, which leads to a splitting of the energy level for $l > 0$. The causes for these deviations and the resulting fine structure of the hydrogen spectrum will be handled in Section 2.5.

Table 2.2: Possible energetic states of the hydrogen electron.

| Shell | s $l = 0$ | p $l = 1$ | d $l = 2$ | f $l = 3$ |
|---|---|---|---|---|
| K $n=1$ | ↑↓ 1s $m = 0$ | | | |
| L $n=2$ | ↑↓ 2s $m = 0$ | ↑↓ ↑↓ ↑↓ 2p $m = -1,0,1$ | | |
| M $n=3$ | ↑↓ 3s $m = 0$ | ↑↓ ↑↓ ↑↓ 3p $m = -1,0,1$ | ↑↓ ↑↓ ↑↓ ↑↓ ↑↓ 3d $m = -2,-1,0,1,2$ | |
| N $n=4$ | ↑↓ 4s $m = 0$ | ↑↓ ↑↓ ↑↓ 4p $m = -1,0,1$ | ↑↓ ↑↓ ↑↓ ↑↓ ↑↓ 4d $m = -2,-1,0,1,2$ | ↑↓ ↑↓ ↑↓ ↑↓ ↑↓ ↑↓ ↑↓ 4f $m = -3,-2,-1,0,1,2,3$ |

Using the Schrödinger equation, it is possible to show that the hydrogen atom cannot always change from an energetic state that can be described by a combination of the quantum numbers $(n, l, m, s_z)$ to another state described by $(n', l', m', s_z')$. This even applies when the quantum number sets both conform to the rules described in Section 2.3. Rather, the following selection criteria must be observed:

- The main quantum number $n$ is allowed to change arbitrarily. Of course, according to the rules in 2.3, it must be a positive integer.
- The orbital angular momentum quantum number $l$ is only allowed to increase or decrease by one and must remain within the value range between 0 and $n - 1$ as defined in Section 2.3.

The allowed transitions for the hydrogen atom are displayed as lines in Figure 2.3. For a transition from high to low n, the transition is accompanied by the emission of a photon. When the transition is upward from low to high n, the same amount of

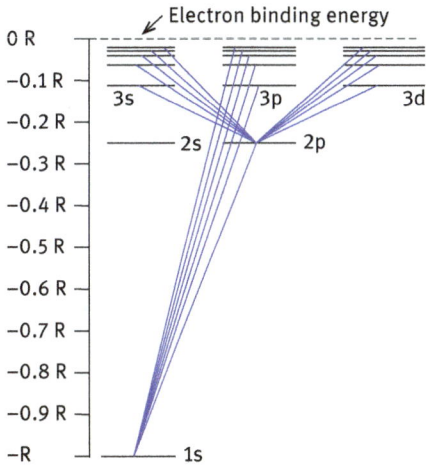

Figure 2.3: Selection of allowed energy transitions in the hydrogen atom.

energy is absorbed. The $Y$-axes are labeled in wave numbers in Figures 2.2, 2.3 and 2.5. The energy $E$ that corresponds to a wave number $\bar{v}$ read from the $Y$-axis is determined by multiplication with Planck's quantum of action and the speed of light: $E = h^*c^*\bar{v}$. $E$ designates the energy with regard to the ionization limit. Thus, the $Y$-axis could also be labeled with energy units, since $h$ and $c$ are constants. The ionization limit or the ground state can be chosen as the zero point.

## 2.5 The fine structure of the hydrogen atom spectrum

The coarse structure of the hydrogen atom spectrum was treated in Section 2.4. However, it was already mentioned that there are small shifts in the energy level for the orbital angular momentum quantum number $l > 0$. Concretely, this means that the levels that appear to be at the same height in Figure 2.3 ($n = 2$, $l = 0$ and $n = 2$, $l = 1$) are not in fact at the same height. The same is true for $n = 3$ and $l = 0, 1, 2$ for $n = 4$ and $l = 0, 1, 2, 3$ etc. The deviations are small, however very high-resolution spectrometers reveal that the lines that correspond to $l > 0$ are actually double lines.

### 2.5.1 Splitting of the energy levels

The double lines are caused by a shift in the energy level that results from the interaction between the orbital angular momentum and the spin.

- The orbital angular momentum $l$ is a vector, which means that it has an absolute value that has a length $|l|$ and a direction. Its magnitude is determined by the value of the quantum number $l$:
$|l| = \sqrt{l(l+1)}\frac{h}{2\pi}$, whereby $h$ is once again Planck's quantum of action.
- The vector $l$ cannot point in an arbitrary direction. Only directions for which the projection on the axis of reference, usually called z, results in values with whole units of $\frac{h}{2\pi}$ are allowed. The length of the vector is $\sqrt{l(l+1)}$, in units of $\frac{h}{2\pi}$. This length is at least $l$, but is always shorter than $l + 1$. The number of possible directions relative to an axis of reference is exactly the value range of the magnetic quantum number $m$, which is an integer between $-l$ and $+l$. This is illustrated in Figure 2.4a.
- The electron has its own angular momentum, the spin. Its quantum number $s$ is always 1/2. The vector $s$ of the spin has an absolute value of $|s| = \sqrt{\frac{1}{2}\left(\frac{1}{2}+1\right)}\frac{h}{2\pi}$ that is, $|s| = \frac{1}{2}\sqrt{3}\frac{h}{2\pi}$. For the spin, half units of $\frac{h}{2\pi}$ are allowed for projection on the axis of reference. The projection $s_z$ of $s$ on the axis of reference z results in the possible values of $\frac{1}{2}\frac{h}{2\pi}$ or $-\frac{1}{2}\frac{h}{2\pi}$. Other half-integer multiples of the unit $\frac{h}{2\pi}$ cannot occur for a vector length of approximately 0.866 units (see Figure 2.4b).

The total angular momentum of an electron $j$ is obtained by adding the vectors $l$ and s. It is reasonable to assume that $|j|$, i.e., the absolute value of the vector $j$, is also quantized and that the relationship $|j| = \sqrt{j(j+1)}\frac{h}{2\pi}$ suffices.

It is then possible to show with simple means that only the following values are possible for $j$:
- If $l = 0$, then $j = \frac{1}{2}$
- If $l > 0$, then $j = l - \frac{1}{2}$ or $j = l + \frac{1}{2}$

Ryder provides an easily understandable derivation of this rule (page 89 et seq. [1]).

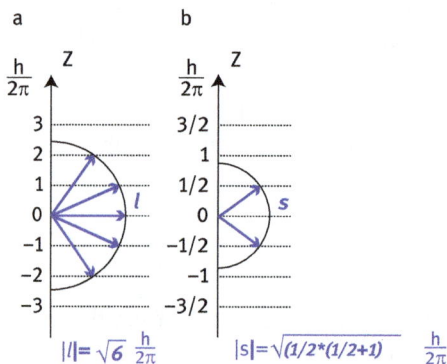

Figure 2.4: Projections of the vectors $l$ and $s$ on the reference axis z.

Thus, there are two energy levels for $l > 0$: $lp_{1/2}$ and $lp_{3/2}$, which leads to the fact that the transition from $lp_{1/2}$ to a lower energy level results in slightly longer-wave radiation than the transition from $lp_{3/2}$ to the same level. Two closely adjacent lines are then observed in the spectrum; a so-called doublet.

Figure 2.5 shows the energy levels for the hydrogen atom up to $n = 4$ and $l = 2$. The splitting into doublets is too small to be able to draw them to scale. Therefore, they are enlarged for three of the doublets. The difference between the 2p levels, which have an excitation energy of 10.199 eV, is only 0.0001 eV. When returning from this level to the ground state, radiation with a wavelength of 121.568 nm is released, whereby the wavelengths of the two doublet lines are separated by only 0.00054 nm, i.e., 0.54 pm. The differences in the energies for the other doublets are even smaller (see the illustrations of the splitting in the enlargements). Spark spectrometers are generally equipped with optics that can, at best, resolve lines that are separated by more that 5 pm. Higher resolution is meaningless when using a spark excitation source as there is a broadening of the spectral lines due to Doppler and Lorentz effects resulting from the excitation source. The doublet structure in the hydrogen atom is, thus, inconsequential for spark spectrometry. The doublet states are indicated by a superscript 2 in Figure 2.5. Please note that the states $1s_{1/2}$, $2s_{1/2}$, $3s_{1/2}$ are also considered as doublet states although no double lines are visible in the spectrum. This terminology is, however, useful as becomes clear after introduction

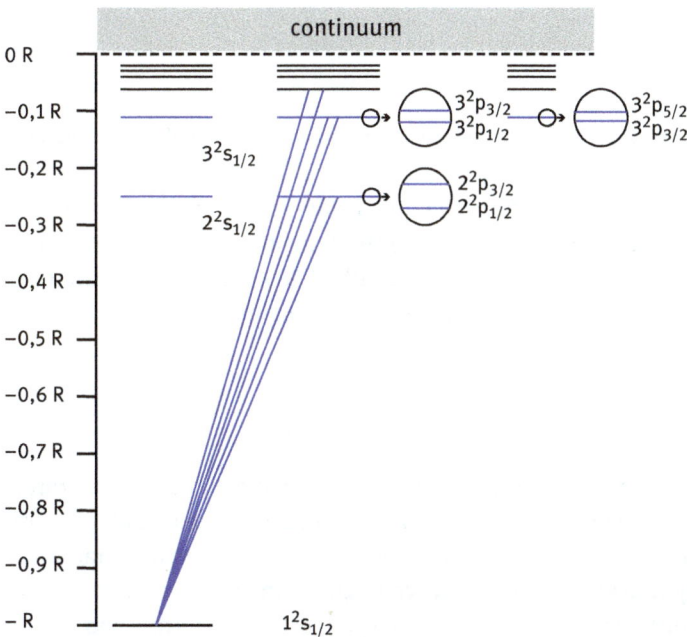

**Figure 2.5:** Energy levels of the hydrogen atom up to $n = 4$ and $l = 2$ with doublet splitting.

of the extended selection rules in section 2.6.3. For atoms that are heavier than hydrogen, the doublet lines are more widely separated. They are frequently found in arc and spark spectra. The yellow sodium doublet, with lines that result from transitions from the $3^2p_{3/2}$ or $3^2p_{1/2}$ state to the $3s_{1/2}$ state, is well known (see Figure 2.6).

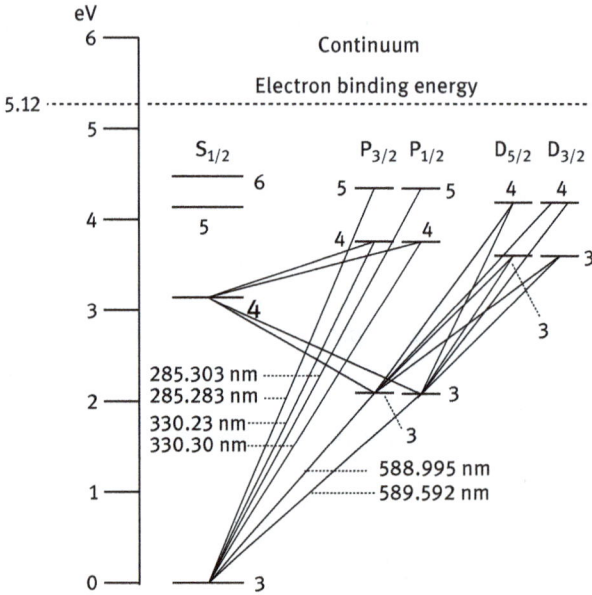

Figure 2.6: Excerpt of the energy level chart for sodium, no transitions for n=1 and n=2 are plotted.

The line generated for the transition from the $3^2p_{3/2}$ state to the $3^2s_{1/2}$ state radiates photons with a wavelength of 588.995 nm, while the transition from the $3^2p_{1/2}$ state to the $3^2s_{1/2}$ state generates photons with a wavelength of 589.592 nm. This difference of 400 pm can be easily observed even with a low-resolution optic system.

## 2.5.2  Origin of the spectral background

Free electrons have energies that lie above the ionization limit, i.e., in the range labelled "Continuum" in Figure 2.5. If such an electron is recombined with a proton, i.e., a positively charged hydrogen nucleus, then the emitted radiation does not belong to any line spectrum. The wavelength of the emitted radiation depends on the energy difference between the energy level the electron had before recombination and the excitation state that it occupies after the recombination.

Of course, the mechanism described also applies to atoms other than hydrogen. The resulting spectral background radiation limits the detection sensitivity in arc and spark spectrometry. Lower temperatures prevail in electrical arcs compared to sparks. As a result, fewer atoms are ionized in arc and therefore less background radiation is produced.

## 2.6 The spectra for atoms with multiple electrons

The electrons for atoms with multiple electrons can also occupy different energetic states that are designated with a set of quantum numbers $n$, $l$, m, and $s_z$. Here too, radiation is emitted during transition to an energetically lower level. However, in this case there is no simple formula like we encountered in eq. (2.8) to calculate wavelengths for the radiated photons.

There are three rules for the configuration of electrons:
- Again, the Pauli principle applies: No two electrons can ever have the same set of four quantum numbers $n$, $l$, m and $s_z$.
- The electrons try to occupy the state with the lowest energy.
- The second Hund's rule applies: If an atom has multiple free states with the same energy, then they are each first occupied by an electron with parallel spin.

Application of these rules leads to an energy level sequence that is valid for most atoms:

$$1s < 2s < 2p < 3s < 3p < 4s < 3d < 4p < 5s < 4d$$

Table 2.3 shows the electron configuration for the first 14 elements (atoms) taking the rules for atomic structure into consideration. Each electron is represented by an arrow that shows the direction of the spin. The electron configuration is given in the last column. The numbers in the header line indicate the main quantum numbers; the letters $s$, p represent $l = 0, 1$. The superscript indicates the number of electrons associated with the preceding combination of the quantum numbers $n$ and $l$.

### 2.6.1 Hydrogen-like atoms and ions

For atoms with a single electron in a single unfilled shell, the relationship is comparatively simple. A fine structure due to the splitting of the spectral lines explained by the effect of the total angular momentum of the electron was described above for the spectrum of the hydrogen atom. It is possible to proceed in the same way for other one-element systems. These are found for the alkali metals in the periodic table: The elements Li, Na, K, Rb and Cs all have a single electron outside of the

**Table 2.3:** Atomic structure and electron configuration of the first 14 elements of the periodic table.

| Z | Element | K 1s | L 2s | L 2p | M 3s | M 3p | Configuration |
|---|---------|------|------|------|------|------|---------------|
| 1 | H | ↑ | | | | | $1s^1$ |
| 2 | He | ↑↓ | | | | | $1s^2$ |
| 3 | Li | ↑↓ | ↑ | | | | $1s^2 2s^1$ |
| 4 | Be | ↑↓ | ↑↓ | | | | $1s^2 2s^2$ |
| 5 | B | ↑↓ | ↑↓ | ↑ | | | $1s^2 2s^2 2p^1$ |
| 6 | C | ↑↓ | ↑↓ | ↑ ↑ | | | $1s^2 2s^2 2p^2$ |
| 7 | N | ↑↓ | ↑↓ | ↑ ↑ ↑ | | | $1s^2 2s^2 2p^3$ |
| 8 | O | ↑↓ | ↑↓ | ↑↓ ↑ ↑ | | | $1s^2 2s^2 2p^4$ |
| 9 | F | ↑↓ | ↑↓ | ↑↓ ↑↓ ↑ | | | $1s^2 2s^2 2p^5$ |
| 10 | Ne | ↑↓ | ↑↓ | ↑↓ ↑↓ ↑↓ | | | $1s^2 2s^2 2p^6$ |
| 11 | Na | ↑↓ | ↑↓ | ↑↓ ↑↓ ↑↓ | ↑ | | $1s^2 2s^2 2p^6 3s^1$ |
| 12 | Mg | ↑↓ | ↑↓ | ↑↓ ↑↓ ↑↓ | ↑↓ | | $1s^2 2s^2 2p^6 3s^2$ |
| 13 | Al | ↑↓ | ↑↓ | ↑↓ ↑↓ ↑↓ | ↑↓ | ↑ | $1s^2 2s^2 2p^6 3s^2 3p^1$ |
| 14 | Si | ↑↓ | ↑↓ | ↑↓ ↑↓ ↑↓ | ↑↓ | ↑ ↑ | $1s^2 2s^2 2p^6 3s^2 3p^2$ |

filled shells, as can be immediately seen with a glance at the rows for Li and Na in Table 2.3. The structure of their spectra is very similar to that of hydrogen.

Figure 2.6 shows the energy level diagram for sodium. For reasons of clarity, not all transitions could be drawn.

Ions for which so many electrons have been removed from the outer shell so that only a single one remains are also one-electron systems (e.g., $He^+$, $Be^+$ and $B^{2+}$).

It is intrinsic to all one-electron systems that two point charges, namely that of the electron and that of the nucleus (where applicable, surrounded by the electrons of filled shells), interact.

The units for the y-axes in Figures 2.6 and 2.8 are given as energies in electron volts (eV). The notation is practical as shown in the following deliberation:

In the case of a transition from a higher energy level $E_H$ to a lower $E_T$, the energy difference is $\Delta E = E_H - E_T$ and a photon with the wavelength $\lambda$ is emitted. The wavelength of such a photon is easy to calculate. By combining the Planck-Einstein equation $E = h\nu$ with $\nu = c/\lambda$ one obtains:

$$\lambda = \frac{hc}{\Delta E} \tag{2.10}$$

However, it is inconvenient that the energy difference $\Delta E$ must be given in J. Working with $\Delta E'$ in eV instead leads to the following simple relationship:

$$\lambda = \frac{1240.7}{\Delta E'} \tag{2.11}$$

This takes into consideration that Planck's quantum of action is $6.626 \times 10^{-34}$ Js and that 1 eV is equivalent to $1.60218 \times 10^{-19}$ J. The unit of the conversion constant is chosen so that the result of the calculation is the wavelength in nanometers; the unit normally used to indicate wavelengths in spark spectrometry. From eq. (2.11), it is immediately apparent there is a minimum excitation energy to each spectral line showing up in the spectrum. If a spectral line appears at, for example, 200 nm, must be a transition with energy difference of around 6.2 eV. This is possible, if electrons are excited to a level of 6.2 eV above the ground state. However, at 200 nm spectral lines resulting from electrons that have a higher excitation level than 6.2 eV with respect to the ground state can occur. This can happen, if the excited state leads to another one that is energetically 6.2 eV lower but still above the ground state.

## 2.6.2 Atoms with multiple outer electrons

The relationships become more difficult when more than one electron occupies the incomplete shells. In this case, it comes to charge shielding effects between the valence electrons. Outer or valence electrons are understood to be those electrons which are not located in completely filled shells or subshells. The charge shielding can be imagined as follows:

Quantum theory makes it possible to determine the probability of the position of the electron in addition to the energy level. Volumes in which electrons are located with a given (high) probability form areas of varying shapes. These areas are called "orbitals." In addition to spheres with different diameters, these orbitals can also be lobes with different orientations or ring forms. The form of the orbital depends on the energetic state of the electron. Because of reciprocal shielding of the outer electrons, which are, on average, located in different positions relative to the atomic nucleus, there is a splitting of energy levels that would otherwise be at the same height.

To take these interferences into consideration, it is necessary to proceed in principally the same way as for the determination of the hydrogen fine structure: Spin and orbital angular momentum must be added. What is new is that the momentums from all the outer electrons must be added. The orbital angular and spin

(a)    (b)

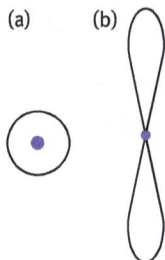

Figure 2.7: Spherical orbital (a) and elongated lobed orbital (b).

momentums from completely filled inner shells do not need to be taken into consideration, since their sum results in zero (first Hund's rule). However, the momentums of all outer electrons must be taken into account.

Addition of the momentums can be performed in two ways:
- First, take the sum of the angular orbital momentums of the outer electrons and then the sum of the outer electron spins. Then add these two subtotals to a total.
- For each of the $a$ outer electrons, add the angular orbital momentum $l_i$ and the spin $s_i$ ($1 \leq i \leq a$) to form $j_i$. Then sum up all $j_i$ to form a total angular momentum $\mathbf{J}$.

Both methods are approximations. Here, only the first method, the so-called Russell-Saunders coupling is to be examined in more detail. It is suitable for light and medium weight elements up to the lanthanides according to Riedel [2]. The second method, the so-called j-j coupling method is more advantageous for elements with higher atomic weights.

Here, a more precise outlining of the Russell-Saunders coupling:

The moments of the spins $s_i$ of all outer electrons ($1 \leq i \leq a$) are added to form a total spin $\mathbf{S}$:

$$\mathbf{S} = \sum_{i=1}^{a} s_i \tag{2.12}$$

The angular orbital momentums $l_i$ of all $a$ outer electrons ($1 \leq i \leq a$) are added to form the total angular orbital momentum $\mathbf{L}$:

$$\mathbf{L} = \sum_{i=1}^{a} l_i \tag{2.13}$$

Finally, $\mathbf{S}$ and $\mathbf{L}$ are added to form a total angular momentum $\mathbf{J}$:

$$\mathbf{J} = \mathbf{S} + \mathbf{L} \tag{2.14}$$

It is possible to define a quantum number $S$ that can assume all combinations of the sum of the $s_z$ projections of the outer electron spins. If $s_s$ is the largest possible sum, then $s_z$ is +1/2 for all electrons. The following values are possible: $s_s, s_s-1, s_s-2, \ldots, -s_s$.

$S$ is regarded as the total spin quantum number.

An additional quantum number can be determined as the sum of the quantum numbers $m$, which is equivalent with the projection $l_z$ in the spatial direction $z$. If $s_m$ is the largest possible sum of the $l_z$ for the outer electrons, then the values for $L$ can be $s_m$, $s_m-1$, $s_m-2$, ..., $- s_m$. $L$ is referred to as the total orbital momentum quantum number.

A quantum number $J$ can also be defined for the combined total angular momentum. $J$ can have the following values:

$$s_m + s_s, s_m + s_s - 1, s_m + s_s - 2, ..., |s_m - s_s|$$

The relation $2S + 1$ is known as a multiplicity, because when $S = 0$ (then $2S + 1 = 1$), single lines usually occur; for $S = \frac{1}{2}$ ($2S + 1 = 2$) double lines; for $S = 1$ ($2S + 1 = 3$) triple lines. Single lines are known as singlets; double lines as doublets and triple lines as triplets.

### 2.6.3 Selection rules and metastable states

The following rules for valid combinations of quantum numbers can be established:
1. Only energy transitions for which the total spin quantum number $S$ does not change are possible; a transition from a singlet state results once again in a singlet state. The same applies to doublet and triplet states.
2. After an energy transition, the orbital angular momentum quantum number $L$ must be different by 1.
3. The total angular momentum quantum number J may only change by –1, 0 or 1.
4. A transition from a state with $J = 0$ to another with $J = 0$ is not possible.

From the selection rules, it follows that there are excitation states from which, due to the selection rules, no return to the ground state is possible. Such states are called metastable. While normal excited states are vacated again after a lifetime on the order of 10 ns, metastable states are occupied for a longer time. Leaving a metastable state can, for example, be triggered as an interaction caused by a collision. Lifetimes in the millisecond range are not uncommon. If the possibility of interaction is prevented, the lifetime can even be extended to seconds (see e.g., [3], p.173). Metastable states are important to understand the function of lasers (see Section 3.2.3).

Figure 2.8 shows the energy level diagram for the helium atom, in which both singlet as well as triplet states occur. Such diagrams are also called Grotrian diagrams (after their inventor Walter Grotrian). Like in Fig. 2.6, s, p and d are replaced by the capital letters S, P and D. These letters represent the sum of the angular orbital momentums of the two outer electrons of the He atom. The superscript numbers right of the capital letters indicate a singlet or a triplet state. Details of this notation will be explained in Section 2.7.

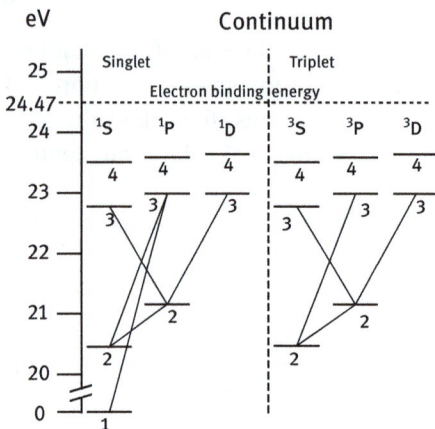

**Figure 2.8:** Term scheme of the helium atom.

## 2.7 Russell-Saunders term symbols

The energetic state of an atom is known when the three quantum numbers $S$, $L$ and $J$ have been identified. It is described in the form:

$$2S + {}^1L_J \qquad (2.15)$$

The quantum number $L$ is not numerically labeled, but is designated with the uppercase letters S, P, D etc.

   Example: The configuration for the magnesium atom is $1s^2 2s^2 2p^6 3s^2$, as can be seen in Table 2.3. Only the two 3s electrons are found outside of the completed shells and subshells and must be considered. To describe the ground state, $l_1$ and $l_2$ for the two outer electrons are both 0 and the sum of the spins is also 0 ($\frac{1}{2} + -\frac{1}{2} = 0$). Thus $J = 0$ and $2S + 1 = 1$. Therefore, the term for the ground state is ${}^1S_0$.

   The rules formulated in Section 2.6 do not always apply to heavy elements. For mercury, for example, the transition occurs between the state ${}^3P_1$ and the ground state ${}^1S_0$ although this transition contradicts rule 1 formulated in 2.6.3. The spectral line 253.65 nm emitted during this transition is even one of the most intense mercury lines.

Sections 2.1 to 2.7 covered the following:
  –  That the excitation of an atom is only possible at certain discrete levels,
  –  that the transition between energy levels is associated with the absorption and emission of certain energies or with the absorption and radiation of photons,
  –  where the energy levels for the excited atoms lie and
  –  that transitions cannot take place between all excitation states.

The formation of atomic spectra could only be briefly outlined here. A more detailed discussion can be found in Banwell and McCash [4]. Other works dealing with this subject can be found in the bibliography [1–3, 5–13].

## 2.8 Characteristics of emitted spectral signals

In emission spectrometry, the electronic transitions generated by spark/arc excitation lead to spectra, which are structured according to the principles discussed in the previous sections. It is not practical to theoretically calculate the positions for the spectral lines suitable for spark spectrometry. Line positions and sensitivities are listed in tables, the so-called spectral atlases. Sensitivity is strongly dependent on the type of the excitation source and of the excitation conditions.

Section 3.5 describes how the radiation from an excitation source is broken down into a spectrum and where in this spectrum the spectral line for a given wavelength can be found.

The line width observed on the focal curve of the optics is also influenced by the construction of the spectrometer. This is discussed in detail in Section 3.5 of this book.

The following influences limit the minimum achievable line widths:
- Collisional broadening
  The movement of the atoms or ions in the gas phase results in collisions between the particles. The electrons are especially affected and deformation of the outer shell occurs. Small changes to the energy levels caused by this result in a slight broadening of the signals for the transitions.
- Doppler broadening
  The excited atoms or ions present in the gas phase are in motion during the emission of radiation. The Doppler effect occurs when this movement is towards or away from the radiation detector. Temperature has a large influence on the amount of the line broadening. Collision and Doppler broadening are dependent on the type of excitation. These are discussed in more detail in Section 3.2.
- The natural line width
  Every photon consists of an oscillation packet of finite length. When a Fourier transformation is applied to the wave packet, a spectrum that consists not only of the central wavelength but also of a wavelength range, albeit narrow, around this central wavelength. Ultimately, the finite line width is a consequence of Heisenberg's uncertainty relationship.

  The natural line width can be calculated with eq. (2.16):

$$\Delta v = \frac{1}{2\pi\tau} \tag{2.16}$$

Here $\Delta\nu$ represents the line width expressed as a frequency and $\tau$ the duration of the photon emission, which is on the order of 10 ns. For a spectral line at 300 nm, the natural line width is approximately 0.005 pm. Thus, it is too small to be observed with an optic of an arc/spark spectrometer. These dependencies are explained by, e.g., Bergmann/Schäfer [14] S. 289 et seq. and in the *Lexikon der Optik* [15] page 430 et seq.

Sections 2.1 to 2.7 discussed which energy levels are possible for an atom. In addition, it was explained that the transition between energy levels is associated with the absorption of given energies or with the radiation of photons. Furthermore, it was shown that transitions cannot take place between all energy levels. However, nothing was said about the intensities of the resulting spectral lines.

The atom and ion line intensities are influenced by three factors:
1. Transition probability.
2. The population density, which in turn depends on the temperature and thus the excitation conditions. Here, it is also important to note that the temperature determines the degree of ionization.
3. The analyte concentration.

The portion of ionized particles $\alpha$ can be calculated using the Saha eq. (2.17). More detailed explanations can be found in Finkelnburg [12] S. 23 and S. 80 et seq. The Saha equation shows that the portion of ionized particles $\alpha, 0 \leq \alpha < 1$ increases with increasing temperature. The following applies:

$$\frac{\alpha^2}{1-\alpha^2}p = \frac{(2\pi m)^{\frac{3}{2}}}{h^3}(kT)^{5/2}e^{-E_i/(k^*T_a)} \tag{2.17}$$

$p$ denotes the pressure, $m$ the atomic mass, $E_i$ the ionization energy, i.e., the energy required to remove an electron from the atomic nucleus. An explanation of the other terms in the equation can be found below eq. (2.19).

The Boltzmann equation is used to calculate the excitation temperature $T_a$:

$$n_a = n_0 \frac{g_a}{Z_0}e^{-E_a/(k^*T_a)} \tag{2.18}$$

Combining the Boltzmann equation with Einstein's equation enables calculation of the spectral radiance $I_\nu$ for a spectral line with the frequency $\nu$:

$$I_\nu = A_{ab}h\nu n_0 \frac{g_a}{Z_0}e^{-E_a/(k^*T_a)} \tag{2.19}$$

The radiance is given in $Wm^{-2}$, $I_v$ is that which is typically known as intensity in spectrometry.

Strictly speaking, eq. (2.19) is only applicable to stationary plasma states. Such conditions do not exist for spark. However, the equation does indicate which factors determine the line intensity.

The variables used in eqs. (2.17) through (2.19) have the following meanings:

$A_{ab}$:   Probability of a transition from state $a$ to state $b$
$h$:        Planck's quantum of action
$n_0$:      Particle density in the ionization state being considered
$n_a$:      Density of particles in the excited state $a$
$g_a$:      statistical weight of the excited state $a$, that is, the number of energetically collapsed energy levels
$Z_0$:      Total state of the ionization level being considered with $Z_0 = \sum_i g_i e^{-Ea/(k*Ta)}$
$T_a$:      Excitation temperature
$E_a$:      Excitation energy
$K$:        Boltzmann constant ($1.380658 \times 10^{-23}$ J $K^{-1}$)

Equations (2.17) and (2.18) are explained by Laqua [16].

The intensities of atom or ion lines for an element are required for quantitative AES as explained in Chapter 3. The fundamental determination of the particle density in the excitation source by means of quantum theory is impractical. Therefore, the relationship between intensities and concentrations is determined by empirical calibration.

Establishing a calibration function is an important part of the analytical work. This is also discussed in Chapter 3.

## Bibliography

[1]   Ryder P. Quantenphysik und statistische Physik. Aaache, Shaker, 2004.
[2]   Riedel E, Janiak C. Anorganische Chemie. Berlin and New York, DeGruyter, 2011.
[3]   Demtröder W. Experimentalphysik 3. Berlin, Springer-Verlag, 2005.
[4]   Banwell Colin N, McCash EM. Molekülspektroskopie: Ein Grundkurs, München, Wien, Oldenbourg, 1999.
[5]   Broekaert JAC. Analytical Atomic Spectrometry with Flames and Plasmas, Weinheim, Wiley-VCH Verlag Chemie GmbH, 2002.
[6]   Ingle JR, James D, Crouch SR. Spectrochemical Analysis. Englewood Cliffs, New Jersey 07632, Prentice-Hall Inc, 1988.
[7]   Hindmarsh R. Atomic Spectra, Pergamon Press, Oxford, 1967.
[8]   Kalvius GM, Luchner K, Vonach H. Physik IV, München, Wien, Oldenbourg Verlag, 1985.
[9]   Heywang, Treiber, Herberg, Neft. Physik, Hamburg, Verlag und Technik, 1992.
[10]  Walker S, Straw H. Spectroscopy Vol. one, Atomic, Microwave and Radio-frequency Spectroscopy, London, Chapman & Hall, 1961.
[11]  White HE. Introduction to Atomic Spectra. New York, McGraw Hill, 1934.

[12]   Finkelnburg W. Einführung in die Atomphysik, Berlin, Springer-Verlag, 1956.

[13]   Johnson RC. Atomic Spectra, London, Methuen & Co Ltd, 1950.

[14]   Niedrig H (editor). Bergmann – Schäfer Lehrbuch der Experimentalphysik, Berlin, New York, Walter de Gruyter, 2004.

[15]   Paul H. Lexikon der Optik. Heidelberg, Berlin, Spektrum Akademischer Verlag GmbH, 2003.

[16]   Laqua K. Ullmanns Encyklopädie der technischen Chemie, vol. 5, Emissionsspektroskopie. Weinheim, Verlag Chemie GmbH, 1984.

# 3 Hardware for spark and arc spectrometers

Emission spectrometers for the direct analysis of solid materials differ depending on the excitation generator used. Here, spark and arc sources are utilized. In the last few years, systems with laser excitation have gone into serial production and been brought onto the market.

The construction of laboratory instruments differs significantly from that of mobile spectrometers for onsite operation.

This chapter describes the state of development that forms the basis for the construction and operation of modern arc and spark spectrometers and that has been achieved through the decades of efforts by numerous experts. It is necessary to note that although the authors describe this state, they do not claim the intellectual authorship of the equipment and procedures presented in this chapter. If the authors themselves have made significant contributions to the state of technology, references to their patent applications and publications are indicated in the relevant locations of the chapter.

## Laboratory spectrometers

Laboratory spectrometers are utilized when analytical results with high precision and good accuracy are required. Detection limits can be achieved that are generally lower than those for mobile spectrometers. The excitation source is almost exclusively the electric spark. With laser sources, it is possible to achieve results that are useful for many applications, but due to the extreme requirements concerning the equipment, laser excitation has only occasionally been applied.

Size and weight play a relatively minor role for the construction of laboratory spectrometers. This is different when considering the measuring time: The analytical results for preliminary samples during smelting processes must be quickly available. Usually samples that have been taken specifically for analytical purposes are analyzed by placing them on a sample stand that is mounted on the main instrument housing. This is an important difference from mobile spectrometers, which have a spark probe for direct analysis of the workpiece.

Figure 3.1 shows laboratory spectrometers from two different manufacturers ((a) GS 1000-II from OBLF and (b) SPECTROLAB LAVM12 from Spectro). In addition to the instruments pictured here, there are comparable spectrometers from other manufacturers on the market.

## Mobile spectrometers

Mobile spectrometers for operation in the warehouse or for production control must be small and easy to transport as the instrument is brought to the sample. There are

https://doi.org/10.1515/9783110529692-003

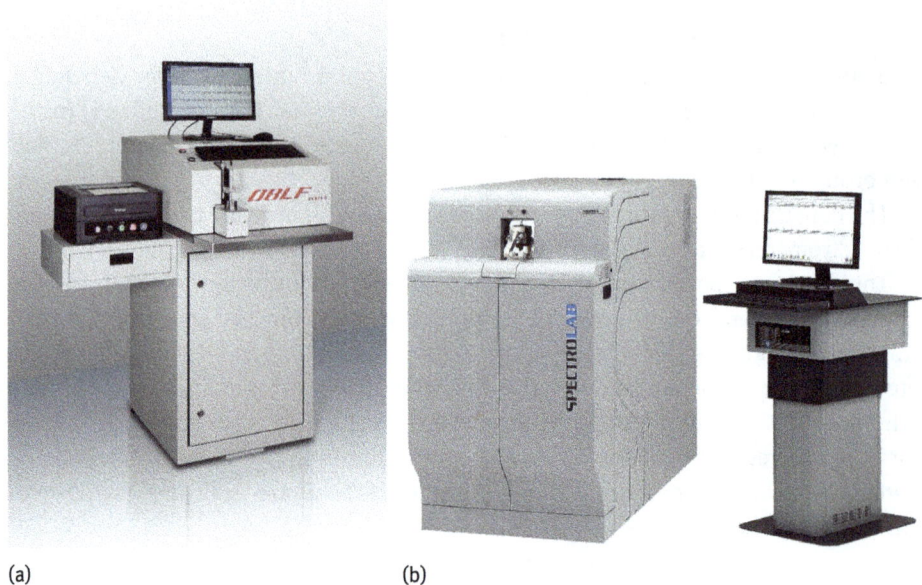

(a)                                        (b)

**Figure 3.1:** Modern spark emission spectrometers. Part (a) printed with the friendly permission of OBLF Gesellschaft für Elektronik und Feinwerktechnik mbH, Salinger Feld 44, 58454 Witten, Germany; Part (b) courtesy of SPECTRO Analytical Instruments GmbH, Boschstr. 10, 47533 Kleve, Germany.

instruments on rollers, portable systems and handheld devices for which the entire hardware is integrated into the test probe. For most mobile spectrometers, both battery and mains powered operation are possible.

The requirements for the precision and accuracy of the results are not as strict as for laboratory spectrometers, because it is usually sufficient to identify the material or to ensure that the correct alloy is present. However, the measuring times must be short in order to achieve a high sample throughput. This is especially important when 100% testing is carried out during incoming or outgoing inspections. In addition, the test probe must be held motionless on the test piece during the measurement, which is not possible for an unlimited length of time.

Compromises must be made for sample preparation. Stationary disc grinding machines, lathes and milling machines are generally used together with laboratory spectrometers. They are impractical for mobile spectrometers, because the measuring system is brought to the test piece. Usually, simple preparation with a handheld tool, for example an angle grinder must suffice. Unlike laboratory spectrometers, excitation sources such as arc and, recently, laser play a significant role. Especially, arc is tolerant of samples that have not been optimally prepared. The range of materials that can be analyzed with mobile spectrometers is very wide, this is for example required in the secondary raw materials industry.

Chapter 7 deals with the constructive peculiarities as well as the analytical possibilities and limits for mobile spectrometers, which are often referred to as industrial spectrometers because they are used outside the laboratory in production.

## 3.1 Construction of emission spectrometers for the analysis of solid materials

Figure 3.2 is a block diagram showing the basic construction of spectrometer systems.

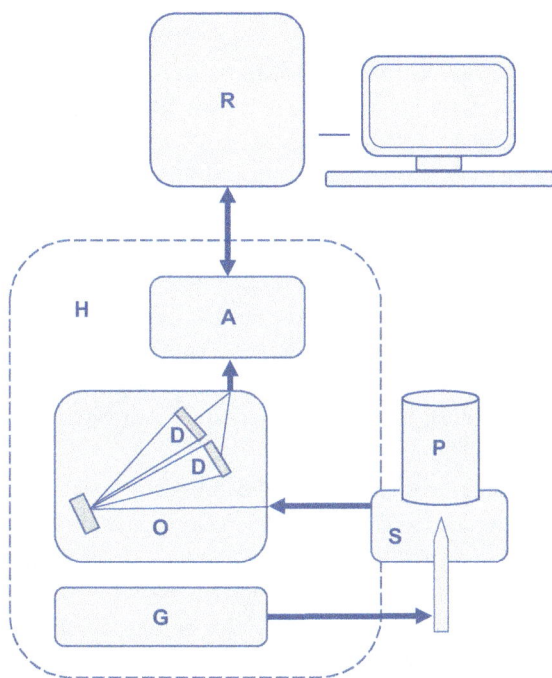

**Figure 3.2:** Block diagram arc/spark spectrometer.

A sample P is on a stand S. There, material from the sample surface is ablated and excited to emit radiation with a line spectrum by an excitation generator G. This process often occurs in an inert gas atmosphere. The emitted radiation is transported to an optical system O. The optics make it possible to separately measure several narrow spectral ranges with the width of a few picometers. Radiation with intensities that are proportional to the elemental concentrations of the analytes occur in these wavelength ranges. The radiation is recorded by detectors D and converted to electrical signals. A readout system A measures these signals, processes them and sends them to a higher-level computer R, where the spectrometer software calculates the element concentrations and displays them or relays them to control other processes. The instrument housing H, in which usually at least the excitation generator, the main optics and the readout system are located, is shown as a dashed line. The main components in Figure 3.2 are described here in more detail.

– *Excitation generators*

When arc or spark generators are used, they are connected on the one hand to the sample to be analyzed and on the other hand to a counter-electrode, the tip of which is positioned close to the sample surface. The arc or spark is generated when a high-voltage pulse ionizes the atmosphere between the tip of the counter-electrode and the sample surface, making it conductive. The gap becomes low impedance, and a stable current flow, which is fed via power electronics decoupled from the ignition, can be established in the range between approximately one and over 100 A.

For arc excitation, this current flow lasts for several seconds. By spark, the current flow is interrupted after a time of 10–1,000 μs and restarted after an interval. Spark frequencies of 50–1,000 Hz are customary.

The sliding spark generator is a special version of spark generator. It is used exclusively for the analysis of nonconducting materials, e.g., for the examination of halogen-containing plastics. Two electrodes are placed several millimeters apart from each other onto the nonconducting sample. The ignition pulse ionizes the sample surface. Again, the ignition is followed by a high-current pulse that may achieve more than 1,000 A. Details about sliding spark generators can be found under T. Seidel [1] and A. Golloch and D. Siegmund [2].

For excitation with laser, the laser beam is focused on the sample surface or just above it. The laser pulses generate temperatures of over 10,000 °C in the focal point. In this way, material is ablated and transformed into a plasma state. After several nanoseconds, during which the plasma generates only a continuous background radiation, it emits a characteristic spectrum, the radiation of which is determined by the atoms and ions present in the plasma. If the laser position were firmly fixed on the sample, then the plasma would always be formed in the same place. For many samples, e.g., metals, a point analysis on an area on the order of a hundred square micrometers is not sufficient for making a true statement about the average analysis of the sample. For this reason, it is necessary to have devices that either move the laser beam over the sample or move the sample relative to the laser beam.

Modern excitation generators usually have a microcontroller that manages the excitation process and communicates with the other components. Section 3.2 describes the construction and function of excitation generators for laboratory emission spectrometers for the direct analysis of solid samples. Some specific aspects are only of interest to mobile spectrometers. They are discussed in Section 7.1.

– *Spark stands and test probes*

Laboratory spectrometers generally use, as mentioned above, spark as the excitation source and are equipped with a spark stand that is firmly attached to the instrument.

High-purity argon (Ar 4.8 with maximum 20 vpm or Ar 5.0 with maximum 10 vpm impurities) is normally used as the excitation atmosphere. The sample is placed onto the spark stand so that the circular spark stand opening is completely covered. A sample clamp presses the sample in place on the opening of the spark stand ensuring tight closure during the spark process. Section 3.3 deals with laboratory spectrometer spark stands.

Figure 3.3 shows examples of the spark stands for two modern laboratory spectrometers ((a) spark stand on the GS 1000-II from OBLF and (b) spark stand on the SPECTROLAB LAVM12 from Spectro).

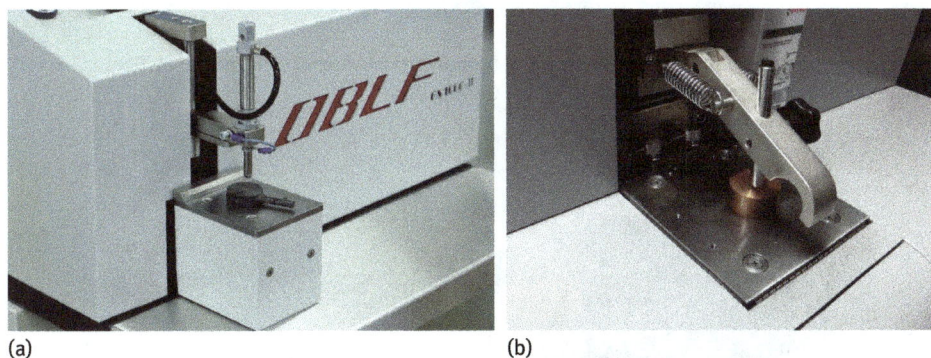

(a)                                                    (b)

**Figure 3.3:** Spark stands of two modern laboratory spectrometers. Part (a) printed with the friendly permission of OBLF Gesellschaft für Elektronik und Feinwerktechnik mbH, Salinger Feld 44, 58454 Witten, Germany; Part (b) printed with the friendly permission of SPECTRO Analytical Instruments GmbH, Boschstr. 10, 47533 Kleve, Germany.

For mobile spectrometers, the measuring probe usually has a practical pistol form (Figure 3.4). The electrical plasma is generated in the spark chamber (metal component at the tip of the test probe). The light is usually transported from the spark chamber to the spectrometer optics via quartz fiber optic. However, fiber optics do not allow radiation under 185 nm to pass through. This is why mini-optics that are only for the spectral lines in the wavelength range between 160 nm and 200 nm are frequently integrated into the spark probe (Figure 3.5). This wavelength range is particularly important for the analysis of steel, as the spectral lines used for important elements such as C, P, S and B are found here. An optical system for the entire spectral range to be measured is built directly into the spark probe for handheld instruments. However, the resolution capacity and, thus, the analytical performance of such optics cannot be compared with those having a longer focal length. Section 7.2

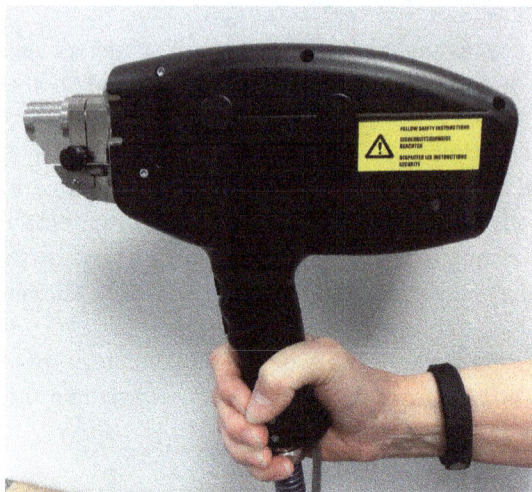

Figure 3.4: Spectrotest probe. Printed with the friendly permission of SPECTRO Analytical Instruments GmbH, Boschstr. 10, 47533 Kleve, Germany.

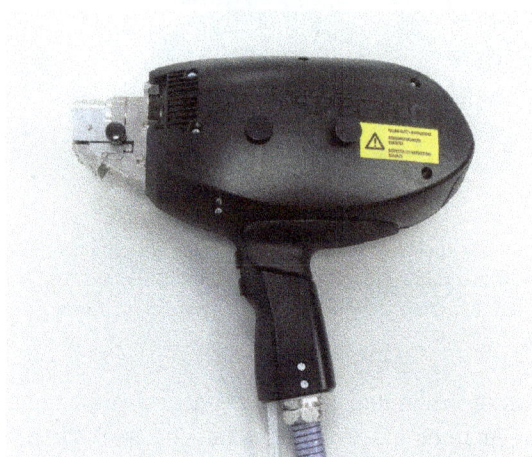

Figure 3.5: Spectrotest probe with UV-minioptics. Printed with the friendly permission of SPECTRO Analytical Instruments GmbH, Boschstr. 10, 47533 Kleve, Germany.

discusses measuring probes for mobile spectrometers. Connection to the instrument with probe hose is described in Section 7.3.

– *Inert gas systems*
Cleanliness and air-tightness of the argon system are important for spectrometers with spark excitation. Just a few ppm oxygen, water vapor or vapor from organic compounds in the argon can severely disrupt the discharge and lead to poor repeatability.

The argon system is usually designed so that, during pauses between measurements, a moderate argon flow of approximately 2–20 liter per hour ensure that penetrating ambient air and humidity are flushed out. A higher gas flow of 100–400 liter per hour is employed during the measurement. With this, the metal condensate created by each spark is removed. The used argon is led to a filter system. This prevents the condensate in the form of particulate matter from contaminating the ambient air. Section 3.4 describes the design of inert gas systems in modern laboratory spectrometers. Argon is also utilized as the inert gas for sparking with mobile spectrometers. If carbon is to be determined using arc, air from which carbon dioxide has been removed is used as the operating gas. Special features of the gas systems in mobile instruments are described in Section 7.6.

*– Optical systems*
Spectrometer optics with longer focal lengths are almost exclusively built using optical systems with concave holographic gratings. The Paschen-Runge mount assembly, for which the spectrum is focused on a circular arc, is the preferred design. Such optics dominate in laboratory spectrometers. This construction is explained in Section 3.5.

In the case of instruments equipped with semiconductor sensor arrays of shorter focal lengths, like those found in smaller mobile spectrometers, a design employing holographically produced so-called "flat field gratings," in which a part of the spectrum is straightened, is preferred. For the commonly utilized sensor length of 30 mm, the circular focus curve of the Paschen-Runge mount would lead to strong defocusing on the sensor edges or in the center of the sensor. Very compact optics can be built using the so-called "crossed Czerny-Turner mount" that employs a plane grating and two concave mirrors instead of a concave grating. These are particularly suitable for handheld instruments. Spectrometer optics for a narrow spectral range frequently use a folding of the light path to save space. Alternative grating arrays, like those according to Wadsworth, may be useful for such applications, as a good resolution for short focal lengths can be achieved here by minimizing image errors. Section 7.4 is dedicated to the optics of mobile spectrometers.

*– Electro-optical sensors*
Semiconductor arrays in line form predominate as sensors. Here, a larger number (usually between 2,000 and 5,000) of light-sensitive elements are lined up in a row. This makes it possible to simultaneously record entire wavelength ranges. Rows in charge coupled devices (CCDs) and complementary metal oxide semiconductor (CMOS) technologies are most frequently used. Photomultiplier tubes (PMT), which dominated until the turn of the millennium, are now only found in large stationary instruments. Their advantage is that the decay behavior of the spectral line is recorded at the end of the spark and the measurement can be limited to the time window most favorable for the detection sensitivity. An exit slit that passes only the

radiation for one wavelength (or to be more exact; a narrowly limited wavelength range) is positioned in front of the photomultiplier on the focal curve.

The surroundings of this range cannot be recorded because exit slits cannot be mounted in arbitrarily small distances. Section 3.6 deals with the electro-optical sensors used in modern arc and spark spectrometers.

– *Measurement electronics*

Instruments equipped with line sensors have logic units to clock the arrays, sometimes analog multiplexer for sensor array selection and A/D converters for conversion of the measurement data. Optics with photo multipliers require an integrator for each PMT as well as analog multiplexers that connect the integrator output with analog-digital converters. Miller integrators with reset circuits are normally used to collect the photomultiplier charges.

Usually, at least one microcontroller records the intensities and then transfers them afterward via interface to a superordinate computer that takes over further processing of the spectra. For state-of-the-art systems, each detector array has its own microcontroller. An additional microcontroller collects the data and organizes transfer to the superordinate computer.

Details about data acquisition are described in Section 3.7.

– *Superordinate computer and peripheries*

Section 3.8 deals with the superordinate computer, usually a PC with a standard operating system. It receives the raw spectra from the semiconductor arrays or the PMT raw intensities from the readout system, executes the spectrometer software and finally presents the results via monitor, printer or into a network. The microcontrollers of the readout system and excitation generator must be able to complete time-critical tasks while maintaining an exact time regime. The PC, however, provides high-level computing power and a user interface in the form of a standard operation system familiar to the operator. It cannot generally offer real-time capabilities.

– *Spectrometer software*

As late as the 1980s, only a few voltage values (one per spectral line) were transmitted to the computer at the end of a measurement. The voltages were picked up from the integration capacitors and converted to digital values. The software calculated the element concentrations from these digital values. At that time, the source program was often only a few thousand lines long. Even today, the calculations performed there are part of the spectrometer software. However, modern software performs additional tasks:

– Before each measurement phase, the microcontroller of the readout electronics and excitation generator are supplied with the parameters necessary for the measurement.

- At the end of the phase, or during single spark evaluation, after every spark, the scalar measurement values or complete spectra are received and stored or processed in real time.
- This is followed by calculation of the element concentrations for the single measurements as well as updating of averages, standard deviations and variation coefficients of the current series of measurements.
- If an error condition has been fulfilled, which would indicate, for example, an unusable sample, the current single measurement is interrupted and discarded and an error message is issued.
- After the series of measurements have been completed, the results are printed, stored in a database and automatically forwarded as programmed. Material specifications can be checked. For samples with unknown contents, a material identification can be conducted.
- In modern instruments, frequently full spectra or at least spectral regions of interest are recorded. In many cases, the software is also able to conduct a complete spectrum recalibration, thus restoring the instrument to its original state at the time of the first calibration.
- The software also enables comfortable control and extension of the calibration.
- There is usually a selection of several languages available for communication with the operator.

This list is not exhaustive but makes it plausible why modern spectrometer operation program can often encompass more than a million lines of source code. The key algorithms used in modern emission spectrometers are discussed in Section 3.9. Additional calculations based on the concentrations found are frequently conducted, especially in mobile spectrometers. Some of these algorithms, e.g., determination of a material grade from the analysis, are presented in Section 7.7.

- *Housing*
When listing the instrument components, the instrument housing must not be forgotten. It is, much like the software, an interface to the operator. Lack of ergonomics, fans, pumps or cooling units that are too loud, or poorly accessible operating components hinder efficient operation of the instrument. Besides, there are comprehensive safety regulations for laboratory instruments, many of which directly or indirectly affect the instrument housing. For Europe, the safety regulations are set out in the DIN EN 61010 [3] standard. A properly designed housing is required in order to comply with the standards for electromagnetic compatibility regarding interference emission and immunity. A declaration of conformity regarding compliance with the standards mentioned is a prerequisite for sales of an analytical instrument within the EU. Similar regulations apply in other regions of the world. It is particularly difficult for mobile spectrometers to achieve compliance with the standards. Section 7.5 examines this issue in detail. The same standards generally apply for laboratory spectrometers.

However, they are easier to meet due to the lower restrictions on space, weight, mobility and environmental conditions.

## 3.2 Excitation generators

The design and performance characteristics of the three most common types of excitation generators for the direct analysis of solid samples are discussed in this chapter.

The *arc generator* has the simplest design and its operating principle is also the easiest to understand. *Spark generators* can be understood from a technical point of view as an extension of the arc generator: The initial phase of a high-current arc is periodically repeated after a pause. The vast majority of the instruments for the direct analysis of metals are equipped with spark sources. Subsequently, *laser sources* are discussed. For decades, systems with laser excitation did not achieve more than experimental status. This has changed in recent years. There are now serially produced systems on the market, especially for simple analytical tasks where the advantages of the laser are important (remote measurement, exact spatial control of the ablation, no electrodes). Handheld instruments for the sorting of aluminum scrap are an example. These are discussed in detail in Section 7.1.4.

### 3.2.1 Arc excitation and arc generators

Electrical arcs are not only used for spectrometric purposes. In principle, carbon arc lamps were invented as early as 1802 by Humphry Davy. The electrodes are horizontally arranged in Davy's construction, which leads to the formation of an arc-shaped plasma due to thermal impetus between the electrodes [4]. The arc owes its name to this fact. Jean Bernard Léon Foucault improved Davy's construction by introducing an automatic regulation of the electrode gap. On page 509 in his book about the development of analytical chemistry [5], Knuth Ohls reported on correspondence between, on the one side, Gabriel Stokes and, on the other side, Kirchhoff and Bunsen after their groundbreaking publication dealing with atomic emission in flames, which led to the founding of spectrometry [6]. Stokes writes that Foucault had already made similar observations with the arc in 1849. The electrical arc first became more widespread in the 1870s, after development of viable power generators. It became a part of instrumental analysis only toward the end of the nineteenth century. In retrospect, the exact timepoint is difficult to determine. In 1875, the arc is briefly mentioned (p. 47) in Zech's textbook [7]; however, it is also noted that the handling is complicated and that the costs are high due to the 40–50 galvanic elements required by the state of technology at that time. In Landauer's work [8], on the other hand, it is called the most suitable excitation source for metals and described in detail. Landauer bases this on works by Liveing and Devar [9, 10] as well as by Kayser and Runge [11].

The timepoint at which the arc became associated with analytics is not surprising: High-current electric arcs provide a bright spectrum compared to the high-voltage spark used back then. They were hence suited to the Rowland gratings, which increasingly supplanted prisms as components for the decomposition of the light at the end of the nineteenth century. In principle, gratings require greater light exposure because the radiation is distributed over several diffraction orders. On the other hand, high resolution is easier to achieve with gratings than with prisms. This enables the meaningful use of an arc burning in air as the excitation source because wide ranges of the visible spectrum are heavily disturbed by molecular bands, such as the radical CN. Molecular bands are groups of spectral lines that have short, regular gaps and increase toward longer wavelengths, only to end abruptly. The gaps between the single lines are so narrow that they cannot always be resolved by smaller optical systems. The bands can still be recognized by the typical sawtooth structure. Figure 3.6 shows sections of the

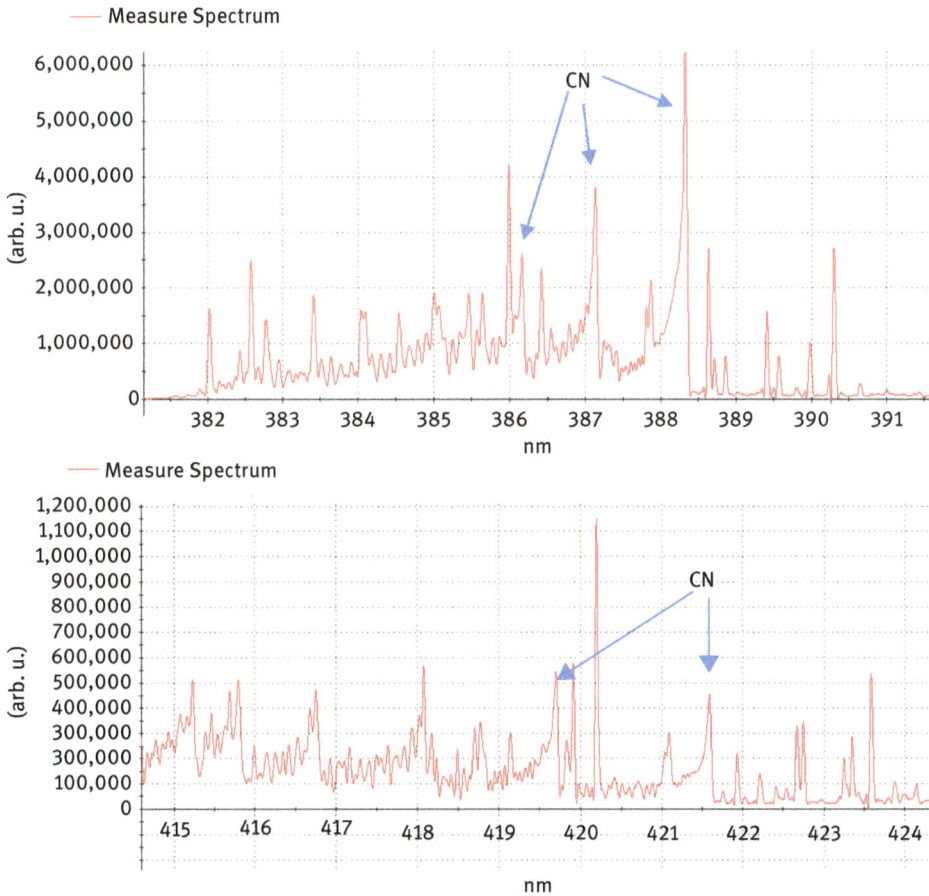

Figure 3.6: Section of the arc spectrum of a graphite sample with cyanide bands.

spectrum for a graphite sample. The strongest cyanide bands are labeled. The spectral lines that belong to the bands can only be separated from the neighboring analytical lines with high resolution. The cyanide bands are particularly strong when a carbon electrode is used.

### 3.2.1.1 The physics of the DC arc

The structure of the electrical arc is shown in Figure 3.7. The geometry is based on that used in modern mobile spectrometers: The arc burns the air between a flat sample and the pointed counter electrode, which usually consists of spectrally pure silver or electrolytic copper. The counter electrode is poled as the anode; the sample functions as the cathode. When the electrode is mentioned further in the text, then it will always mean the counter electrode, although, of course, the sample also has a function as an electrode. The chosen distance between the electrode and the sample surface (gap) is generally 1–2 mm.

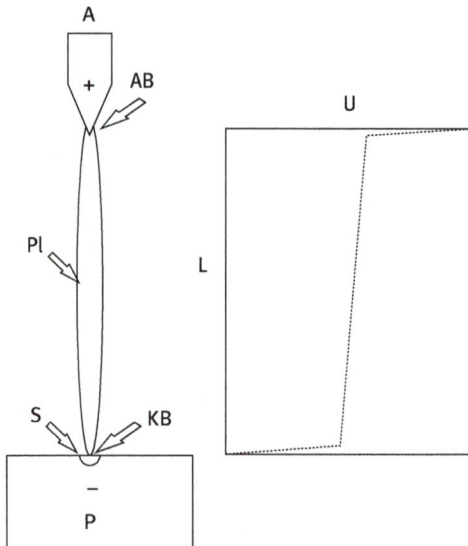

Figure 3.7: Structure of a DC arc.

Explanation of the symbols in Figure 3.7:

*left*

A    Counter electrode (anode)
P    Sample (cathode)
Pl   Plasma, (area between anode and cathode drop area),
     the plasma extends over almost the entire gap between the electrodes.
S    Melt at the arc base

A     Counter electrode (anode)

KB    Cathode burn spot (starting point for the arc on the sample), just above it is
      the cathode drop area as a thin film over the burn spot.

AB    Anode burn spot (starting surface on the tip of the electrode for the arc), the drop
      area for the anode is found under this as a thin film on the tip of the electrode.

*right*
Qualitative course of the voltage drop (U) over the distance (L) between the tip of
the counter electrode and the sample surface. In the drop areas, the voltage increase is even steeper than drawn here.

## The mechanisms of the ignition process

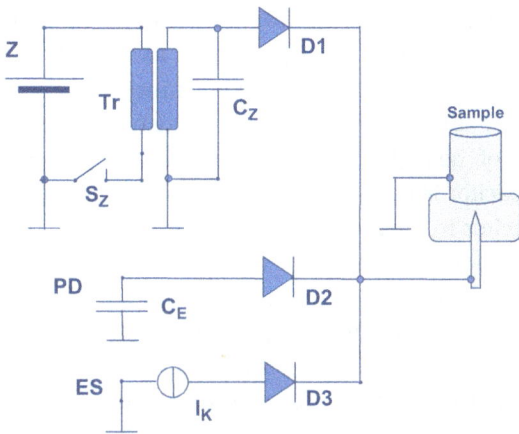

Figure 3.8: Equivalent circuit diagram of the DC arc.

Figure 3.8 shows an equivalent circuit for a DC arc with circuit components for ignition
(Z), plasma development (PD) and energy supply in the stationary arc phase (ES).

– Ignition of the arc occurs through contact or a high-voltage impulse.
  Here, ignition is to be understood as the achievement of a state in which the
  gap between the electrode and the sample becomes conductive. Modern instruments usually work with fixed electrode gaps and high-voltage ignition, which
  is why this type of ignition mechanism is considered here.
      Electron/ion pairs are constantly formed by ionizing radiation in the gap, i.e.,
  the gas-filled area between the sample and the tip of the electrode. Even in complete darkness, cosmic radiation leads to the formation of such charge carrier
  pairs. According to Demtröder [12], cosmic radiation causes an ion pair concentration of $10^6$ per liter. Küpfmüller and Kohn [13] cite $10^{-18}$ A as the magnitude of the
  current that is generated by cosmic-radiation-induced charge carrier pairs between

two surface electrodes each with a size of 1 cm² and being 1 cm apart from each other. If high voltage is applied to the electrodes, the charge carriers already present, are accelerated in the electrical field (see Figure 3.9). This is the case even if the high-voltage is only present for a short period of time. The accelerated particles collide with neutral atoms after traveling their free path lengths. If they have absorbed so much energy in this way that it is sufficient to ionize the gas particles with which they collide, then these are also transformed into charge carrier pairs, whereby the electrons produced are accelerated toward the electrode and the ions toward the sample. Thus, the number of charge carrier pairs increases with a snowball effect (see Figure 3.10).

The ignition voltage pulse is usually generated by interrupting the primary circuit of a high-voltage transformer (Tr in Figure 3.8). In most cases, a capacitor $C_Z$ of 100–300 pF, which transfers energy into the gap after the breakdown, is connected to the secondary side of the high-voltage transformer.

On the primary side of the transformer Tr, the electronic switch $S_Z$ is initially switched on to magnetize the high-voltage transformer. As soon as enough energy has been stored in the core, Tr is switched off. The voltage increases until there is either a breakthrough or the entire energy stored in the core fully charges the capacitor on the secondary side of the high-voltage transformer. This capacity is always available, even if no capacitor is connected. For example, the secondary winding forms a parasite capacitor to ground and the primary winding. The ignition voltage pulse should be able to reach a height of 10–15 kV for the gap geometries and materials mentioned at the beginning of this section.

The ignition voltage cannot be maintained for long, as it inevitably collapses under the strain of the continuously decreasing resistance of the gap. After $C_Z$ is emptied, the circuit for the ignition stops performing due to decoupling effected by the diode D1.

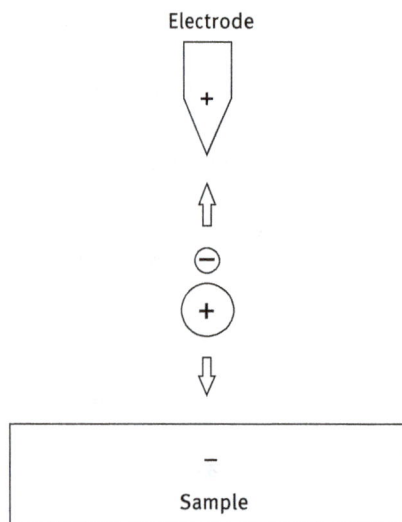

Figure 3.9: Generation and acceleration of charge carriers in the electric field between sample and electrode.

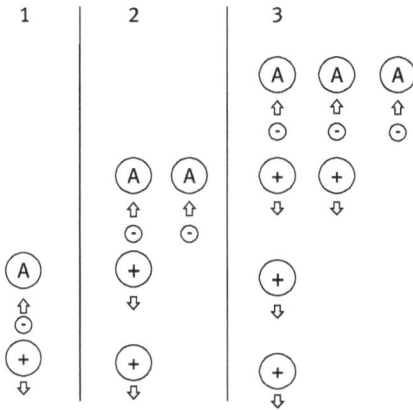

Figure 3.10: Duplication of charge carriers.

- If the ignition voltage is only present for even a few 100 ns, so many charge carrier pairs are generated that a voltage of 200 V between the electrodes is sufficient to maintain and even increase the charge carrier concentration in the gap. Even now, the plasma is not fully developed. However, the concentration of the charge carrier pairs is even lower than that in the stationary phase of the arc, i.e., it still has a relatively high resistance. The circuit for development of the plasma (PD in Figure 3.8) ensures that the arc is transferred to its stationary phase. To this end, additional energy must be brought into the gap. It is sufficient to apply a capacitor $C_E$ charged to 200 V with a capacity of one microfarad to the gap. The capacitor then discharges and the voltage falls quickly, but the gap becomes less resistant during the discharge process. After the discharge of $C_E$ a relatively low voltage between the electrodes is sufficient to prevent a decay of the charge carrier concentration.
- The circuit for the introduction of energy in the stationary arc phase (ES in Figure 3.8) consists, in addition to diode D3 for decoupling, of a constant current source $I_K$. There is a voltage across this current source. $U_{Qmax}$ is defined as the largest voltage drop that can be generated across $I_K$. To regulate a falling current, it must be able to rise a few 10 volts above the burn voltage of the discharge gap in the stationary arc state. In the initial phase of the arc, the current is lower than the set current; therefore, $U_{Qmax}$ drops over $I_K$. As soon as the voltage across $C_E$ falls below $U_{Qmax}$, $C_E$ loses its function due to the decoupling by D2 and stabilization of the arc by the current source $I_K$ becomes effective.
- The circuit for plasma development and the power source for maintaining the arc can be combined into one unit if $U_{Qmax}$ can become large enough. If the conditions mentioned at the beginning of this sections are assumed, $U_{Qmax}$ must be able to achieve a value of at least 200 V. In the past, this was easily realized by connecting a DC voltage $U_V$ of about 300 V to the gap via a resistor $R_V$. Choosing a voltage of about 300 V had practical reasons. The mains voltage (in Europe 220 - 240 V) was

decoupled by a separator tranformer, rectified and smoothed by a capacitor of a few 100 µF. In Figure 3.11 this circuit is represented by the voltage source $U_V$. The gap is high resistance at the beginning of the plasma development phase. It is assumed that it has a resistance R on the order of 300 ohm. If $R_{PE}$ = 130 ohm is selected and a breakdown voltage of 40 V in the stationary state is assumed, then at the beginning of the plasma development there is a voltage of (300 × 300)/(300 + 130) volts or 209 V over the gap. The plasma quickly becomes low resistance and the voltage drop over the gap is reduced to approximately 40 V. Then, an arc current of (300 − 40)/130 amps, i.e., 2 A flows. If, for whatever reason, there is a reduction in the charge carriers in the spark gap, their resistance increases and the voltage over the gap also increases, so that the original charge carrier concentration is quickly restored; there is a current readjustment. The disadvantage of this simple circuit as shown in Figure 3.11a is its poor efficiency: Based on the example data, a power of 80 W is introduced into the plasma; the losses in resistance, however, are 520 W. The efficiency is only about 13.3%, whereby additional losses have not yet been taken into account. The simple circuit according to Figure 3.11a is therefore out of the question for modern arc generators that frequently need to be supplied with power using batteries. Section 3.2.1.2 presents more modern concepts.

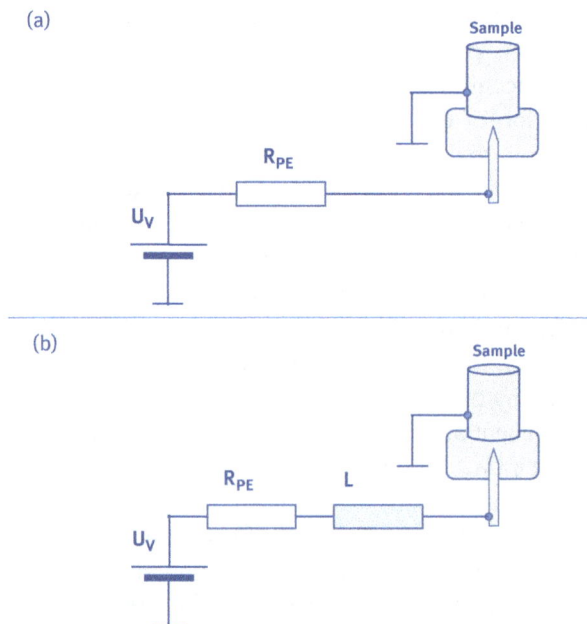

Figure 3.11: Simplest arc source.

### Electrical conditions of the arc in a stationary state

The burn voltage of the arc for steel, copper or nickel alloy samples using a counter electrode made of spectrally pure silver or electrolytical copper is on the order of 35–40 V, if the electrode gap is between 1.5 and 2 mm and the arc current is approximately 2 amps. In this context, burn voltage is the voltage that can be measured between electrode and sample while the arc is burning. It does not react strongly to changes in the arc current. This is easy to understand: With increasing current, more atoms are ionized and the power line is available to the charge carrier pairs generated. One should, therefore, not think that the arc plasma has a constant electrical resistance. In the early days of arc spectroscopy, it had already been noticed that the arc burned with more stability with an increasing input voltage ($U_V$ in Figure 3.11a) and an increasing resistance ($R_{PE}$ in Fig. 3.11a). If the input voltage is too low, then the arc can spontaneously extinguish, which is easy to explain: Suppose the arc is in a stable state and has the momentary resistance $R_P$. The differential resistance here is the ratio between the burn voltage and the arc current. If the arc becomes a little colder due to a disturbance, a draft for example, the plasma temperature decreases. Fewer electron/ion pairs are formed than recombined. The momentary resistance of the plasma $R_P'$ increases. The arc current I, that was previously $U_V/(R_{PE} + R_P)$, is reduced to $U_E/(R_{PE} + R_P')$. If $R_{PE}$ is much larger than $R_P'$, this is not really tragic. If, however, $R_P'$ is of the same order of magnitude as $R_{PE}$, then the current decreases and, for this reason, even fewer charge carrier pairs are formed. The resistance of the plasma continues to increase, which once again reduces the current. The arc extinguishes within milliseconds. If inductance L is placed in series to the ohmic resistance (see Figure 3.11b), the voltage $U_V$ can be reduced to about twice the burn voltage. Inductances are known to resist sudden changes in current; the following relationship applies:

$$U_L = L*(dI/dt). \tag{3.1}$$

This means that if the arc current decreases by dI in the time dt, a voltage of $U_L$, which is added to the input voltage $U_E$, drops through the inductance. The higher overall voltage compensates for the increase in the resistance of the plasma and stabilizes the formation rate of the electron/ion pairs.

The assumption of a stable arc state is certainly hypothetical and the mechanism described above works for a free burning arc only insofar as that it is prevented from being extinguished. It is simple to maintain a stable arc current, but this does not apply to the burn voltage. It is frequently observed that these remain for a second in a voltage range with the width of about one volt. Then the voltage can break out upward or downward and either remain stable at this new level or quickly switch between higher and lower levels. These changes in the arc radiation can frequently be seen with the naked eye: Switching between bluish and yellowish arc radiation can be observed in irregular time periods on the magnitude of a few seconds. This switching can also be heard: The hissing that accompanies the arc can suddenly change.

The difference between the arc and the glow discharge is the high cathode temperature. For the glow discharge, relatively few electrons are released from the cathode by the Auger effect and accelerated toward the positively charged anode. The plasma remains high resistance, so that voltage drops of several hundred volts occur between the electrodes. In contrast, the electron emission from the arc cathode is due to the high temperatures prevailing at the contact point on the sample, which functions as the cathode. The plasma temperature is able to create enough heat at the arc burn spot only with high melting point samples such as tungsten, molybdenum, tantalum and graphite. For materials with lower vaporization temperatures, the vaporization cools the arc's contact point. The energy input from the plasma alone is no longer enough to generate a sufficiently high temperature. There is, however, an effect that ensures a sufficiently high temperature in the arc burn spot: There is a very high field strength directly above the cathode. It's easy to see. It was explained above that the number of charge carrier pairs multiplies with a snowball effect. Then, the ions are accelerated toward the cathode; the electrons toward the anode. Directly above the cathode, a comparatively small number of thermally emitted electrons is in contrast to the very large number of all the ions in the plasma that have been generated and not recombined. The zone with a positive space charge is known as the cathode drop area. It is thin, but a large part of the burn voltage is dropped here.

In order to become an idea of the dimensions of the cathode drop area, it is worthwhile to concern oneself with works dealing with the physics of electrical contacts. Here, the understanding of arc occurrence is especially important as the arc can develop when the distance between the contacts is increased and interruption of the circuit is prevented, which can lead to erosion of the electrodes – it is therefore a disturbing effect. Holm [14] develops a physical model and performs a sample calculation for a 20-amp arc between silver electrodes and compares it with experimental data. Holm determines a gap of only $4.8 \times 10^{-8}$ m between cathode surface and center of the cathode drop area. The current density at the arc contact surface on the electrode is determined to be $0.5 \times 10^8$, which leads to an area of only $0.02$ mm$^2$ for the given current strength at the contact surface, i.e., burn spot. Surface limitation of the cathode burn spot to a very small area follows from Holm's model and coincides with the experimental data.

The potential difference at the cathode, i.e., the cathode drop, is just a few volts, but because it is applied over a small distance on an order of $10^{-8}$–$10^{-7}$ m, the field strength is very large. For the example of the 20-amp arc, the thickness of the cathode drop area layer has an order of magnitude of the mean free path for the electrons. The free path for ions is shorter due to their larger cross sections. The ions are strongly accelerated above the cathode just before they hit there. This greatly heats up the cathode. Holm calculated a surface temperature on the order of 4,000 K, which is in line with the experimental data.

As the distance from the cathode surface increases, the excess of ions decreases and the number of ions and electrons is balanced. This is the beginning of the area of the plasma that is also occasionally referred to as the positive column of the arc. The electrons are accelerated in the electrical field, and they can collide with neutral atoms. If the electrons have a sufficiently high energy, then excitation or ionization of the given atom occurs. Particles that are already ionized can be excited or even doubly ionized. Excitation or ionization by collisions between positive ions and neutral particles can also take place. However, because of the smaller free path of the ions, the electrons are mainly responsible for ionization and excitation.

The relationships that cause line spectra to appear are discussed in detail in Chapter 2 of this book. Figure 3.14 serves as a reminder. The energy levels to which the atoms of an element can be brought by collision are indicated by dotted horizontal lines. Only exactly defined energy levels, represented by dashed lines, can be reached. Being raised from one level to a higher level is also possible. After a short time, the excitation level is vacated by either returning to the ground state or to an energetically lower excitation level (indicated by dotted lines in Figure 3.14). A photon is emitted, the wavelength of which is given directly by the difference in the energy levels. The energy levels are usually recorded in the unit electron volts (eV). An eV is the kinetic energy that an electron absorbs through an electrical field with an acceleration voltage of one volt. This is a very small energy of about $1.6 \times 10^{-19}$ J. When dropping to a lower excitation level, the differences in energy levels $\Delta E$ (expressed in eV) can be directly converted to the photon wavelengths $\lambda$ emitted. The equation below applies:

$$\lambda = (h^{\star}c)/\Delta E. \tag{3.2}$$

Here, h represents Planck's quantum of action and c the speed of light (see also eq. (2.10) and (2.11)).

In addition to excitation by collision, excitation can result from the absorption of a matching photon. This effect is the subject of further discussion below about the self-reversal of lines.

If the collision supplies the atom with energy above the upper dashed line, it is sufficient to remove an electron from the ion. The associated energy is called the "ionization energy." A scheme similar to that in Figure 3.14 applies to the newly created ion; excitation and, as long as there are still electrons in the ion, a higher ionization state are possible.

Like at the cathode, there is again an increased field strength at the anode due to the prevailing surplus of electrons. All the electrons resulting from impact ionization arrive here unless they have previously been eliminated by recombination with ions. The effect is like that at the cathode, but with the opposite sign. The area with a negative space charge is called the "anode drop area." The anode is also heated by bombardment with electrons that are accelerated by the increased field strength above the anode. The energy input at the counter electrode is considerable. There is also a surface limitation of the anode burn spot. However, Holm ([14], p. 305) points out that the anode burn spot becomes larger and the cooling is improved as the electrode gap increases. A sufficiently wide gap can therefore minimize material degradation from the counter electrode.

The voltage curve in the plasma is shown on the right side of Figure 3.7. The change in voltage is largest at the electrodes in the cathode drop area (lowest segment of dotted line) and the anode drop area (highest segment of dotted line). The voltage drop in the area of the plasma (middle segment of dotted line) is comparatively small.

### 3.2.1.2 Construction of arc generators

Figure 3.11a is a schematic diagram of a very simple arc generator; the arc is ignited when the counter electrode comes into contact with the sample. By triggering a spring or electromechanically using a lifting solenoid, the electrode is drawn away from the sample surface, thus "pulling" an arc. The electrode tip typically remains in its final position 1.5 mm above the sample surface. A plasma is formed and persists until $U_V$ is switched off. For the reasons explained in section 3.2.1.1, the sample is poled as the cathode as shown in Figure 3.8 and 3.11, i.e., it is connected to the minus pole of the voltage supply.

The advantage to arc generators with contact ignition is that no high-voltage ignition pulse is required to generate the plasma. This can be extremely beneficial for handheld instruments in which the plasma is located close to sensitive electronic components. Figure 3.12a shows an example of a handheld instrument with a contact ignition.

In modern arc generators, however, fixed gaps between the sample surface and the counter electrode, as described in the previous section and shown in Figure 3.8, are more common for reasons of easier maintainability. Here, instead of by contact, the ignition is performed using a superimposed voltage that ionizes the gap between the tip of the electrode and the sample surface.

In Section 3.2.1.1, it was already calculated that the simple stabilization of an arc by using a voltage supply with several 100 V in combination with a power resistor is unfavorable because of the low efficiency.

For this reason, switching regulators are used to set the required current by which most of the ohmic losses are avoided. Figure 3.13a shows the principle: The electronic switch S, usually a MOSFET or IGBT, connects the plus pole of the voltage supply U with a coil L; the inductivity of which is known to prevent sudden changes

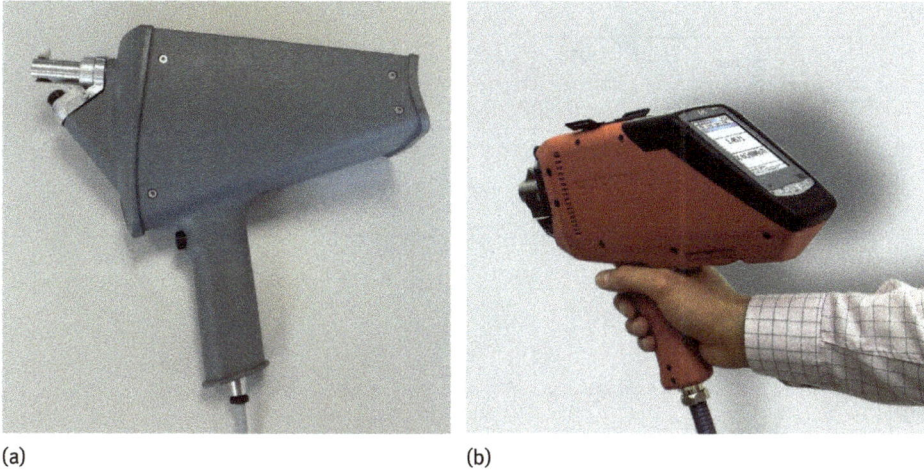

(a)                                                    (b)

Figure 3.12: left or (a): SpectroSort handheld arc spectrometer with contact ignition; right or (b): Spectro iSort handheld arc spectrometer with high-voltage ignition. Printed with the friendly permission of SPECTRO Analytical Instruments GmbH, Boschstr. 10, 47533 Kleve, Germany.

in current. The increase in current follows eq. (3.3) and reaches the target arc current with a delay. If the target arc current upper limit is reached, S is switched off.

$$I(t) = \frac{U}{R} \times \left(1 - e^{-(t*R)/L}\right), \tag{3.3}$$

Whereby
I(t) Function of the current depending on the time since closing the switch S (A)
U   Supply voltage (V)
R   Sum of the ohmic resistances in the circuit including the plasma resistance (Ohm)
L   Sum of the inductivity in the circuit, essentially the discrete inductivity in the circuit diagram Fig. 3.13 (in units of H)

Since the inductivity L resists sudden changes in the current, the current flow runs through L and the diode D4 (dashed arrow). The energy stored in the coil's magnetic field dissipates and the current drops slowly. When it falls below a lower limit, S is switched on and the cycle repeats itself. Figure 3.13b shows the current flow. At the times indicated by upright solid lines, the switch S is closed; it is opened again when the dashed lines are reached.

The circuit principle corresponds to that of the secondary step-down converter from the switching power supply technology. Details concerning the calculation can be found, for example, in Tietze/Schenk ([15], p. 944 et seq.).

Today, arc generators are found mainly in mobile spectrometers. The arc burns in ambient air or, if the element carbon is to be determined, in air from which $CO_2$

(a)

(b)

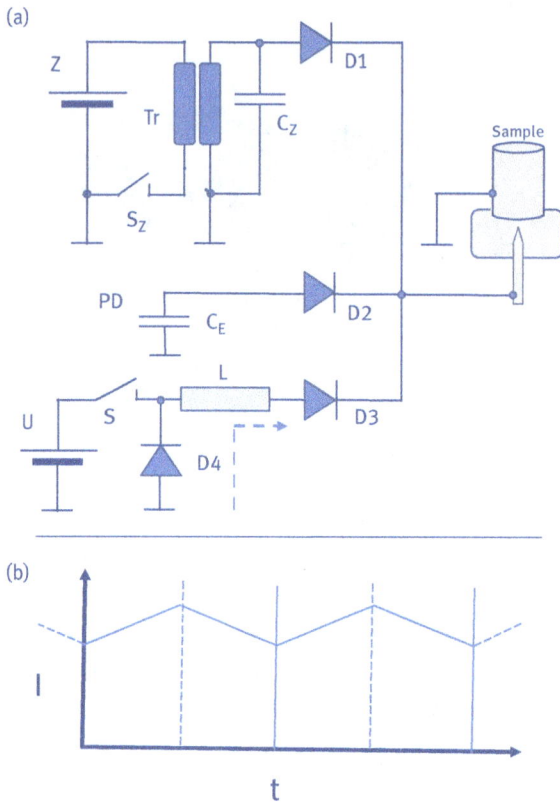

Figure 3.13: Arc source with switch mode current regulator.

has been removed. Several hundred such systems are manufactured worldwide every year. Their design and characteristics are discussed in detail in Chapter 7.

Laboratory spectrometers with arc excitation are relatively rare. These are used only for special applications such as the determination of impurities in pure copper or graphite. The authors estimate that a small two-digit number of these systems are manufactured worldwide each year.

In contrast, a large number of systems in the lower three-digit range are produced for the analysis of oil. The oil to be analyzed is transported by a small graphite wheel from a container to the counter electrode, which is also made of graphite. These systems do not work with DC arcs, but with interrupted arcs, by which the arc only burns for a maximum of a few milliseconds and then it is interrupted. They are actually spark devices which belong in Section 3.2.2. The associated stand and its operation are described in Section 3.3.3.

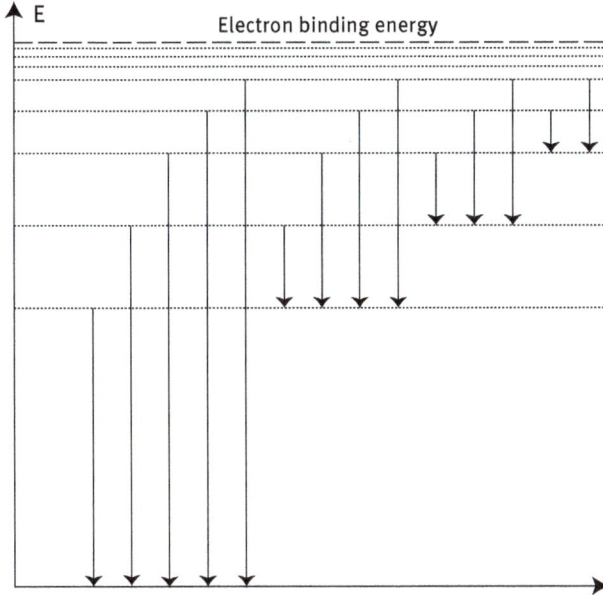

Figure 3.14: Energy levels for excitation and ionization.

Meaning of the symbols and signs in Fig. 3.14:

E is the energy axis.

The dotted lines indicate which energetic states are possible for an excited atom after emission of a photon.

### 3.2.1.3 Characteristics of the electric arc as an excitation source

If the direct current arc is to be used as the excitation source, it is worth evaluating its analytical properties. This makes it possible to judge whether the arc is suited to a particular task.

#### Character of the generated spectrum

Although the lines from ions can also be found in the arc spectrum, atomic lines dominate. This dominance is so pronounced that the term "arc lines" is used as a synonym for atomic lines.

#### Precision

As mentioned above, even small irregularities in the measurement conditions lead to changes in the position of the plasma, to changes in the gap resistance and thus to temperature fluctuations. Relative reproducibility for intensities of about 5% to

10% can be achieved. To improve the repeatability, the intensity of an analytical line is divided by the intensity of a line for the main element, e.g., the intensity of an Fe line in iron base. This reference line is called an "internal standard." Precisions of about 1 to 5% can be achieved for these intensity ratios, i.e., the quotients from analytical line and appropriate internal standard.

### Detection limits

On the one hand, arc generators provide a low background spectrum; on the other hand, the precision is much worse than that for spark. The detection limits achievable with direct current arcs are, therefore, not necessarily better than those that can be obtained with sparks. This becomes clear when comparing the typical detection limits for the spark mode for a mobile spectrometer (Table 7.6) with those in arc mode for instruments with similar optical layouts (Table 7.7).

### Shape and scattering of calibration curves

The calibration function is often unsatisfactory. Individual samples deviate from the calibration curves. The scattering of the calibration curves is significantly worse than that for spark excitation. Relative deviation between the measured and true values of 10% and more are not uncommon. The cause for interelement interferences and also the moderate success in calibrating for arc excitation is the fact that material ablation is essentially achieved through ions that are accelerated in the electrical field in the cathode drop area and then they hit the surface of the sample. These are argon ions when working with spark, at least when the discharge is short-lived. In contrast, metal ions are accelerated with arc. With arc, the light nitrogen and oxygen ions only play a role directly after ignition. The composition of the analyzed sample influences the ablation process, the plasma temperature, the ionization rate and, thus, the measured intensities measured for analyte and associated internal standard. Modern spectrometer software make it possible to take interelement interferences into consideration. The algorithms are discussed in Section 3.9. However, the influence of the sample composition on ablation, ionization and excitation is very complex.

This is illustrated in an example:
The higher concentration of a third element may lead to an increased energy input and, thus, to stronger material ablation, which tends to increase the signals for the analyte and the internal standard. The plasma temperature can, however, decrease anyway, because a higher proportion of the total energy available is required for material vaporization. This is an effect that can lead to signal reduction. If the plasma temperature falls, the possible partial ionization of the atoms for the analyte and internal standard can be reduced. There are, then, more atoms available for excitation, which in turn may cause an increase in intensity of atom lines.

The common algorithms for correction of interelement interferences assume simple, linear interactions. Their use offers only marginal improvement to the scattering of the calibration curves.

The *self-absorption* and *distillation effects* are two additional effects that complicate arc calibration. If a set of standards that contain an element with increasing concentration is measured, the following effect, that results in calibration curves like that shown in Figure 3.15, may be observed for several lines: At first, the intensities increase to approximately the same extent as the concentration increases, then the increase in intensities slows down until, despite growing concentration, the intensities no longer increase; finally the intensities even decrease. This effect can be observed for many sensitive lines, e.g., the copper lines at 324.7 and 327.3 nm. Figure 3.16 explains the effect. Radiation from the arc plasma P that lies within the spatial angel α can be measured by the spectrometer optic. α is drawn as an angle in Figure 3.16 but has the form of a cone in three-dimensional space. If the arc is ignited, then after a short amount of time, a layer forms around the plasma consisting of such atoms as can be found in the plasma. The photons from the interior of the plasma encounter atoms from this outer layer, excite them and release the radiation after a short time. Emission occurs in all spatial directions and not just within the cone visible to the optic. So, the major portion of the radiation is lost. In Figure 3.16, only the radiation represented by a bold solid arrow enters the optic. The probability that the described effect occurs grows with increasing analyte concentration. The radiation entering the optic no longer increases proportionally with increasing concentration and, in end effect, even decreases.

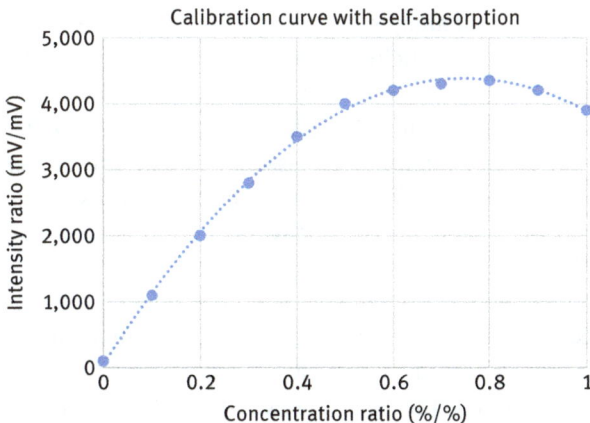

Figure 3.15: Calibration curve in the case of self-absorption.

Electrode

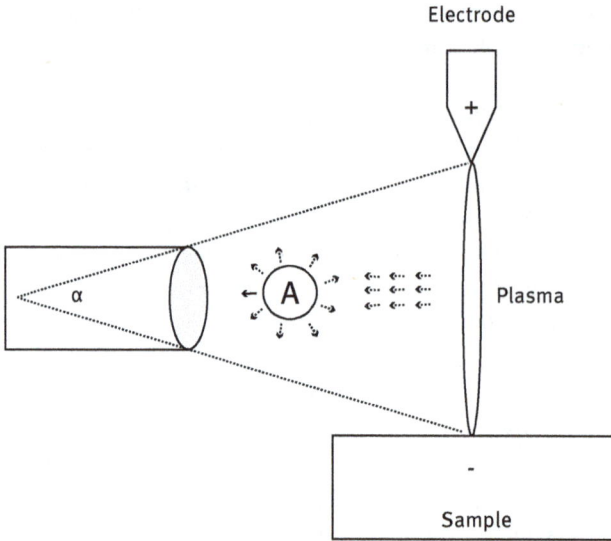

Plasma

Sample

**Figure 3.16:** Absorption and reemission as the cause of self-absorption.

The *distillation effect* occurs because the elements with the lowest boiling points are vaporized first in the melt at the arc base. The elements with higher melting points remain in the melt and the concentrations grow with the length of the arc burning duration; the calibration functions change with the length of the measuring time.

Due to the complications for calibration with arc as discussed above, use of so-called "fingerprint algorithms" makes sense. With these algorithms, a large number of materials with varying contents are stored together with their spectra and elemental concentrations. In addition, a set of analyte and internal standard lines with the associated calibration functions are included for each master sample. A spectrum is recorded for the sample to be determined and compared to the stored spectra. The master sample with the best match is determined. Then, the elemental contents are interpolated using the calibration functions associated with the master sample. First-degree polynomials generally suffice as calibration functions, since the master sample and the material to be determined have similar compositions. Section 7.7.2 details the fingerprint algorithm.

Despite all the obstacles described above, it should be noted that when the analysis of materials with similar concentrations is concerned, determination with arc mode leads to the most effective results. This is illustrated in Table 7.9, where the analyses of several low alloy steel samples are listed together with the expected values. It can be seen that even the results for austenitic steels and high-speed steels are usable.

### Measuring time

The required measuring times are short. A single measurement typically lasts three seconds. Simple testing applications can be conducted with measuring times of under one second. This is the case, for example, when sorting materials of clearly distinguishable compositions.

### Sample preparation

Measurements are even possible on dirty or slightly corroded surfaces. The arc makes contact at a point on the surface sample and burns through most existing contamination.

### Instrumentation

Arc generators are simply constructed and therefore inexpensive to manufacture. As already mentioned, air is normally used as the excitation atmosphere for mobile spectrometers with arc. There are no costs for an inert gas. There is also no need to carry a compressed gas tank to the measurement site (a 10 l argon cylinder weighs more than some portable spectrometers).

### Handling

Working with a direct current arc requires care. The counter electrode becomes very hot in an air atmosphere. This leads to the so-called "memory effect:" If a sample with a high concentration of an element E is measured, an alloy consisting of E and the electrode material forms on the counter electrode. In subsequent measurements, there is a signal for E even when the samples do not contain it. The counter electrode must be either ground or replaced. The thermal load also leads to burning-off of the counter electrode. The gap must be regularly reset and the tip sharpened to maintain constant parameters for electrode form and electrode gap.

Section 7.8.3 describes numerous measurement applications that can be conducted with mobile spectrometers in arc mode.

### 3.2.2 Spark excitation and spark generators

Most spectrometer systems used today for the analysis of metals utilize spark generators instead of the arc generators described above. The arc was discussed before the spark because each individual spark can be thought of as the initial phase of an arc as explained in the previous section.

Spark generators were used very early in spectrometry. The year 1859 is considered to be the year in which spectral analysis was born due to the groundbreaking publication by Kirchhoff and Bunsen. Read about developments in the first years in Görlich [16] and Junkes [17]. Both books were written on the occasion of the 100th birthday of spectrometry. Görlich reports that in the academy communication from December 11, 1859, Emil Du

Bois-Reymond wrote that lines could be seen in a spark spectrum that "are dependent on the nature of the metals between which the spark jumps." In the early days, however, the reliability and reproducible operation were problematic. Practice-oriented methods for quantitative spectral analysis were presented for the first time in the groundbreaking work *Die chemische Spektralanalyse* (Chemical Spectral Analysis) by Walther Gerlach and Eugen Schweitzer [18]. These methods use the spark as the excitation source. The textbook discusses aspects of the spark generator hardware in detail (Chapter 3, p. 25 et seq.).

Feussner's spark generator [19, 20], relieved analysts of solving electrotechnical problems. For a long time, wireless communications were based on transmitters that used spark gaps as a way to generate high-frequency oscillations. So, when in about 1930, a reliable spark generator was required for quantitative analysis, it was borrowed from radio technology. Long after Feussner's generator was introduced, there was still discussion as to the exact way this circuit works. It was only years later, after it was possible to observe voltage flows with oscillators, that it was fully understood. A depiction of its function was published by Heinrich Kaiser in 1938 [21]. By then, Otto Feussner had already died. Therefore, it makes sense to explain the function of this generator with an equivalent circuit diagram. The original circuit diagram does not necessarily show how it works.

### 3.2.2.1 The physics of the spark

Figure 3.17 shows the schematic structure of Feussner's high-voltage spark generator.

The high-voltage transformer Tr charges the capacitor C to a voltage $U_C$ of several kilovolts. A rotating spark gap $GAP_{AUX}$ periodically minimizes the auxiliary air gaps between the output of resistor R and the electrode. The minimum overall auxiliary gap distances are reached, when the rotating electrode is in an upright position (Figure 3.17). The movement of the rotary electrode is synchronized to the charge process of C in a way, that C is fully charged just before the minimum auxiliary gap distances are reached.

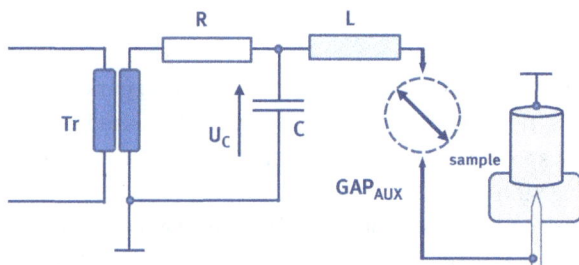

Figure 3.17: Spark generator according to Feussner.

As soon as the rotary electrode approaches the upright position, there is a breakthrough of the auxiliary and analytical spark gaps. The capacitor C and the coil L form an oscillating circuit dampened by the spark gap. Eq. 3.4 indicates the duration t of such a period of oscillation.

The peak voltage at C is reduced from period to period so that the discharge extinguishes after a few periods, as soon as $U_C$ falls below the spark gap burn voltage.

The maximum value of the discharge current during the first discharge period is achieved after:

$$t = 2 \times \text{pi} \times \sqrt{\frac{C}{L}} \qquad (3.4)$$

The peak current is:

$$I_{max} = (U_c - U_b) \times \sqrt{\frac{C}{L}}. \qquad (3.5)$$

Table 3.1 provides a numerical example with realistic dimensioning.

**Table 3.1:** Dimensioning example for a high-voltage spark source.

| | | |
|---|---|---|
| Capacity | C | 10,000 pF |
| Inductance | L | 10 μH |
| Voltage of C before spark ignition | $U_{CO}$ | 20,000 V |
| Sum of all burn voltages (analytical and auxiliary gaps) | $U_B$ | 200 V |

In the example above, the time between breakthrough of the spark gaps, i.e., formation of a low-resistance ionized plasma channel and reaching the maximum discharge current is less than 0.5 μs. The mechanisms that lead to breakthrough and plasma development have already been described in connection with the electrical arc (see Section 3.2.1). The stationary state reached after a few milliseconds by the arc is not achieved by the spark. Current flows only through the ionized discharge thread; its densities are up to 1,000 A mm$^{-2}$. Kipsch [22] and de Galan [23] report peak temperatures of up to 40,000 K.

As with the arc, a pointed heating zone is formed on the cathode end of the discharge channel, where the boiling point of the cathode material is reached. The sample material vaporizes at these spots, leaving craters with a diameter from 20 to 40 μm [24]. A cloud of vaporized metals forms around the impact point. Due to the high temperature, the cloud expands away from the crater in all directions of the discharge atmosphere with speeds of up to 1,000 m s$^{-1}$ [24]. Thus, the major portion of the excitation of the sample material takes place outside of the current conducting discharge channel. Figure 3.18 schematically shows the explosive expansion of the metal vapor cloud during one single spark.

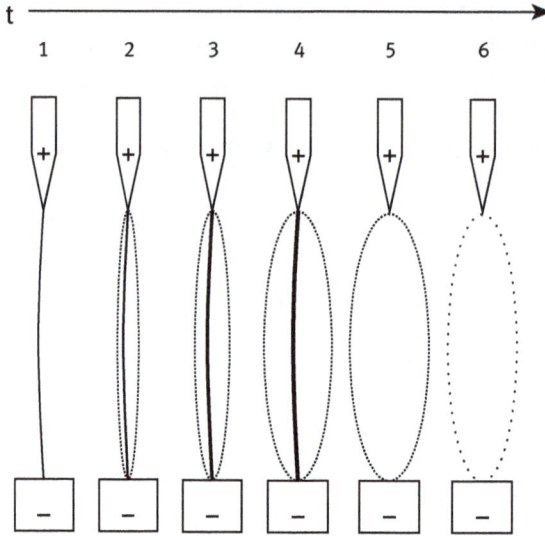

Figure 3.18: Expansion of the metal vapor cloud after spark impact.

Meanings of the terms in Figure 3.18:

t:      Timeline

t1:     Ignition

t2–t4: Snapshots during the current conducting phase. Dotted lines symbolize the area, where radiation is emitted. The conductive channel is represented by the solid line.

t5:     Radiation emitting area immediately after the end of the current flow

t6:     Snapshot of several tens of microseconds after the end of the conductive phase.

Depending on the spark energy and the discharge atmosphere, emission of radiation can be detected more than 100 μs after the end of the current flow. However, the intensity of the emission decreases.

With spark excitation, even elements with high boiling points are vaporized in the (ion) impact crater; therefore, distillation effects like those in arc do not occur. The pressure caused by the high temperatures at the beginning of a spark discharge increases the cross-section of the discharge channel explosively. As a result, the peak temperature decreases; the conditions become similar to those of the direct current arc. However, the spark discharge does not last long enough to form a temporally constant temperature distribution over the cross-section of the current conducting plasma channel. If the discharge is very short, the excitation process itself occurs mainly after the electrical discharge has been completed [25]. There is a high electron density in the still-narrow plasma channel in the first phase of the spark discharge; it results in a strong continuum due to electron bremsstrahlung.

As shown in Section 2.8 of this book, spectral lines have a finite, albeit small, minimum width on the order the magnitude of $10^{-2}$ pm (see also Skoog/Leary [26], p. 219). The spark discharge's high pressure and the high temperatures lead to broadening of the line profiles. Doppler broadening and pressure broadening are the two most important effects to observe.

Doppler broadening arises due to the fact that the atoms in the metal vapor cloud (see Figure 3.19) move toward the observation point or away from the observation point (window W in Figure 3.41). The full width at half maximum $\Delta v$ of the broadening can be calculated with eq. (3.6). The intensity distribution results in a Gaussian profile around the center frequency $v$.

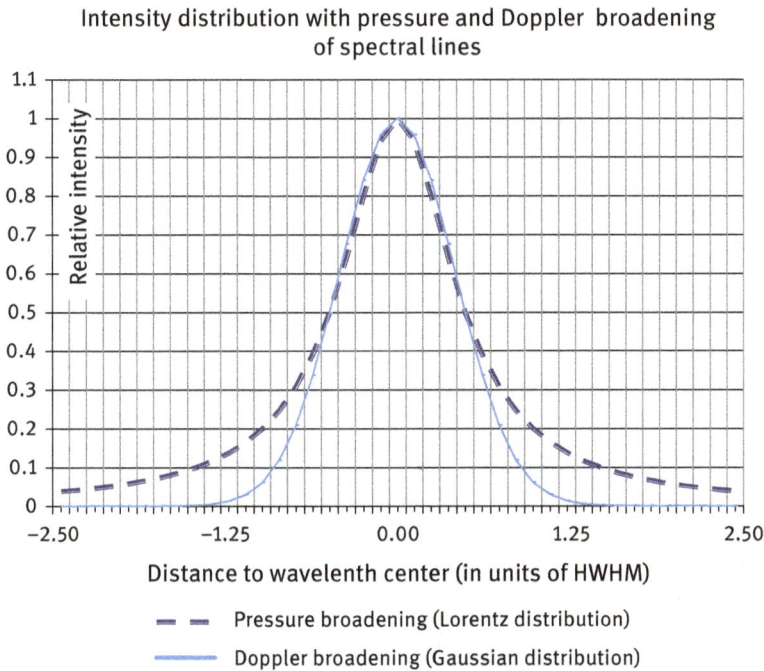

Figure 3.19: Intensity distribution with pressure and Doppler broadening.

$$\Delta v_D = \frac{2^* v}{c} \sqrt{\frac{2^* k^* T^* \ln(2)}{M^* u}} \tag{3.6}$$

Meanings of the terms:

$\Delta v_D$ Full width at half maximum of the line broadening due to the Doppler effect $(s^{-1})$

c  Speed of light (m/s)
k  Boltzmann constant ($1.380658 \times 10^{-23}$ J K$^{-1}$)
T  Absolute temperature (K)
v  Center frequency of the spectral line (s$^{-1}$)
M  Relative atomic mass
u  Atomic mass unit ($1.6605402 \times 10^{-27}$ kg)

Derivation of this relationship can be found in Kneubühl [27] (p. 62 f).

Equation (3.6) shows that the Doppler effect is particularly pronounced for high temperatures. The spectral lines for light elements are more strongly broadened than those belonging to elements with large atomic masses.

Line broadening occurs on the one hand due to the Doppler effect and on the other hand because the particles collide in the plasma. This leads to a shift of the energy level at the ground state. Kneubühl [27], p. 60 et seq. derives a formula for the approximate calculation of pressure broadening.

$$v_P \approx \sqrt{\frac{3}{4 \times M \times u \times k \times T}} \times d^2 \times p \tag{3.7}$$

Whereby:
$\Delta v_P$ Full width at half maximum of the pressure line broadening (s$^{-1}$)
c    Speed of light (m/s)
k    Boltzmann constant ($1.380658 \times 10^{-23}$ JK$^{-1}$)
T    Absolute temperature (K)
M    Relative Atomic mass
u    Atomic mass unit ($1.6605402 \times 10^{-27}$ kg)
p    Pressure (Pa)
d    Diameter of the colliding particles (m)

Pressure broadening leads to a widening of the profile according to the Lorentz function. The flanks of the Lorentz function fall off more slowly than those for a Gauss function with the same full width at half maximum. The differences in the forms of the line profiles are sketched in Figure 3.19. Pressure broadening tends to influence wavelengths that are further away more strongly than Doppler broadening with the same full width at half maximum.

Pressure broadening is relevant in the first phase of the spark discharge when the current density and temperature are at a maximum. The pressure is then high. As the discharge progresses, the plasma expands explosively, which leads to a strongly dropping pressure. Emission phenomena can still be measured even when the discharge current no longer flows and pressure and temperature have dropped to a great extent. The lines then have a minimal width.

Time-resolved spark spectrometry makes use of this circumstance. Integration does not take place throughout the entire discharge, but only during a window in time in which the influence of neighboring interfering lines and background radiation caused by bremsstrahlung is excluded. The beginning of the discharge is ignored because then only (collision broadened) spectral lines from the discharge atmosphere and background radiation (see Section 2.5.2) are produced.

Of course, eqs. (3.6) and (3.7) can also be applied to direct current arcs. The temperatures there are, however, lower and there is not such a high pressure in the plasma as for spark. For these reasons, arc spectra have narrower lines than non-time-resolved spark spectra.

An exact description of the high-voltage discharge can be found, for example, in Kaiser and Walraff [28]. Detailed theoretical contemplation about the processes in arc and spark were published by Weizel and Rompe [29].

### 3.2.2.2 Construction of a modern spark generator

To combine the advantages of arc and spark excitation, modern emission spectrometers are equipped with a hybrid form, the so-called externally ignited medium voltage spark generator. The current sources $I_E$ and $I_Z$ are switched on and off clocked by the spark repetition frequency (Figure 3.20a). As in the arc generator, circuit components for the ignition (Z), plasma development (PD) and for formation of the current flow (ES) are present. Assuming suitable dimensioning, it is possible to conduct plasma development and current flow formation using the same hardware components. However, in addition to better energy efficiency, separation has the advantage that the voltage drop through the power source for pulse formation (ES) can be limited to a low value, which can be beneficial from a safety point of view. Timing and proportional currents are given in Figure 3.20b. In Figure 3.20b ES produces rectangular current characteristics. Other shapes can be generated here. The representation on the timeline is not to scale. With a pulse duration of 100 µs and a spark frequency of 400 Hz, the pauses are 24 times as long as the pulse.

C-L-R networks were used in the past to obtain a defined discharge current flow (Figure 3.21). By briefly closing switch S, the capacitor C is charged before each spark. Then a high-voltage pulse is generated using $S_Z$, thus initiating the discharge, whereby the resistance R, the capacitor C and the coil L form the discharge current curve. The diode D3 switched in the blocking direction over capacitor C prevents oscillation and thus material degradation of the counter electrode. D1 and D2 serve to prevent the ignition voltage from flowing off over the power circuit.

Today, it is more common to build power sources with semiconductor components that enable power sequences in any shape and frequency. This makes it

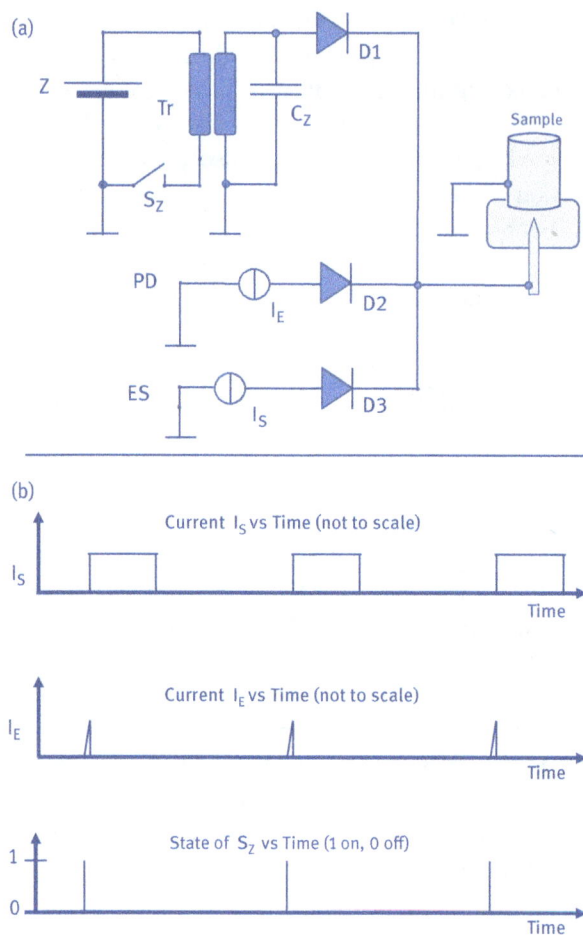

Figure 3.20: Block diagram of a medium-voltage spark source.

possible to produce on the one hand short, high-current "spark-like" and on the other hand long-lasting, low-current "arc-like" discharges using a medium-voltage spark source. The switching principle was already presented (see Figure 3.13) and explained in Section 3.2.1.2, which deals with arc generators. This switching circuit makes it possible to create every desired spark shape. The spark is completed when S is not closed again after falling below the target current. The switch S remains open in the intervals between two sparks. With suitable dimensioning, the circuit according to Figure 3.13a enables arc and spark operation.

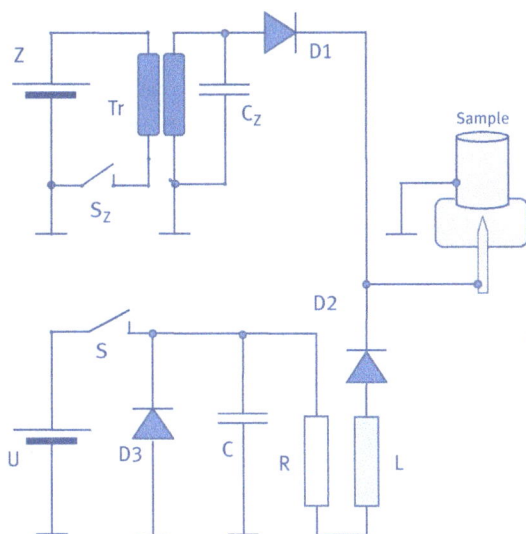

Figure 3.21: Medium-voltage spark generator with C-L-R pulse shaping.

### 3.2.2.3 Characteristics of the spark as an excitation source

A great advantage to the spark is the fact that the individual sparks represent statistically independent events.

When intensities resulting from successive sparks for single spectral lines are compared, there are large deviations from spark to spark. Coefficients of variation of 40% are not unusual here. These fluctuations are caused by the fact that even on a perfectly homogenous sample it is difficult to achieve similar plasma conditions for successive discharges. Also, metallic samples almost always have a grain structure, which can be easily recognized when, after metallographic grinding and polishing, the surface is viewed through a microscope. Therefore, different sparks may ablate different compositions of metal. In addition, the samples often have inclusions that are preferably ablated by the sparks (see Chapter 6). In practice, the material that enters the plasma varies in the course of the spark sequence.

Integrating a large number of individual spark occurrences leads to good repeatability. Assuming that the remaining hardware in the spectrometer system is free of errors, the deviation of the single intensities follows a normal distribution and increasing the number of sparks by a factor of $n^2$ leads to a factor $n$ improvement of the repeatability.

### Statistics on the number of sparks

Sparks enable a larger sample surface to be included in the analysis than in the case with arc. The sparks do not always hit the sample at the spot directly opposite from the electrode tip. After several sparks, charge carrier pairs consisting of metastable excited argon atoms form in the atmosphere around the electrodes. The energy required to achieve the first ionization level for argon is 11.76 eV; the metastable 4s state lies 11.55 eV above the ground state. A small energy input, e.g., from collisions, is enough to ionize the metastable excited atoms. The breakthrough of the gap requires a lower voltage when it occurs over bridges consisting of charge carriers (see Figure 3.22).

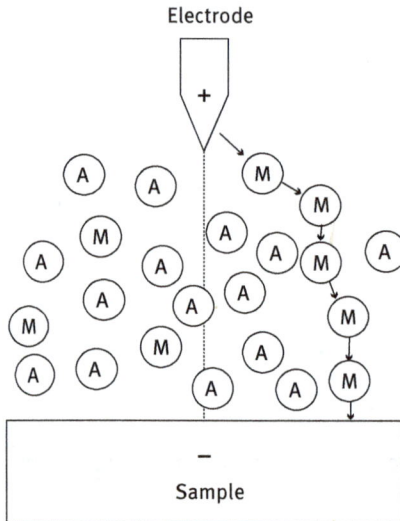

Figure 3.22: Formation of a planar burn spot caused by charge carrier bridges.

Whereby in Figure 3.22:

A:   Neutral atoms

M:   Metastable excited atoms from which charge carrier pairs can be formed

## Burn spot size

Several factors can lead to an increase in the size of the burn spot:

- An increase in the electrode gap leads to an increase in the burn spot diameter: The larger the electrode gap becomes, the longer is the path that can lead over charge bridges and the further the discharge contact point can move away from the point opposite the electrode tip.
- The burn spot becomes larger with increasing spark frequency. The time between two sparks, during which removal of metastable excited argon can take place, becomes shorter as the frequency increases.
- Energy-rich discharges also increase the size of the burn spot. Then more charge carriers and, when argon is used as the discharge atmosphere, more of the burn spot enlarging metastable argon atoms are generated (more about metastable excitation states can be found in Sections 2.6.3 and 3.2.3).

- The length of the single spark also plays a role: If the energy is introduced in a short time, the plasma expands very quickly, which rapidly removes the ions and metastable atoms from the gap. For a longer, lower-current discharge with the same energy, this effect does not occur to the same extent. The burn spot becomes larger as result.
- Occasionally, a pin electrode, consisting of a millimeter-thick tungsten wire, is used instead of a compact electrode. Such electrodes get so hot at their tips that they glow. The burn spots resulting from pin electrodes are usually significantly larger than those produced from the use of compact electrodes.

### Common geometries

Gaps between 3 and 4 mm are usual. Tungsten counter electrodes with a diameter of 6 mm that are sharpened with an angle of 90° are generally used. Smaller electrode diameters of about 4 mm are used in mobile spectrometers. Here, the electrode gap is also usually somewhat smaller (typically 2.5 mm).

After every single spark, the material previously vaporized from the sample condenses and forms finest metal particles that are largely transported away by the argon flow. A portion of this condensate, however, settles in the spark stand or on the upper surface of the electrode. The surface of the compact electrode must therefore be regularly cleaned with a wire brush – after each measurement or after every sample. Pin electrodes made of thin tungsten wire, mentioned in the paragraph above, are an alternative. It is not necessary to brush these. However, pin electrodes have a limited lifetime, which does not exceed several thousand measurements. In contrast, compact electrodes last practically indefinitely.

The amount of sample degraded by a single energy-rich spark was determined for pure aluminum. For the test, a parameter with about 20 mJ spark energy was chosen. The amount degraded was determined to be 32 ng per spark. This corresponds to a crater diameter of 35 µm, if this has the shape of a hemisphere. These values were determined by weighing a sample, sparking it and, after a large number of sparks, weighing it again. The test was conducted independently by two spectrometer manufacturers and produced similar results in both cases.

### Concentrated and diffuse discharges

With dirty samples, samples that are porous have cracks or holes, or if contaminated argon is used, then frequently several parallel channels are formed instead of one single discharge channel per spark event (see Figure 3.23, burn spots bottom left and top right). Such discharges are called "diffuse." The discharges by which a single discharge channel is formed is, in contrast, called "concentrated" (see Figure 3.23, lower row from burn spot 2). Only with concentrated discharges is there

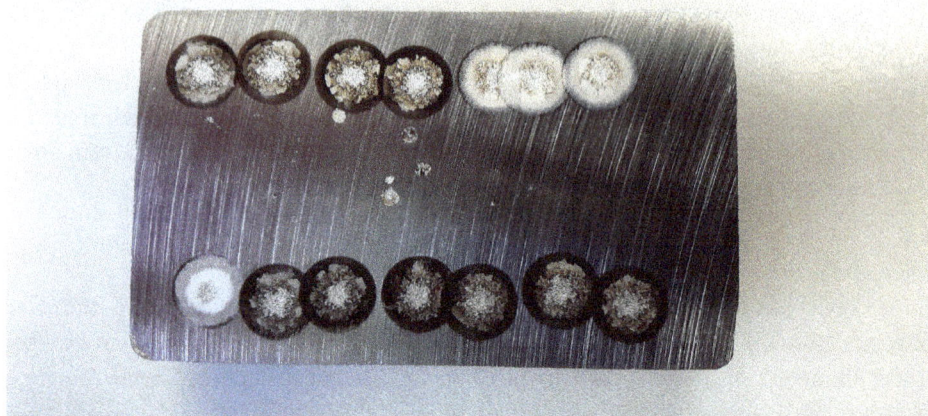

**Figure 3.23:** Burn spots with high (lower left) and low (lower right) fractions of diffuse discharges. Printed with the friendly permission of SPECTRO Analytical Instruments GmbH, Boschstr. 10, 47533 Kleve, Germany.

any significant material ablation. Diffuse discharges leave numerous very low-volume contact points that lead to whitish spots on the surface. During a spark, such unusable sparks can be identified: The burn voltage of concentrated discharges in an argon atmosphere lies, assuming common materials and geometries, on the order of between 20 and 40 V. The burn voltage drops several volts for diffuse discharges. This information can be recorded during the spark and is used to interrupt the measurement when unusable sparks accumulate; a new sample is then requested. Additionally, the length of the pre-spark phase, discussed in the next section, can be dynamically adjusted. It is completed when a predetermined number of concentrated discharges is reached.

### High energy pre-spark

It has been proven that it is useful to spark the sample surface with a series of high-energy sparks before the actual measurement. Caused by the high energy of the single sparks, it is ensured that during the actual measuring phase only parts of the surface are hit that had already been hit during the pre-spark.

This has two advantages:
- The pre-spark ensures that dirt that was on the sample surface before the measurement is vaporized. Contamination on the surface cannot be avoided; this also includes moisture on the surface.

- The area on the surface that is intended to be measured is homogenized by re-melting. Slickers [24] reports a 40–100 μm diameter remelted area on the sample for a crater diameter of 20 μm.

The energy of the single sparks in the pre-spark phase must not be too high. Before the start of each single spark, every point on the sample surface has to be solidified. Reproducibility is worsened if a melt forms in the burn spot. There may be distillation effects as discussed in connection with arc, which also negatively influence the analytical accuracy. For the same reason, the spark frequency chosen must not be too high. The maximum spark energies and frequencies selected in a method depend on the thermal conductivity of the material group to be analyzed and must be experimentally determined before the method is calibrated.

Typical parameters for the pre-spark time are currents of 50–70 amps for times of about 80 μs. Spark frequencies of 400 Hz are common. For many materials, it is possible to increase the frequency to 1,000 Hz, whereby the length of the single sparks is then usually shortened to 40–50 μs.

## Using different measuring parameters

It was already discussed in the last section that the measuring time is preceded by a pre-spark phase with energy-rich parameters. The measuring time itself is usually divided into several phases:

1. Sparks with a short duration (40–60 μs) and medium current strengths (10–20 A) with rectangular current course are suitable for the determination of the metal alloying elements. Spark frequencies from 300 to 600 Hz are common.
2. Trace elements are determined with longer sparks with durations up to a millisecond and lower current than those discussed in point 1. These are called "arc-like discharges." Due to the longer spark times, the conditions in the later spark phases are similar to those in arc: The plasma is expanded and there are low pressures and temperatures, which, because of the likewise low ion density (see Saha eq. (2.16)), lead to a reduction of the background radiation. But because it is still a spark phase, in which the statistics for the single sparks improve the reproducibility, the repeatability is good compared to the arc. However, in the later phases of the single sparks, not only argon ions but also the atoms of the alloying elements now present in the plasma are accelerated toward the sample surface. Ablation and excitation processes depend on third-element concentrations in the sample. This leads to more scattering in the calibration curves than for the parameters described in point 1. Due to the less dynamic plasma expansion and the longer discharge time, a layer of cold atoms forms around the plasma. It absorbs radiation and emits in all spatial directions. The result is a nonlinear calibration curve for which the slope increases with higher concentrations like the one shown in Fig. 3.15. This effect has already been described in Section 3.2.1.3.

3. The most sensitive detection lines for some analytes lie below 180 nm. These are mainly gases such as oxygen and nitrogen, but also lines from metalloids and a few metals. According to eq. (3.2), lines with wavelengths below 180 nm have an excitation energy of at least 6.9 eV. Energy-rich spark parameters with currents of 40 amps and more are required to generate a sufficient number of excited atoms. The duration of the single sparks must then be about 100 μs.

4. The next section and the entire Chapter 6 discuss single spark analysis as a tool for detection and characterization of nonconducting inclusions in the metallic matrix. If small inclusions are to be detected, it is often advantageous to use very low-energy sparks. Currents below 10 amps and spark times of 20 μs are then commonly used.

5. Oils and fuels can be analyzed with the so-called "rotrode technology." With this technology, a small wheel made of spectrally pure graphite carries the liquid to be analyzed into the plasma, which burns between a graphite counter electrode and the graphite wheel (rotrode). A depiction of such a spark stand can be found in Section 3.3. The parameters as described in point 2 are generally used for this technology, as the focus is on detection sensitivity. Frequently, the spark duration must be limited to prevent ignition of the sample. Then the parameters are set between the parameters described in point 2 and point 4.

### Burn-Off curves

As early as 1930, Gerlach and Schweitzer [18], p. 110 et seq., observed the following phenomenon:

When a gold sample containing lead is broken and the break is sparked, then the measured signals obtained at the beginning of the spark phase are different from those at a later point in the spark phase. At the beginning, the Pb signal was excessively high. After several seconds, it decreased and remained largely stable. Gerlach and Schweitzer explained this phenomenon by saying that the lead preferred to settle at the grain boundaries. Since the breakage also occurs along these boundaries, the first sparks form a thin lead layer on the surface. In the later course of the spark process, the sparks hit mainly in the grain interior, where the lead content is lower.

A similar effect can be observed in gray-solidified cast iron. Here, graphite inclusions are enclosed in the iron matrix. After a short initial spark phase, high signals are obtained for carbon. The sparks hit the graphite inclusions that sublime and disappear from the matrix. Figure 3.24 shows a grinding pattern for a metal sample after 10 s pre-spark with 400 Hz and subsequent grinding and polishing. In the middle of the sample (dark zone in the center of the image), a deeper ablation due to the higher probability of being hit by a spark can be observed. Simple polishing was not possible here. This area should be ignored. The area immediately bordering on the dark zone is interesting. It is clearly

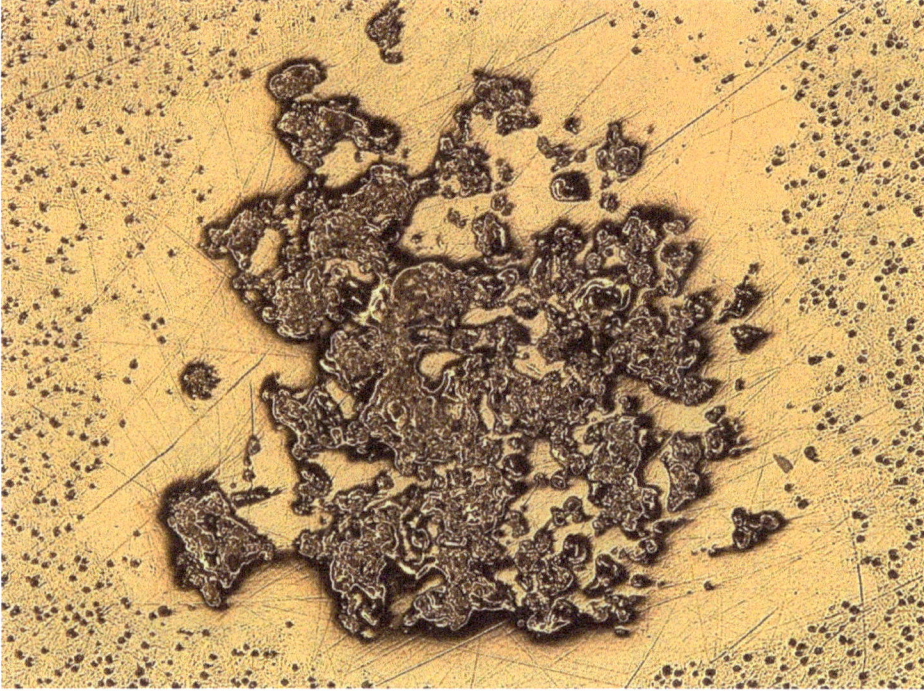

**Figure 3.24:** Micrograph of a gray cast sample with graphite inclusions after a 10 s pre-spark time. Printed with the friendly permission of SPECTRO Analytical Instruments GmbH, Boschstr. 10, 47533 Kleve, Germany.

visible that the graphite inclusions (small dark dots, still present on the left and right side of Fig. 3.24) have disappeared here.

Gerlach and Schweitzer's measuring capabilities were limited. They could only record the first seconds of a spark phase with photographic imaging, make a further image of the rest of the spark process and compare the two photographs. Today, measuring technology is more advanced. Direct registration of the spark signal with electro-optical sensors makes it possible to record the intensity of each single spark. These sensors and the associated measuring electronics are described in Sections 3.6 and 3.7.

Figure 3.25 shows the burn off curves for the elements iron and carbon measuring a cast iron sample with graphite inclusions. The structure of the sample is similar to that shown in Figure 3.24. Each curve point represents a spark packet consisting of the sum of five consecutive sparks. The first spark packets deliver only low intensities, as in the beginning there is still moisture or other types of contamination on the

(a) Spark package intensities vs. Spark package numbers for Fe in partial grey solidified cast iron

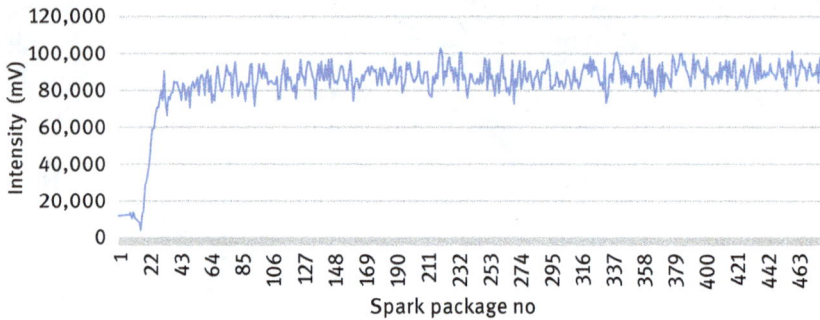

Intensity (mV)

120,000
100,000
80,000
60,000
40,000
20,000
0

Spark package no

(b) Spark package intensities vs. Spark package numbers for C in partial grey solidified cast iron

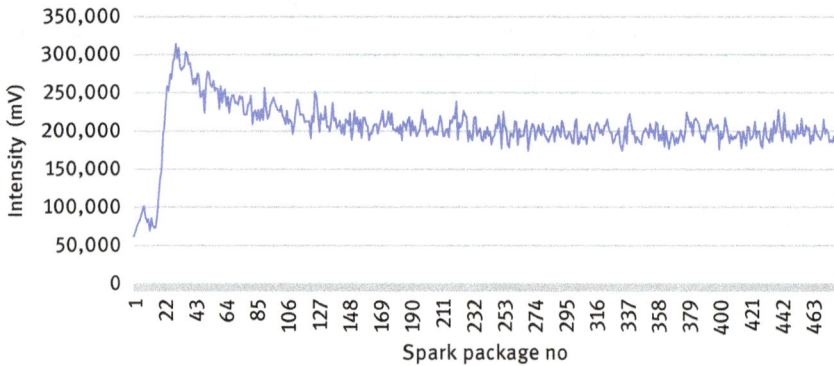

Intensity (mV)

350,000
300,000
250,000
200,000
150,000
100,000
50,000
0

Spark package no

**Figure 3.25:** Burn-Off curves of Fe and C for a partially gray-solidified cast iron sample. Printed with the friendly permission of SPECTRO Analytical Instruments GmbH, Boschstr. 10, 47533 Kleve, Germany.

surface of the sample, which lead to diffuse discharges. After about 20 spark packets, the intensities increase to a higher level. The so-called "stationary spark state" has been achieved.

If an element preferably hit at the beginning of the spark process is observed, then a burn-off curve like that in Figure 3.25b is obtained (graphite inclusions were hit in Fig. 3.25b). The excessive intensity at the beginning does not necessarily apply to every spark. The probability that a new spark hits a spot that was already sparked becomes larger with an increasing number of sparks. In the "graphite in cast iron" example, in the beginning phase, only a part of the surface consists of graphite inclusions. If an element is homogeneously distributed within the metallic matrix (here Fe), burn off curves similar to Figure 3.25a are obtained.

## Single spark analysis

In the last section, it was seen that the spark largely removes graphite inclusions at the beginning of the pre-spark time. If there are nonconductive inclusions in a metallic matrix, an effect very similar to that described above can be observed. It occurs, e.g., for $Al_2O_3$ inclusions in steel. The spark prefers to hit on the boundaries between the metallic matrix and nonconductive inclusion. A small ridge forms at the transition between conductor and the nonconductor. This leads to an increase in the field strength there and increases the probability of a spark over at this spot (for details, see chapter 6.) Figure 3.26 shows a cross section of a sample

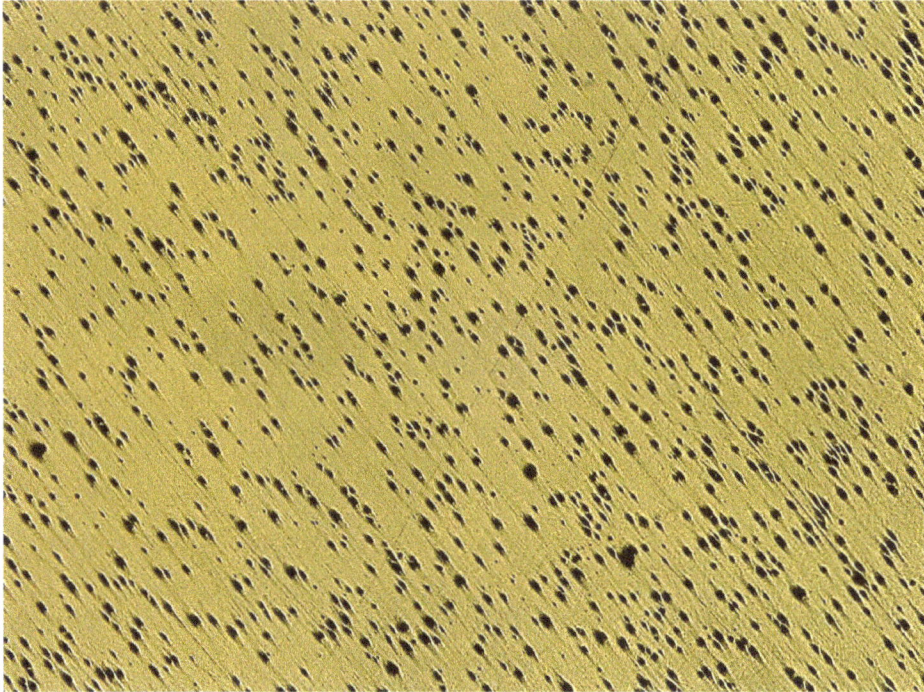

Figure 3.26: Micrograph of a sample with inclusions, scale 1:200.

containing inclusions. The black spots are inclusions. A spark can then either degrade only the metallic matrix or hit on the edge of an inclusion.

Measuring every spark individually results in two different signal levels for the analyte elements in the inclusion:

1.  If only the metallic matrix without inclusions is hit, the intensity depending on the analyte content in the metallic matrix is obtained.
2.  The analyte intensity is usually significantly higher when the spark attacks at the edge between inclusion and metal. The reason for this lies in the

fact that the analyte concentration in the inclusion is always high. It can be easily calculated from the molecular formula of the inclusions. The Al content of $Al_2O_3$ is 53 mass percent, while the aluminum concentration in steel is usually less than 0.1% (there are, however, steel alloys with aluminum concentrations in the percent range to improve behavior at high temperatures).

The intensities of the lines from the matrix elements (i.e., the signals from iron lines in steels) behave in an opposite manner. Their signal is reduced when an inclusion is hit. It is also possible to make conclusions about the composition of the inclusion: If, during a single spark, there is a high aluminum intensity with simultaneous high oxygen signal, it is an indication that an aluminum oxygen compound was hit. The experience practitioner knows what compounds may occur in the materials analyzed. Chapter 6 deals extensively with the possibilities and limitations of single spark analysis for compact metal samples.

Single spark detection also plays a role in the examination of oils with the rotrode technique. The determination of the number and size of metal particles is of interest here. Two types of information can be gained from this:
–   It is possible to deduct from which component wear particles originate from the element combination of simultaneous signals. Particles that are large or numerous may indicate a defect. Maintenance may prevent downtime or malfunction of the system.
–   Even when contamination of the lubricant is within normal limits, an oil change may be necessary when a maximum limit is reached. For large machines, e.g., marine diesel engines, this can be a cost-intensive measure that is only performed when it is really required.

## Time-Resolved Integration
It was already mentioned in Section 3.2.2.1 that it may be convenient to integrate only parts of the spark for use in calculating the analyte content:
–   Background is caused by bremsstrahlung during the current conducting phase of the spark. This phase is often suppressed to increase the detection sensitivity.
–   Atom and ion lines arise during the current conducting part of the spark. After the end of the current flow, in the so-called "afterglow," the ion lines disappear. The background radiation is also greatly reduced. The remaining atom lines, while comparatively faint, can be measured with only a few interferences.
–   The risk of radiation absorption by a layer of cold atoms increases with the duration of the single spark while the collision broadening decreases.

It may be useful to optimize the measurement for individual analyte lines by selecting a fitting section of the spark. In this way, it is possible to improve the detection sensitivity, linearity, scattering, and usable concentration range of the calibration curve or to reduce line interferences. However, an advantageous result is not always obtained compared to integration of the entire spark, because the process of splitting the single spark intensities introduces an additional source of error. The hardware required for integration of partial sparks is discussed in Section 3.7.

### 3.2.3 The laser as an excitation source

Efforts to use the laser as a source of excitation is almost as old as the laser itself and was first realized in 1960. After 1963, there were many publications. Horst Moenke and Lieselotte Moenke-Blankenburg published a textbook for laser micro-emission spectral analysis in 1966 [30]. The laser as an excitation source for atomic spectroscopy is discussed in depth in the textbook *Analytical Emission Spectroscopy* by Jozsef Mika and Tibor Török [31] from 1973. The acronym "LIBS," an abbreviation for laser-induced breakdown spectroscopy, has become accepted as the term for the technique in which the laser is used as the excitation source for ablation, ionization and excitation.

#### 3.2.3.1 Basic operation of a solid-state laser

In connection with arc excitation, it was determined that the atoms are brought into an excited state (one of the levels above the ground state in Figure 3.14) through the introduction of energy, collisions or radiation. Normally, the atom falls to an energetically lower level within a very short time of about $10^{-8}$–$10^{-7}$ s after the excitation; a photon is emitted (see Chapter 2 of this book). As already shown in eq. (3.2), the wavelength $\Lambda$ of the photon is $(h \times c)/\Delta E$, whereby $\Delta E$ is the difference between initial and final energy levels. However, there are excited states that have a longer lifetime, so-called metastable states. Metastable states remain unchanged as long as there is no interaction of the excited atoms with other particles (see Section 2.6.3). But because these interactions always occur, e.g., due to collisions, the atoms also leave these states after times that are significantly longer, e.g., in the millisecond range.

Assume that an atom is in an excited state, below which there is a lower energy level with a difference of E. If a photon with the wavelength $\Lambda$, for which $\Lambda = (h \times c)/E$ applies, encounters the excited atom, the return of the atom to the lower level is triggered and a second photon with exactly the same radiation direction, wavelength and phase position is emitted. This way of emitting a photon is called "stimulated emission." Both photons can now cause the emission of additional photons in the same way and so on. A laser beam that is very easy to focus is formed.

The portrayal above is, in one way, too simplified: Usually, so-called "solid-state lasers" are used for excitation purposes. The active medium of the laser is a crystal doped with impurities. In these, the energy levels are not as sharply defined as Figure 3.14 suggests; rather, the energies can vary in so-called "bands." Here bands refer to wide ranges around the drawn energy levels.

To trigger laser radiation, there must be enough atoms in an excited state. If the energy transition from $E_{High}$ to $E_{Low}$ is intended, there must even be a very large number of atoms in the state $E_{High}$. Conversely, only very few atoms may be found in the state $E_{Low}$, otherwise the stimulated radiation would be absorbed by the atoms that would then be found in the state $E_{High}$. For this reason, it is usually unfavorable to choose the ground state for $E_{Low}$ (baseline in Figure 3.14). The atoms are raised to a higher-energy level by so-called "pumping." Pumping can be caused by irradiation with an external source, e.g., a flash lamp, or with a laser diode. It is sufficient when pumping brings atoms in the laser rod to an energetically higher-level $E_{Pump}$, which lies above $E_{High}$. This can be a broad energy band. $E_{High}$ is reached from $E_{Pump}$ by releasing energy to the crystal. The left side of Figure 3.27a shows the energy transitions for a laser where $E_{Low}$ is the ground state. This scheme is called the "three-level-system" because the three levels $E_{Pump}$, $E_{High}$ and $E_{Low}$ are represented, whereby $E_{Low}$ is identical with the ground state. This scheme applies to the ruby laser. A four-level system where $E_{Low}$ is energetically above the ground state is shown in Figure 3.27b. This system is used by the Nd:YAG laser that is frequently used as an LIBS excitation source. The fiber laser is another type of solid-state laser that has recently become popular especially for handheld instruments. The way in which is works is explained in Section 3.2.3.3.

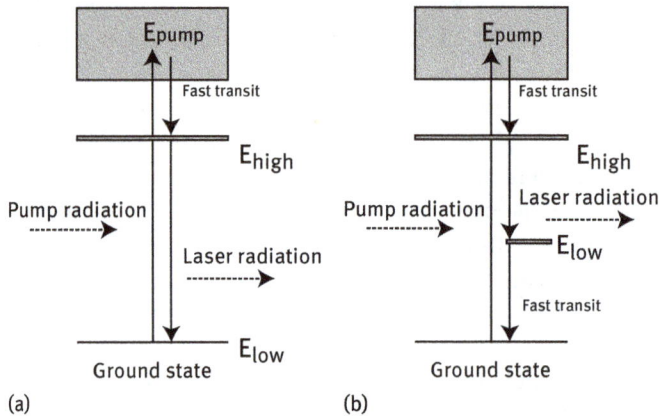

Figure 3.27: Energy transitions in the laser.

Figure 3.28 shows the basic construction of a solid-state laser. The laser crystal is a cylindrical rod that is partially silvered on the front side and fully mirrored on the back. Parallel to the laser rod is a flash lamp, whose radiation is focused onto the laser crystal by a parabolic cylindrical mirror. The process of creating a laser pulse

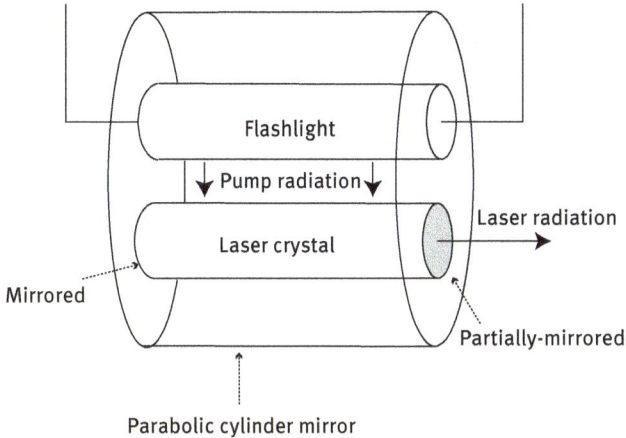

Figure 3.28: Basic structure of a solid-state laser.

starts by switching on the flash lamp: The energy of the flash lamp raises atoms in the laser crystal to a higher, metastable energy level. They usually remain in this state for a duration in the millisecond range. Externally supplied photons with an energy of $E_{High} - E_{Low}$ or photons with the same energy emitted by the interaction of metastable atoms with other crystal atoms can collide with other excited atoms in metastable states. These are stimulated in this way to release energy and strike other atoms in an excited state, which in turn are triggered to stimulated emission. Laser radiation with photons of the same direction, phase position and frequency is created. The mirrored surfaces on both ends of the laser ensure that only a portion of the radiation exits through the front side. The reflected rays return to the crystal and can induce metastable excited atoms to emission. On the back side of the crystal, they are again reflected into the direction of the front side together with the emission they stimulated. When the wave properties of the generated radiation are observed, it is possible to consider the optical path $l_O$ separated by the mirrored surfaces as a resonator, between which a standing wave forms. The optical segment is obtained by multiplying the geometric gap with the refractive index of the crystal. If the amplitude of the radiation at the mirrors is 0, there is a resonance. It then follows that for this kind of resonance $2 * l_O$ must be a multiple of the wavelength, when remaining by the wave view of the photons, the standing waves overlap and radiation with a high total amplitude is produced.

### 3.2.3.2 Lasers with quality switches

The design according to Figure 3.28 has a major disadvantage for laser spectroscopy: The energy is generally not high enough to ignite a plasma. According to Cremers and Radziemski [32], a power density with a magnitude from $10^8$ to $10^{10}$-W cm$^{-2}$ is required. The laser output must be increased in order to exceed the indicated power density threshold. For a given pulse energy, this can be accomplished by increasing the speed of the pulse energy. If, for the same pulse energy (which is measured in Joule, i.e., Watt × seconds), it is possible to shorten the pulse duration by a factor n, then the power is increased by the factor n. According to Kneubühl and Sigrist [27] (p. 364), it is possible to generate peak powers of 10 kW with flash lamp pumped lasers of the type described above. The pulse duration is 1–5 ms, which corresponds to the length of the pump lamp pulse. To achieve shorter laser pulses, the design was therefore modified in the way shown in Figure 3.29.

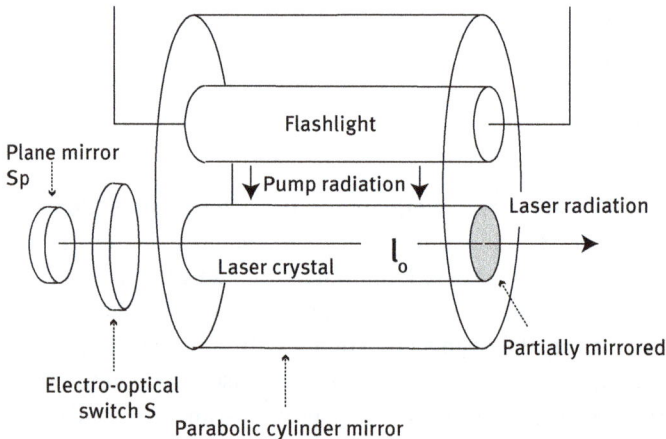

Figure 3.29: Q-switched laser.

There is an electro-optical switch S, for example a Pockels cell, behind the laser rod, and behind this is a front surface mirror Sp. Generation of the shortened laser pulse proceeds as follows: The flash lamp pulse pumps the atoms in the laser crystal into the energy band $E_{Pump}$, by giving up energy to the crystal, their energies fall to $E_{High}$. They remain at this energy level because this state is metastable. The optical switch S does not allow any radiation to pass through. Rather, the radiation emitted from the laser rod is absorbed in S. When viewing photons as particles, this means that the photons exit the laser crystal without, as in Figure 3.28, remaining in the crystal and being reflected back and forth and, thus, moving excited atoms to stimulated emission. When viewing photons as waves, no resonance occurs since this requires two mirrored surfaces. As long as S does not allow radiation to pass, the pump radiation is only used to bring atoms to the state $E_{High}$. If the switch S is abruptly switched to transmit, then the conditions become like those in Figure 3.28. The only difference is that there are now many atoms in the state $E_{High}$. These are abruptly triggered to stimulate emission in a very short time period. Of course, the distance of the mirror Sp behind the optical switch must be correctly chosen. Once again, $2 \times l_0$ (path between the partially mirrored front side of the laser crystal and the mirror Sp) must be a multiple of the laser wavelength. The so-generated laser pulses have durations in the nanosecond range and achieve powers of up to 50 MW [27] (p. 364). Pulses with this output can be easily focused to the $10^9$ W $^*$cm$^{-2}$, mentioned at the beginning of this section, as being required to generate a LIBS plasma.

### 3.2.3.3 Fiber laser

Solid-state lasers, by which the laser beam is generated with a doped crystal rod, have been described in Sections 3.2.3.1 to 3.2.3.2. In recent years, however, another type of solid-state laser has been gaining in importance, that is, the so-called "fiber laser." Fiber lasers were initially used for material processing, such as for engraving or for cutting metal sheets. But, because of their numerous advantages, they are being increasingly used for other purposes, e.g., LIBS applications.

The advantages of fiber lasers are as follows:
- High beam quality that enables the beam to be focused on a small surface
- Efficient conversion of the pump power into the laser power
- Long lifetime and reliability
- Compact design
- Resistant to misalignment

The fiber laser technology is constantly evolving, driven by mass markets such as telecommunications. The fiber laser is characterized by the fact that the coherent laser beam is generated within a fiber-optic cable. A flexible fiber optic takes over the function of the crystal rod described in Sections 3.2.3.1 and 3.2.3.2. The fiber-optic cable consists of an elongated cylindrical fiber core surrounded by a layer of a medium, the cladding, which is also optically transparent. The cladding is optically less dense than the fiber core, i.e., its refractive index is smaller than that of the core. As a result, the radiation that enters the fiber optic within a defined range of angles first exits it at the end of the fiber optic. When the radiation reaches the boundary between the core and the cladding, total internal reflection occurs and the radiation bounces back into the fiber-optic interior. The way in which fiber optics work is described in Section 3.5.5.2. Figure 3.62 shows the path of the radiation through a fiber optic. The total internal reflection makes it possible for the radiation to pass through the fiber optic even when it is bent. The core and cladding of the fiber laser are made of quartz glass, whereby the refractive increase of the core is achieved with the addition of aluminum [33]. Just like the crystal rod of an Nd:YAG laser, the core of the fiber laser optic is doped with foreign atoms so that laser emission can occur. Erbium, thulium and ytterbium were mentioned as possible doping elements by Dong and Samson [34]. Simultaneous doping with erbium and ytterbium is also common according to [34]. The fiber laser also has reflecting elements at both ends, of which at least one is partially transparent so that the generated laser beam can be released. The cross section of the core must be selected so that it is small enough to ensure a defined length of the optical light path $l_0$ for the radiation traveling through the core (see Section 3.2.3.1). As described there, $2 \times l_0$ must correspond to a multiple of the wavelength emitted by the laser. I.e., it must be prevented that the radiation can travel through the fiber both along a direct path parallel to the core axis as well as a zigzag course with many total internal reflections. This is achieved by reducing the diameter of the laser-active fiber core to an order of magnitude of 10 µm and less.

The fiber laser contains all the elements that can also be found in the diagram of the solid-state laser in Figure 3.28. The configurations of these elements are, however, different than those in the solid-state laser. Especially, the type and position of the pump source and the design of the reflective surfaces on the ends of the resonator are usually configured in a different way.

Coupling of the pump energy is carried out through one of the end faces of the core. Laser diodes are suitable as pump sources. Unlike the Nd:YAG solid-state lasers presented in the previous section, only a relatively small area, namely the end face of the fiber core, is available for energy coupling. This area has, as mentioned above, a diameter that is only on the order of 10 µm. Although laser diodes generate a high radiation density on a small surface, a fiber end face of 80 µm² limits the pump power and, thus, the output power of the laser.

Various measures can be taken to avoid this limitation:
– Combination of the radiation from multiple laser diodes
– Cladding pumping using double clad fibers

A fiber-optic cable with n inputs and one output is necessary to combine the radiation from n laser diodes. Each laser diode irradiates one input, the output in turn illuminates the pump end of the laser fiber. Such fiber optics are known as "tapered fused bundles" (TFBs). Figure 3.30 shows this principle for a Y-formed split fiber optic and two pump diodes. Unfortunately, n cannot be increased infinitely because the requirements for the focusability of the laser diode radiation increase with increasing n.

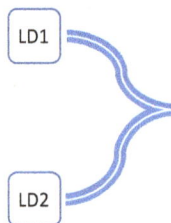

Figure 3.30: 2:1 Tapered Fused Bundle.

Another measure is suited to getting more pump energy into the fiber core: A three-phase fiber structure is chosen. Instead of surrounding the laser fiber with just one cladding, the first cladding C1 with a diameter of $d_{C1}$ is surrounded by a second with a diameter of $d_{C2}$. The second cladding C2 is less dense than the first, i.e., it has an even smaller refraction index than the first. Thus, C1 and C2 form a second fiber-optic system with cladding C1 as the core, in the larger end face of which more radiation can be coupled. Figure 3.31 shows the principle. The restriction that limits the diameter of the laser-active inner core to 10 µm does not apply to C1. C1 can have a diameter from several 100 to several 1,000 µm. There is a much larger surface available for coupling of the pump energy.

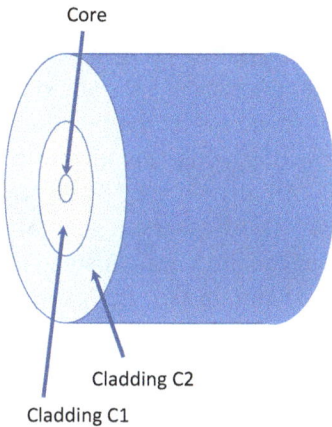

Core

Cladding C2

Cladding C1

Figure 3.31: Double Clad Fiber.

The problem here is that a concentric arrangement of core, C1 and C2 results in a limited number of total internal reflections and, thus, "traversing" of the fiber optic. The probability that a photon of the pump radiation strikes the core is low. To increase the number of total internal reflections, the C1 cladding is given a different shape; instead of being cylindrical, C1 can, for example, be hexagonal. Other outer forms for C1 or a nonconcentric placement of the laser-active core within C1 lead to comparable results. See Dong and Samson [34] (p. 90), for examples.

The C1 cladding has such a large cross-section that the energy from several diode pump lasers can easily be injected through a TFB with inputs for each pump diode and an output that is connected to the laser-active fiber. An informative instructive video from the company Nufern [35] reports that six pump diodes each with 600 W can be combined. It is also noted that the requirements on the pump diodes with respect to focusability are low, enabling cost-effective implementation.

In addition to the different pumping methods, it makes sense to use a different type of reflective surface on the ends of the resonator than those illustrated in Figure 3.28. Of course, the pump end of the fiber could be a mirror. However, this would have to be limited to the area of the core itself. The C1 cladding would have to remain nonreflective to enable unhindered coupling of the pump energy. Mirroring only the core requires exact positioning of the mirror coating and carries the risk of thermal overload, because a high-power density may occur here due to the small mirror surface.

These drawbacks can be avoided by using a so-called "fiber Bragg grating" (FBG). In this case, the core is treated on the ends in such a way that there is a periodic alternation of the refraction indices between two values $n_{K1}$ and $n_{K2}$. Of course, $n_{K1}$ and $n_{K2}$ must both be larger than the refractive index $n_{C1}$ of the C1 cladding to ensure the light guiding function. Figure 3.32 shows the result schematically. The periodic alternation of the refractive index results in a grating that works like a mirror for some wavelengths. The wavelength that is reflected can be adjusted with suitable choice of the periodic interval of the FBG-layers (sketch of layers see Figure 3.32). The grating structure is achieved by doping the fiber core with germanium and exposing it to the interference pattern of two lasers. Both lasers periodically generate a doubling and an extinction of the radiation that result in the desired changes in the course of the fiber core at the positions of intense irradiation. Section 3.5.3 in this book deals with gratings for spectrometry, including their holographic production methods. An arrangement is shown there that can be used for the exposure of grating blanks with lasers to generate a grating structure. (see Figure 3.58). Picture the end of the fiber optic instead of the grating blank in this image; this is the arrangement suitable for the generation of a fiber Bragg grating in the fiber core. A simplified, but intuitively easy to understand illustration of the arrangement specifically for the exposure of laser fibers can be found in Meschede [36], in Section 7.4.4.

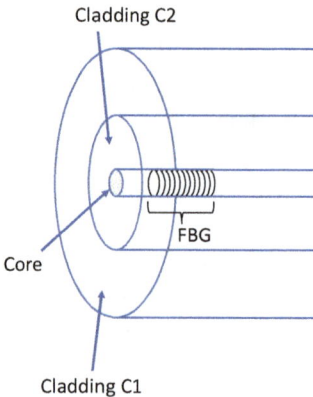

Figure 3.32: Fiber Bragg Grating (FBG) at the end of a laser fiber.

The ratio between the transmission and reflection of an FBG can also be controlled. It is also possible to generate semitransparent reflecting elements that are only reflective for the desired laser wavelengths. Various possibilities for the implementation of

FBGs in different fibers are listed on the Homepage of the company Femto Fiber Tec GmbH [37]. There, it is also noted that there are procedures that function without germanium doping of the fiber core.

### Fiber lasers with quality switches

As already stated in Section 3.2.1.2, very short pulses of a maximum of a few nanoseconds are desirable for LIBS applications. One step in this direction is to equip the fiber laser with a quality switch. The design is somewhat more complicated than for Nd: YAG lasers as pumping must take place over the end face of the fiber. This complication can be solved by using a dichroic mirror as an additional component. A dichroic mirror is characterized by the fact that it is transparent for certain wavelength ranges but reflects radiation from other ranges.

The Q-switched fiber laser can, then, be built as sketched in Figure 3.33:

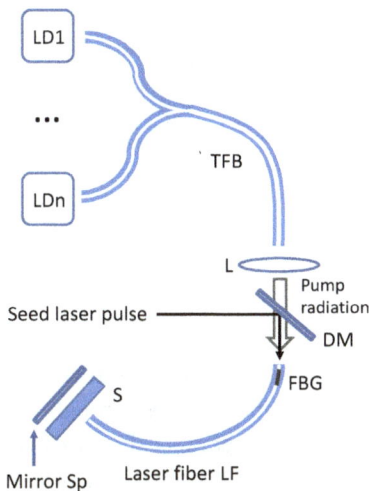

Figure 3.33: Sketch of a fiber laser.

The pump diodes LD1 .. LDn irradiate the n fiber bundle TFB inlets where the radiation for all n pump sources is brought together. After exiting the TFB, the pump radiation hits a lens L, which focuses them onto the end face of the laser fiber LF. Then it passes through the dichroic mirror DM, which is transparent for the pump radiation, and hits the cladding C1. The pump process takes place in the core of the fiber optic, as described on the previous pages, as long as the optical switch S remains closed. If S is made transparent, multiple reflections occur between the mirror Sp and the fiber Bragg grating and, thus, stimulated emission of the laser radiation can exit the system through the semitransparent Bragg grating FBG. The laser radiation, which has a significantly longer wavelength than the pump radiation is reflected by the dichroic mirror DM and exits the system in this way.

The switch S can be designed so that it can be placed between two partial fibers. In this case, Sp can also be implemented as a Bragg grating resulting in a mechanically compact and robust arrangement. More information can be found in Dong and Samson [34] in the "High Power Isolators" section of Chapter 6.

Acousto-optical modulators (AOM) can also take over the function of the optical switch S. Such AOMs are based on the following principle (see Figure 3.34):

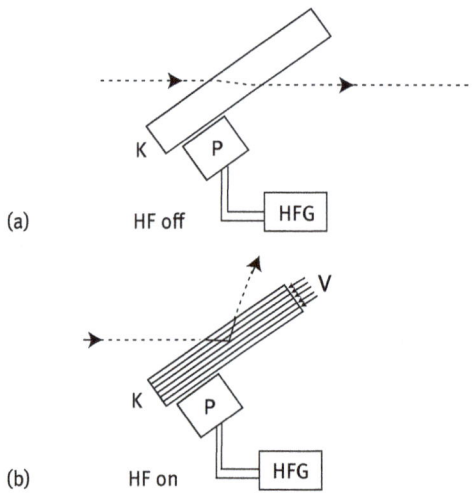

Figure 3.34: Principle of an Acousto-Optical Modulator (AOM).

A crystal K is coupled with a piezo element P that is connected to a high-frequency generator HF. When the high-frequency is switched off, radiation passes through the crystal virtually unhindered (Figure 3.34a). There is only a slight parallel shift because the crystal now acts as a refractor. Refractors are discussed in Section 3.5.5.4 of this book, and the path of radiation through a refractor is shown in Figure 3.66. The high-frequency generator (HFG in Figure 3.34) can generate vibrations with frequencies on the order of 10 MHz up to the GHz range. The piezo elements cause mechanical vibrations in the crystal. This leads to periodic compression V in the crystal, which increases the refractive index in the compressed areas. This creates a structure similar to that of a fiber Bragg grating. The radiation is diffracted at this structure and takes a different path than when the high-frequency is switched off.

## Master Oscillator Power Amplifier (MOPA) laser systems

MOPA laser systems consist of two components:
1. a low-power excitation laser (seed laser) and
2. a power amplifier.

The radiation from a seed laser is optically increased by a power amplifier. The seed laser can be a Q-switched solid state or fiber laser. This laser does not need high power and can thus be optimized for other parameters, e.g., high beam quality. A laser fiber with a doped core, which we know from fiber lasers, is used as the power amplifier. It is pumped by laser diodes that are fed into the amplifier laser fiber through a tapered fused bundle. The amplification process is now triggered by injecting laser radiation from the seed laser into the core of the amplifier fiber. This can be done in a similar way to that shown in Figure 3.33. A dichroic mirror transmits the pump radiation and directs the longer wavelength radiation from the seed laser at a right angle toward the amplifier fiber core. Here, stimulated emission of the excited atoms occurs abruptly as soon as the seed laser emits a radiation pulse.

The duration of the pulses from MOPA lasers can be varied in a range of from several hundred nanoseconds down to a few nanoseconds. Particularly short pulses are frequently desirable for LIBS applications.

### 3.2.3.4 Ablation and excitation using a laser with a quality switch

The LIBS plasma develops as sketched in Figure 3.35:

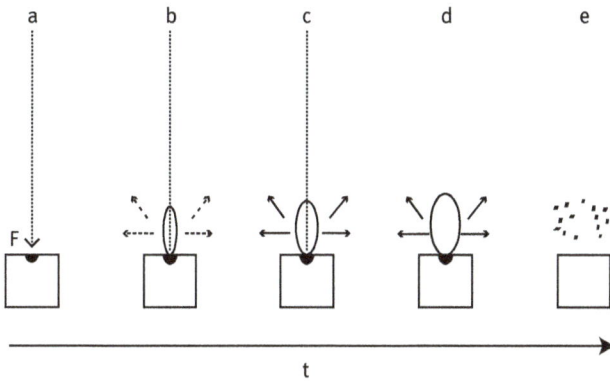

Figure 3.35: Development of an LIBS-plasma.

At the beginning of the pulse, the radiation acts on a small portion of the sample surface F on which the laser beam (dotted vertical arrow) was focused (Figure 3.35a). Immediately afterward, the material begins to vaporize (Figure 3.35b). After a very short time, a plasma that initially emits only thermal radiation (short dotted arrows) forms. After a few nanoseconds, radiation from the elements ablated from the sample is also emitted (solid arrows). The laser radiation input lasts until the end of the pulse (between c and d). The pulse duration is on the order of 10 ns. The plasma expands between ignition and the end of the pulse. When the pulse ends, the plasma expands further at first and the emission of spectral lines also continues. However, it becomes ever colder. The emitted radiation weakens steadily (Figure 3.35d), until it finally goes out completely. The ion line radiation disappears very quickly after the laser pulse, but the atom lines first completely subside after several microseconds. Finally, the temperature falls below the melting point of the ablated material; condensate particles form that, because of the plasma expansion, move away from the location of the plasma from which they originated (Figure 3.35e).

Figure 3.36 shows the spectra of the radiation generated during the different phases of a single pulse. Immediately after plasma ignition, almost exclusively continuum radiation is emitted (Figure 3.36a). In the following, approximately 500 ns, the spectral lines of the analytes, which are at first very intense but also strongly broadened, develop (Figure 3.36b). The causes of this broadening have already been discussed in the context of spark discharge (see Section 3.2.2.1 and Figure 3.19).

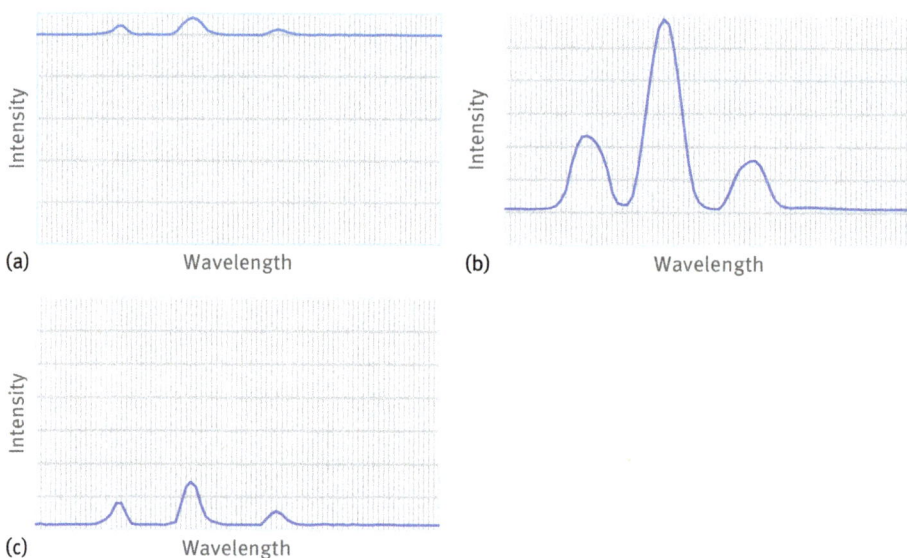

Figure 3.36: Spectra during development of a LIBS-plasma.

However, it is much more pronounced in LIBS plasmas. The line width decreases in the course of the next microseconds, but the intensities also decrease (Figure 3.36c) and finally disappear altogether.

In the report "Development of techniques for the production of glassy standard material for the analysis of slags by spectroscopic methods" [38], there are example spectra on pages 28 and 29 that were recorded in seven steps each with an integration duration of 0.5 µs between 0 and 3.5 µs after the laser pulse. If the spectrum from the first 0.5 µs is compared with that from the time between 1.5 to 2 µs, it can be seen immediately after the laser pulse that the laser pulse lines can take on a width of 2 nm. For an integration beginning at 1.5 µs, the lines are much narrower. Their widths are reduced to fractions of a nanometer, but they are also less intense. In the time window between 3 and 3.5 µs, no lines are visible. The further the starting point of the integration is after the end of the pulse, the lower is the background. However, the effective radiation also decreases. The intensities for the analyte lines are significantly higher when an argon atmosphere is used. The effective intensities can increase by a factor of five or more. The disturbing band spectra from the ambient air are also avoided. Wong, Bolshakov and Russo [39] name the high achievable plasma temperature, the easier plasma ignition compared to air and a comparatively higher electron density as the reasons for the intense spectra in an argon atmosphere. A further increase in intensities can be achieved by utilizing multiple pulses. The first pulse is used for material ablation. Once its continuum radiation has subsided, additional pulses can be used for further excitation of the generated plasma. This can be accomplished with flash lamp pumping, which were described above, without additional hardware. If the pump pulse has a length of, for example, 200 µs, then the ablation pulse can be enabled by opening the switch S in Figure 3.29 after e.g., 70 µs. It is closed again immediately after this first pulse and the pump process starts again. Additional laser pulses are possible in the period between several microseconds after the first pulse and the end of the pump pulse.

A comprehensive representation of the multiple pulse technique together with numerous other subjects concerning LIBS can be found in Noll [40]. The monograph from Cremers and Radziemski [32] provides a wide range of information about LIBS.

For very simple applications, the data gained from a single laser pulse supplies sufficient information about the test piece. Usually, however, the procedures described above are cyclically repeated over a measuring time in the seconds range to increase the overall signal and, with the sum of numerous pulses, to obtain a more statistically secured total spectrum. Depending on the design of the system, the pulse repetition rate can be in a range from just a few hertz to 10 kHz.

The advantage of the LIBS method over spark is that it is possible to control exactly where ablation of the material is to occur on the sample. This is also possible when there is a larger distance between the laser and the sample. This advantage is useful in several respects:

- Only a small part of the sample surface needs to be prepared
- The sample preparation can also be done with laser pulses
- Analysis can be done from a distance of several meters

Simplification of the sample preparation usually also means a shorter time until the analysis is available. Even small time savings bring significant cost savings in the steel producing process.

The EU study "Fast analysis of production control samples without preparation" [41] was performed in the light of this. Four different LIBS systems in combination with various types of sample preparation were examined.

– The first LIBS system was equipped for triple pulses generated by two laser systems. Here, the rolling skin is removed from the surface with the first pulse. The following pulses then strike the metal in the sample interior. The system is targeted for various process steps in steel production. An accelerated analysis of pig iron was also investigated. Sampling was optimized to obtain as thin a scale layer as possible.
– The second system was designed to analyze pin samples with a diameter of 9 mm. Pin samples can be prepared by cutting through them and using the end faces for analysis. This type of sample preparation, developed by the Institut de Recherche de la Sidérurgie (IRSID, French Research Insititute of Siderurgy), is faster than conventional milling or grinding.
– Unprepared and conventionally ground steel and pig iron samples were to be analyzed with a third system.
– The fourth system was designed for application areas different from those mentioned above. It was intended for sorting control from distances on the order of 40 m.

All four LIBS instruments operated with Nd:YAG lasers with pulse energies of several 100 mJ with pulse frequencies between 10 and 20 Hz.

To discover how the analytical performance of these LIBS systems is compared to spark-OES, they were calibrated with conventionally ground standards and examined with regard to the achievable detection limits.

It showed, that, similar to spark-OES, many laser pulses are needed to get the required reproducibility. Furthermore, it is necessary to move the sample with respect to the laser beam, to increase the treated sample surface (see Figure 3.37). Neglecting this action leads to steep, deep craters, shadowing the emitted radiation and reducing signal intensities.

The best detection limits achieved for carbon were 8 ppm, for Si 22 ppm, for Mn 10 ppm, for Cr 26 ppm and for Mo 3 ppm (for details, see [41], p. 69). The detection sensitivities for the first three instruments were quite close to each other. However, the detection limits mentioned here are all worse than those achievable with spark spectrometry (see Table 5.1); still they should be sufficient in many cases.

Sample preparation with high-energy laser pulses has a negative impact on the sensitivity; the detection limits worsened for all elements except carbon by an average factor of about 3. This worsening was found for both low alloy steel ([41], p. 48) and cast iron ([41], p. 50 et seq.). However, in practice, there are applications for which the achieved analytical performance is sufficient. The determination of C in pig iron was problematic. This is due to the fact that carbon is present in the samples partly as graphite and partly as cementite. This problem also occurs for the analysis with spark OES as described in Section 5.1.2 in this book.

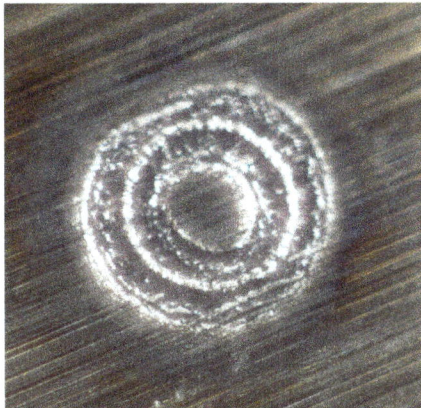

Figure 3.37: Steel sample with laser burn spots.

The IRSID pin sample method proved to be powerful and feasible. The attempts at grade identification from a larger distance with the fourth laser system tested were also positive. The transition zone between two grades in continuous casting, for example, could be monitored with such a system.

The above-mentioned EU study "Development of techniques for the production of glassy standard material for the analysis of slags by spectroscopic methods" [38] was also instigated to shorten the time required for the analysis. According to the state of the art, the slag sample, e.g., taken with a ladle, is ground up and then either pressed to a tablet together with a binder or fused with a flux to a glassy bead. The sample prepared in this way is then analyzed with X-ray fluorescence instruments. The elements of special interest in slags are: Si, Ca, Fe, Ti, Al, P, S, Mg, Mn, V, Cr, Na and K. These elements are usually present as oxides. Direct analysis of slags with LIBS could shorten the time to analysis. Figure 3.38 shows a slag sample with laser impacts. Since LIBS, like spark spectrometry, is a comparative measurement method, calibration samples are required to create methods. The production and examination of such samples was a key objective of this project. The production of samples with adequate homogeneity and stability was successful, which could be verified with X-ray fluorescence (XRF) analysis. The XRF analyses of these samples conducted by five laboratories showed good agreement. The same samples were analyzed by four different laboratories with LIBS. Here, the spread between the laboratories was significantly higher. Also the variation coefficients within the series, which were between 0.1 and 1% with X-ray fluorescence instruments, were considerably higher for the LIBS measurements – frequently in the range between 1 and 10%.

The investigated slags were glassy solidified, partially transparent specimens. The laser beam penetrates deeper here compared to metallic materials. The energy introduced not only leads to the formation of the plasma but also to the ablation of vitreous particles.

**Figure 3.38:** Slag sample with laser impacts. Printed with the friendly permission of Rubikon Verlag Rolf Kickuth, Bammentaler Straße 6–8, 69251 Gaiberg, Germany.

The study shows that direct LIBS analysis of slag samples is possible. Further development is, however, required in order to approach the performance of XRF instruments.

### 3.2.3.5 Commercially available LIBS systems

The devices on the market today (2018) usually fall into one of the following two categories:

–  Handheld instruments that are used mainly for the sorting of light metals. Such instruments usually operate with lasers with small pulse energies and pulse repetition rates in the kilohertz range. They are good supplementary devices to handheld X-ray instruments by which the determination of light elements is problematic. A detailed description of handheld laser instruments can be found in Section 7.1.4. Recently, such handheld systems are also offered for other metal bases, for example, for steels.

–  Spectrometer systems permanently installed in recycling plants for the identification of smaller metal pieces, e.g., for the sorting of shredder scrap. In a typical scenario, the parts are transported on a conveyer belt above which a portal with the laser spectrometer is mounted. The system captures every piece of metal and shoots it with a laser pulse, which is, when necessary, preceded by a pulse to clean the surface. The spectrum is recorded by an optical system and evaluated; the analytical requirements are not very high. Frequently, it is enough to determine the base metal and make a rough grade identification. For the aluminum industry, for example, shredder scraps have to be separated in the main alloy groups (see Section 7.1.4 and Table 7.1). Some distinctions within the main groups can also be made if there are large differences in the concentrations. In this way, the alloys which contain lithium, can be distinguished from the lithium-free alloys as either lithium is present in the percent range or only traces of the element are contained. If the scrap pieces are arranged one after the other on the conveyor belt, they can be blown into containers with compressed air after identification. The containers then contain sorted material fractions.

Werheit, Fricke-Begemann, Gesing and Noll described a system for sorting aluminum shredder scrap that operates with pulse rates of 40 flashes per second [42]. The pulses are generated by a Nd:YAG laser. The double pulse technique described above is used. The energy of each double pulse is 200 mJ. The system targets pieces within a measuring volume of $600 \times 600 \times 100$ mm$^3$ and moving with a belt speed of three meters per second. It can separate the goods to be sorted in eight fractions with a correctness of over 95%.

Similar equipment for aluminum scrap recycling is offered by SECOPTA in Teltow near Berlin. Figure 3.39 shows the arrangement of the system. This system was discussed at the 141$^{th}$ session of the *Chemikerausschuss* (Chemistry Committee) [43].

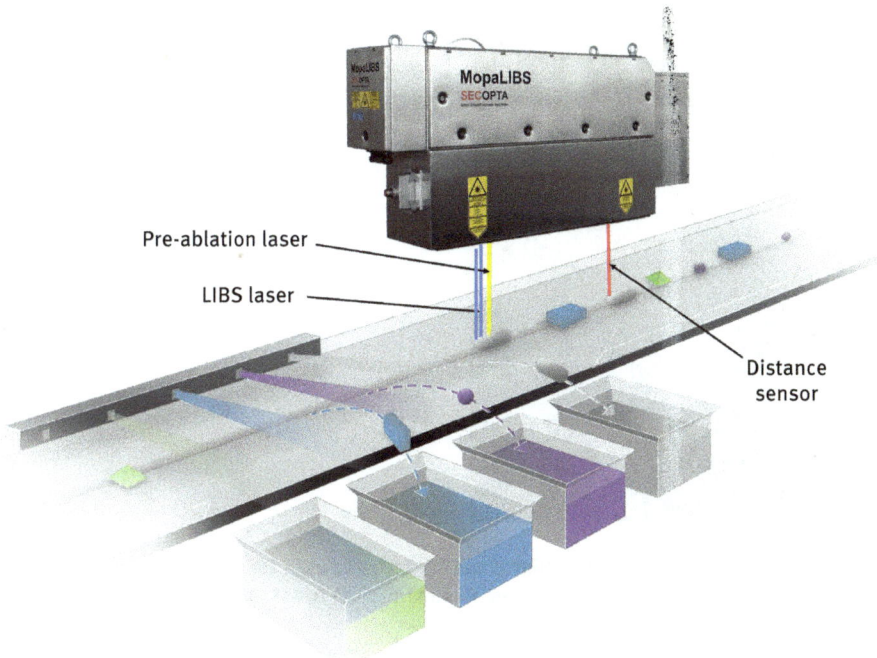

Figure 3.39: High-speed measurement principle for metal scrap sorting by LIBS, with friendly permission of Secopta analytics GmbH, Rheinstrasse 15 B, 14513 Teltow, Germany.

The system's task is to sort irregularly shaped metal pieces depending on their Si, Cu, Zn, Mg and Mn contents. A conveyor belt, on which the pieces are found, moves in Figure 3.39 from right to left with a speed of up to several meters per second. The pieces first pass the distance sensor labeled in Figure 3.39. The laser systems are correctly focused based on the distance measured. The focal point can vary by up to 140 mm here. When the targeted point on the sample is under the pre-ablation laser, it emits one or more pulses that clean the sample surface, whereby the several micrometers of surface is ablated. Shortly thereafter, the cleaned sample surface is below the analytical laser and analytical pulses are emitted onto the pre-cleaned spot. The LIBS radiation is dispersed and recorded by an optical system and then evaluated. Complete spectra that are processed with multivariate methods are recorded. The lasers used are fast; they achieve pulse frequencies of more than a kilohertz.

Figure 3.40 shows the pre-ablation pulse on the right side and the analytical pulse on the left. The result of the analysis determines how the metal piece is further processed. It can, for example, be blown into a container with a blast of compressed air.

**Figure 3.40:** Photograph of pre-ablation (right) and LIBS plasma used for the analysis (left), with friendly permission of Secopta analytics GmbH, Rheinstrasse 15 B, 14513 Teltow, Germany.

## 3.3 Arc and spark spectrometer sample stands

The term "sample stand" refers to the part of the spectrometer system where the plasma is generated.

The sample stand must fulfill the following general requirements:

- It must be possible to position the sample simply, quickly and with minimal risk of error on the sample stand, but, above all, the size of the gap, i.e., the distance between the electrodes, must be reproduced as accurately as possible.
- It is necessary to make the radiation generated in the gap available to the optical system. The optics can be in the direct light path. They are, then, only separated from the sample stand by a window and/or a lens. Alternatively, the radiation is transmitted, using windows and lenses, into a fiber optic that transports the radiation to the optic. Sample stands frequently enable a combination where one optic is in the direct light path and additional optical systems are connected via fiber optics.
- The windows and lenses for transmission of the radiation must be positioned so that the condensate that is formed during the spark does not, as far as possible, settle on them. This can be achieved with a sufficient distance from the plasma, or if available, through suitable direction of the operating gas flow.
- When using an operating gas flush, the interior of the sample stand should be designed so that the protective gas flow removes the residues formed during the measurement as completely as possible from the interior of the sample stand.
- Maintenance and cleaning work should be simple and error-tolerant and it should be able to be conducted quickly with a minimum on tools.
- When argon is used as the operating gas, the spark stand should be highly gas-tight, as contamination of the argon in even the ppm range can seriously disturb the excitation process.

### 3.3.1 Spark stands

The sample stand of a spark emission spectrometer is referred to as a spark stand. Figure 3.41 shows a cross section of such a spark stand. The sample S is placed on the spark stand plate P, in which there is an opening H. The sample is pressed onto P by a sample clamp N, thus ensuring a gas-tight seal and a reproducible electrode gap.

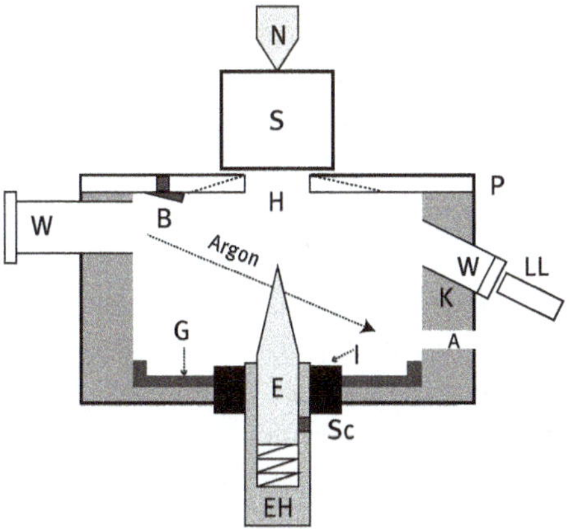

Figure 3.41: Cross-section of a modern spark stand.

The spark stand plate is fastened onto the spark stand housing K, on the bottom of which an insulating body I and, in its center, the electrode holder EH is found. EH holds the electrode E. E is pre-loaded with a spring and is fixed in place with a locking screw Sc. The electrode gap is set by loosening Sc, placing a distance gauge, which protrudes through the spark stand opening H, onto P, allowing the spring to work to adjust the electrode and then tightening Sc again. Such a distance gauge (labeled Sp) can be seen in Figure 3.42. These distance gauges are also known as spacers.

**Figure 3.42:** Spacer, V-gauge and gas-tight bell. Printed with the friendly permission of SPECTRO Analytical Instruments GmbH, Boschstr. 10, 47533 Kleve, Germany.

The radiation generated is made available to the spectrometer optics either directly (through the left window W) or by transmission (through the right window W) through a fiberoptic LL. The spark stand opening H is beveled on the bottom (indicated by dotted lines in Figure 3.41) to prevent undesirable shading of the radiation generated near the sample surface. In this way, the fiber optic can view the entire plasma. While the fiber optic transmits only radiation above 185 nm, even shorter wavelength radiation enters the spectrometer via direct light path. In the wavelength range below 200 nm, radiation generated in the cathode drop area directly below the sample surface leads to an increased background signal and limits the detection sensitivity. An adjustable diaphragm B prevents this background radiation from entering the optic. Often, it is useful to measure partially with and partially without the shutter, because using the diaphragm can limit the repeatability and increase scattering in the calibration functions. For this reason, B is frequently designed to be switchable; the diaphragm can be moved out of the light path. The diaphragm does not necessarily have to be located inside the spark stand. It may be found in other places along the light path, for example, inside the optical system.

The argon stream is usually directed from the windows W into the spark stand where it then flows through in the direction of the argon outlet A. This type of argon flow protects the windows from contamination and removes the metal condensate generated by the spark. This is not completely achievable. A portion settles on the underside of the spark stand plate or in the interior of the spark stand. This is why there is a glass, boron nitride or mica insert G on the insulator K. It often has the shape of a pot to take up the condensate not transported by the argon and to increase the electrical leakage path. This prevents a high-resistance short circuit of the ignition pulse. The levels of oxygen, moisture and hydrocarbons in the argon must not be significantly higher in the gap as in the container from which the argon comes. Therefore, every window W as well as the spark stand plate is sealed against the spark stand housing with O-rings.

If small parts that do not completely cover the spark stand opening are to be measured, they can be covered with a gas-tight adapter. Such an adapter (a gas-tight bell), labeled with G, is shown in Figure 3.42. The plunger at the top ensures that current flows through the piece to be analyzed. A wire adapter can also be seen (Figure 3.42, D) along with a copper block that is placed on foils to be analyzed to ensure planar placement and heat dissipation. There is a V-shaped gauge (Figure 3.42, V) on the plate to facilitate positioning of the adapters. There is also a screw (Figure 3.42, S) with which the bell, copper block or wire adapter can be fixed. The screw replaces the sample clamp, which would otherwise be used. The sample holder can be folded out of the way for small part analysis.

Spectrometer systems are usually intended for the analysis of alloys of different metal bases. It can happen that one of the base metals is to be determined as a trace element in another base. If such a trace element is to be determined after previously measuring samples that contain this element as the main element, contamination by the condensate in the spark stand may hinder trace element analysis.

The following measures can be taken against this:
–   The entire spark stand can be designed so that it is exchangeable. This method has the disadvantage that moisture usually adheres to the surfaces of the newly mounted spark stand even when stored under dry conditions. There is also air in the inevitable dead volumes that hinder the discharge and, above all, the already-critical analysis of oxygen and nitrogen. The spark stand could be stored in an argon atmosphere, but this would be an additional effort.
–   It is usually more practical to change only the spark stand plate, electrode and spark stand inserts. There is then no air containing dead space. Storage in a desiccator is recommended to avoid moisture adherence. The electrode cleaning brush should be used for a single metal base, because condensate particles from the brush could get on the electrode tip when cleaning the electrode.

Contamination can also occur if the sample is moved over the surface of the spark stand plate when being positioned. For this reason, it should also be cleaned regularly if there is a risk of contamination. This is the case when, for example, carbon steel is to be measured after the analysis of high-alloy steels. Figure 3.3 shows examples of the spark stands from modern laboratory spectrometers.

An interesting spark stand variation is described in the German patent application DE102009018253 A1 [44]. Here, the spark stand has two electrodes that alternately spark the sample. This can greatly reduce the measuring time. One might think that the same effect could be achieved by increasing the spark repetition frequency, but this is not the case. When the frequency is increased, the locations on the sample that are directly opposite the electrode tip are excessively heated. The temperature increase influences the analytical performance, particularly the reproducibility. The solution described in this invention [44] does not have this disadvantage, as the energy of the sparks is distributed over twice the area.

The geometry of the spark stand chamber and the level of the gas flow determine how well the condensate produced by sparking is transported out of the spark stand. High gas flows lead to higher costs and are, therefore, undesirable. A laminar argon stream, with as little swirling as possible, has proven itself to be advantageous for keeping the spark stand chamber free of condensate. Spark stands constructed favorably in this respect are described in the European patent specification EP2612133 B1 [45].

### 3.3.2 Sample stands for arc instruments

Currently (2018), the electrical arc is used as an excitation source only in mobile spectrometers in significant quantities. The sample stand for mobile spectrometers is usually constructed so that the electrode holder, radiation output and exhaust are used for both arc and spark operation. Switching between arc and spark modes is performed by connecting the appropriate adapter. Figure 3.44 shows a probe head without an adapter to the left. The same probe with the spark adapter is shown in the middle. It can be seen with the arc adapter on the right. Probes for mobile spectrometers are discussed in detail in Section 7.2.

Arc and spark stands differ in the following aspects:
- The requirements for gas-tightness are lower for arc than for spark stands that operate with an argon atmosphere. The reduced requirements for arc also apply when the arc is used with air from which $CO_2$ has been removed to enable the analysis of carbon. Several 10 vpm carbon dioxide are unproblematic.
- In arc mode, the counter electrode becomes hotter than in spark mode. The electrode holder must, therefore, be cooled. In most cases, passive cooling via heatsink is not sufficient. Active ventilation of the electrode holder is especially

necessary for mobile spectrometers. Because of size and weight limitations the probe cannot be equipped with a high-volume heatsink.

– The electrode degradation is larger in arc mode than in spark mode. Thus, readjusting of the electrode gap must be quick and simple.

– Contamination of the counter electrode is more persistent for arc mode than spark, where it is only necessary to remove a little condensate dust from the tip of the electrode. Therefore, a sturdy brush (as shown in Figure 3.43 above) is required. The opening of the stand must be large enough to enable cleaning.

– There is no removal of the condensate with a gas flow in arc mode. This applies even when flushing with purified air. The flushing flow may only be low here as higher flows lead to dispersion, i.e., there is a change in the position of the plasma. Such dispersion must be avoided because the change in position may vary from measurement to measurement. The contamination can be stronger when there is an air tight seal between sample and spark stand opening. For these reasons, it is necessary to regularly clean the arc sample stand. The construction of the stand must enable fast and simple cleaning. If a proper cleaning is neglected, there can be a (high-resistance) short circuit between the counter electrode (anode) and the spark stand housing (cathode, connected to the sample). The risk of contamination of the window and lens facing the plasma is greater in the arc mode. These surfaces also need to be quick and easy to clean.

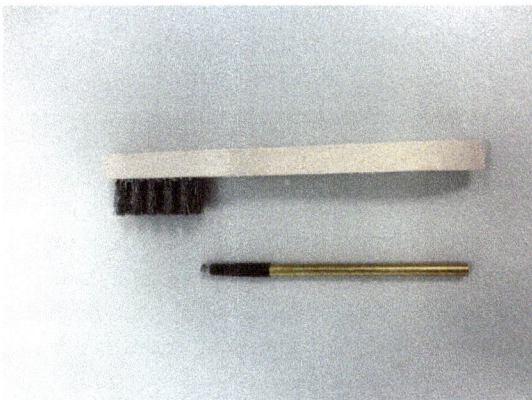

Figure 3.43: Electrode brush for arc (top) and spark (bottom). Printed with the friendly permission of SPECTRO Analytical Instruments GmbH, Boschstr. 10, 47533 Kleve, Germany.

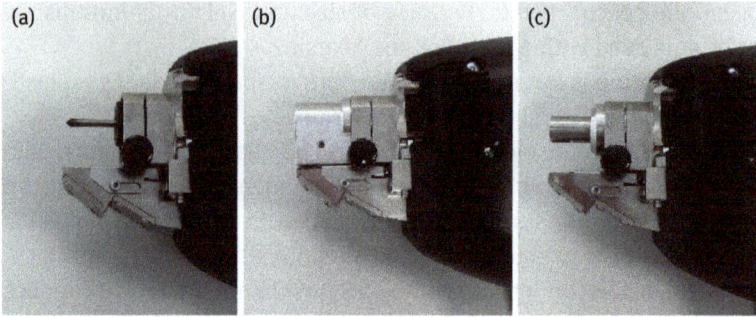

Figure 3.44: Probe without adapter (a), with spark adapter, (b) and with arc adapter (c). Printed with the friendly permission of SPECTRO Analytical Instruments GmbH, Boschstr. 10, 47533 Kleve, Germany.

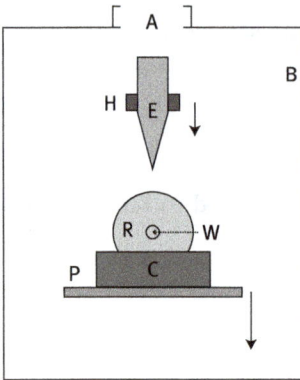

Figure 3.45: Basic structure of a rotrode sample stand.

### 3.3.3 Sample stands for the rotrode

While there is great similarity between spark and arc sample stands, the rotrode sample stands have a completely different construction. Figure 3.45 shows the basic design.

A small container C holds approximately 2 ml of the liquid to be analyzed. The container is placed on a platform P. P can be lowered by about a centimeter for sample placement and is then brought into measuring position by pressing a lever. In the measuring position, the rotrode R, a small wheel made of spectrally pure graphite about 12 mm in diameter and 5 mm thick, is immersed in the liquid to be measured. There is a sharpened counter electrode E usually 6 mm in diameter fastened above the rotrode. It is also made of spectrally pure graphite. In most cases, graphite material with a carbon content greater than 99.9995% (the sum of all impurities is less than five ppm) is used for rotrodes and electrodes.

The electrode holder H can be fastened in one of two vertical positions. The electrode is placed into the holder in the lower position. Then it is raised into the desired gap distance with a lever. The rotrode rotates slowly before beginning the measurement. The spark process can start when the section of the rotrode opposite the counter electrode is wet. It is usually conducted with a rather low frequency of around 100 Hz and lasts between 10 s and 1 min. Like with the analysis of metal samples, there is a pre-spark before the measurement itself.

The following is required of a rotrode sample stand:
- The rotrode must be changed after every sample. This must be designed to be as simple as possible.
- The counter electrode must also be removed after every sample. It can be sharpened like a pencil and then be reused. When being reinserted, the tension on the spring of the electrode clamp should be loosened without tools by pulling on a knob so that the electrode can fall onto the rotrode surface. When the knob is released, the spring fixes the counter electrode in place.
- The chamber B around the rotrode stand should be air-tight. A door is opened to work on the stand. Sparking can only be started when the door is closed. Encapsulation is necessary because gases that are hazardous can form during the spark process. There is a connector A at the top of the chamber B to install an extractor hood.
- The interior of the chamber should be made of a corrosion resistant, easy-to-clean material, as contamination, e.g., through spillage of the liquids to be analyzed or from residue fall out, is inevitable. The materials for the interior should be fire resistant because generally flammable substances are analyzed. Stainless steels have proven their effectiveness.
- The shaft W, on which the rotrode is attached, should also be made from a corrosion resistant material. Preferably, a material that is not to be analyzed should be used, e.g., silver. It may be sufficient to select acid-resistant stainless steel as the shaft material when only less aggressive materials are to be analyzed.

Principally, the method is suited to the analysis of all elements. Carbon and components in the ambient air are excluded. However, the detection sensitivity is limited for the elements that have their most sensitive lines below 200 nm. Nadkarni [46] reports 0.01 ppm as the detection limit for magnesium, while that for tin is 0.88 ppm using the less sensitive line of 317.5 nm. The more sensitive line for tin, at 189.98 nm is strongly absorbed by the atmosphere.

## 3.4 Operating gas systems

Excitation by arc and laser occasionally takes place in an air atmosphere. Even the analysis of lubricants with the rotrode technique does not require an externally supplied operating gas. In most other cases, however, the sample stand is flushed.

### 3.4.1 Commercially available operating gases

By far, the most widely used auxiliary gas is argon. It is almost always used in conjunction with spark and frequently when using laser as a source. That is why Section 3.4 is concerned mainly with argon and argon-hydrogen mixtures. Purified air is occasionally used as an operating gas in mobile spectrometers.

Table 3.2 displays the argon grades commonly used in spectrometry together with maximum oxygen, water vapor, nitrogen and carbon contamination. In many cases, the grade "argon for spectrometry" (argon 4.8), a variety with a maximum of 20 vpm total contamination, is sufficient. This grade can, however, contain up to 10 vpm nitrogen, which is problematic when this element is to be determined. Nitrogen in steel must often be monitored at levels below 100 ppm, whereby it is necessary to distinguish between nitrogen dissolved in the metallic matrix and nitrogen bound to other elements. See for this, e.g., Niederstraßer [47] or Schriever [48]. What is particularly disturbing is that every vpm nitrogen in the gas simulates an intensity increase of around 20–50 ppm in the sample. The multiplier is dependent on the spark parameters. In these cases, the use of argon 6.0, which has a maximum of 0.5 vpm nitrogen, is strongly recommended. This grade is, of course, much more expensive than argon 4.8. Argon 5.0 is the compromise to be considered, while it is somewhat more expensive than argon 4.8, it contains only half as much nitrogen.

Table 3.2: Argon grades for spark spectroscopy.

| | Argon for spectroscopy (Ar 4.8) | Ar 5.0 | Ar 6.0 | Argon/Hydrogen mixture for spectroscopy |
|---|---|---|---|---|
| Argon (Vol.%) | 99.998 | 99.999 | 99.9999 | balance |
| $H_2$ (Vol.%) | -,- | -,- | -,- | 2–5% |
| Oxygen (vpm) | 3 | 2 | ≤0.5 | 2 |
| Nitrogen (vpm) | 10 | 5 | ≤0.5–1* | 10 |
| Water vapor (vpm) | 5 | 3 | ≤0.5 | 3 |
| Carbon compounds (vpm) | ≤0.5 | ≤0.2 | ≤0.2 | ≤0.5 |

*: Manufacturer specific

Sometimes, when the requirements are not too high, it is possible to achieve surprisingly good results with welding argon. This is, however, not always the case. Then, the burn spots are whitish due to diffuse discharges (see Section 3.2.2.3), and there is too little and irregular ablation. The reason for the varying results is that while the welding argon cylinders are often filled with the same grade of argon as those filled with the grade "for spectrometry," the preparations are, however different: Argon cylinders for spectrometer argon are evacuated before being filled; thus, the resulting purity of the full welding argon cylinders depends on their history.

Argon hydrogen mixtures consist of argon 4.8 and 2–5% hydrogen 5.0, i.e., hydrogen with a purity of 99.999%. In most cases, an addition of about 2% is chosen. The hydrogen binds the residual oxygen content of the argon. This is advantageous when small oxygen levels in the sample need to be determined, for example, during analysis of electrolytic copper. Slickers ([24], p. 334) established that while argon-hydrogen is beneficial for the analysis of alloys of iron, nickel and cobalt bases, it can lead to diffuse discharges when larger amounts of B, Al, Mg, Zn and Ti are present. Modern spectrometer systems are frequently designed for the determination of the alloys of several bases. Many spectrometer manufacturers use exclusively argon as the operating gas so that users are not expected to purchase different operating gases.

### 3.4.2 Argon supply forms and transport of the gas to the spectrometer

In Europe, operating gases are commercially available in cylinders with volumes of 2, 5, 10, 20 or 50 L. Only the 50 L cylinders are suitable for laboratory spectrometers. Mobile spectrometers, on the other hand, also use smaller containers, because the total weight of the system is often important. The cylinders are delivered with a filling pressure of 200 bar. The 50 L bottles are also available with a 300 bar filling pressure, which corresponds to an argon quantity of about 15 $m^3$.

In addition to single cylinders, it is also possible to get bottles in bundles. In this case, there are usually 12 bottles connected in parallel. Here, 200 or 300 bar filling pressures are available.

Using a bundle has the advantage that changing the container is required less frequently, because this is associated with an interruption of the argon supply. This procedure can lead to leaks or the entry of nitrogen, moisture, oxygen and hydrocarbons. Then it is necessary to introduce a flush phase to reinstate the analytical performance of the system. In view of this, the use of a tank with liquid argon is a better choice.

The argon is delivered with a tanker truck and filled into a tank with a size between 200 and 10,000 L. These tanks resemble huge thermos flasks, in which the temperature must be kept low to keep the argon in a liquid state. The necessary cold arises from expansion of the gas removed. Figure 3.46 shows such a tank.

Liquid argon for spectrometry purposes is delivered in the varieties 5.0 and 6.0. Technically, operation of spectrometers with liquid argon from a tank is advantageous because the supply line never needs to be interrupted. However, this solution is only economical when there is a certain minimum consumption.

The operating gas container is not usually located directly next to the instrument, but even in this case, a supply line between the spectrometer and the argon

**Figure 3.46:** Tank for liquid argon. Printed with the friendly permission of SPECTRO Analytical Instruments GmbH, Boschstr. 10, 47533 Kleve, Germany.

container is required. Care is necessary for the conception and implementation of this connection.

Stainless-steel tubing and those made of copper in accordance with DIN EN 13348 [49] are suitable.

The argon supply lines must be gas-tight, because where gas leaks out, ambient air can enter the system. Once the supply lines have been installed, they must be pressurized. There should be no significant fall in pressure. It is recommended to leave the installation of the argon line to an experienced specialist. The gas suppliers can generally help in finding the right company.

### 3.4.3 Argon purifying systems

If there is no sufficiently pure argon available or the cleanliness of the feed line is questionable, then the argon should be purified right before the spectrometer. This can be done in two different ways:
- The argon stream can be passed through cleaning cartridges in which water vapor, oxygen and hydrocarbons are removed. Combined cartridges that remove

all the mentioned impurities are also available. The oxygen is removed via chemisorption, i.e., removed by a chemical reaction with a suitable reagent of chromium, nickel, titanium or copper base. Moisture and hydrocarbons are normally removed by passing the argon through a molecular sieve made of suitable zeolites or through an active carbon filter. The impurities remain in the pores of the molecular sieve, the argon is allowed through. Molecular sieves of zeolites can be regenerated by heating. Some of the reagents for oxygen removal can also be recycled. This is usually done by flushing them with hydrogen at high temperatures (several 100 °C).

It is often possible to easily distinguish between used and fresh cleaning media. This is made possible by the transparent cartridge housing and the addition of suitable indicator substances.

The color change generated by the indicator makes it possible to estimate how long the capacity of the cartridge will last. If the spectrometer system is to be operated in the European Union, the restrictions applicable here for the use of hexavalent chromium must be observed. These are set out in the so-called "ROHS Directive" [50]. For this reason, some of the indicator reagents cannot be used.

- Argon cleaning with passive cartridges leaves the nitrogen content in the argon unchanged. If the nitrogen must also be removed, active purification systems are available. Tried and tested systems are offered by the company Sircal Instruments, for example. The argon is passed over titanium and copper reagents that are heated to 680 °C and 450 °C, respectively, during operation [51].

### 3.4.4 The spectrometer's internal argon system

Argon is used for the following purposes in the spectrometer:
1. Argon is flushed through the spark stand during the measurement. The functions in the excitation process were discussed in detail in Section 3.2. It has also already been explained that the argon stream transports the metal condensate generated during sparking out of the spark stand. Typical argon flows during measurement are about 2 liters per minute.
2. The argon stream is switched on up to three seconds before the measurement begins to flush out any oxygen introduced into the spark stand when changing the sample. The flow rates are usually the same as those used during the measurement, but occasionally, a short strong burst of argon is triggered.
3. A reduced gas flow remains between the measurements. This keeps the light paths in the spark stand transparent for radiation between 115 and 190 nm. In this way, the instrument is also ready for operation even after a pause in sparking. The flush rate usually is between 1 and 30 liters per hour.
4. Smaller laboratory spectrometers and the UV optics in mobile spectrometers are often kept transparent with argon flushing. This gas flow is always switched on

whenever the flushing gas flow described in point 3 is turned on. Occasionally, the argon is first passed through the optic and then reused to flush the spark stand. In this case, intermediate cleaning of the gas is recommended to at least remove the moisture flushed out of the optic.

5. For reasons of cost, it is desirable to completely turn off the argon supply during prolonged inactivity of the spectrometer, e.g., over the weekend. When the argon is switched back on, however, the instrument must then be flushed for a relatively long time (e.g., 30 min) with a higher flow rate. This reactivation can be controlled with the software program so that the spectrometer is ready to measure at the start of the workday.

6. Shutters to block the light paths and the switchable shutter mentioned in Section 3.3.1 are preferably operated with pneumatic cylinders for reasons of speed and reliability. The argon present in the spectrometer is also utilized for this purpose.

7. Large spectrometer optics are often filled with argon. This argon is then constantly recirculated by pumping in a closed system and is passed through a purifying system like that described in Section 3.4.3.

It is technically safer and, due to the simpler installation, usually more economical to realize the argon connections with holes drilled in a so-called "argon block" instead of working with individual copper tubes.

Figure 3.47 shows such a block.

Figure 3.47: Argon block of a laboratory spectrometer. Printed with the friendly permission of SPECTRO Analytical Instruments GmbH, Boschstr. 10, 47533 Kleve, Germany.

### 3.4.5 Exhaust system

The argon that leaves the spark stand through the outlet cannot be directly released into the ambient air because it contains condensate. This condensate consists of particulate matter that is undesirable in the workplace. Figure 3.48 shows a possible design for a waste gas cleaning system. The waste gas hose S coming from the spark stand is immersed in a wash bottle W. The wash bottle is filled to about 20 cm over the end of the hose with water. The argon flows out of the hose and rises to the surface of the water. A portion of the condensate is washed out in this way. Large gas bubbles are disadvantageous as they reduce the washing effect and can cause pressure fluctuations in the spark stand. For this reason, there is a perforated inset E at the end of the hose. The holes in this inset must not be too small, otherwise there is a risk of clogging. The water in the wash bottle not only removes condensate, but also prevents oxygen from entering the spark stand. However, it is detrimental that moisture diffuses back into the hose. This influences not only the discharge, it can also lead to clumping of the condensate and, thus, clogging of the exhaust hose. The hose should, therefore, be cleaned in regular intervals. It must be easily detachable from the spark stand to perform this cleaning quickly and efficiently. The pre-cleaned argon then enters a filter housing G that contains a filter cartridge F. The argon passes through the filter that consists of fibrous material. After this, the argon which is now fully free of condensate is either released directly into the surrounding environment or connected to an exhaust system. The latter is especially recommended if toxic materials are measured. In this case, an extractor should be directly attached to the spark stand plate to prevent condensate from exiting the spark stand opening. The filter cartridge must be replaced in regular intervals. It is important to be aware that the condensate contained in it can spontaneously ignite. The condensate is very finely grained, which

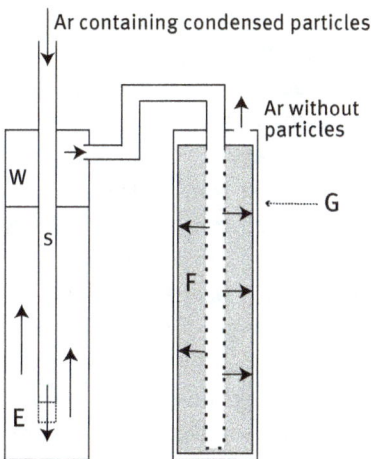

Figure 3.48: Exhaust gas cleaning system.

leads to a large total surface area. Care must be taken especially when opening the filter housing because in this moment, the filter is exposed to air.

## 3.5 Spectrometer optics

In the early years of spectrometry, spectral apparatuses equipped with prisms were used as optical systems. Figure 3.49 shows the design of such a spectral apparatus in a version that is suitable for photographic spectrum recording. It works as follows:

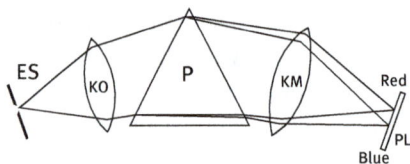

Figure 3.49: Principle of construction of a prism spectral apparatus.

The combined light emitted by the plasma passes through the optical system entrance slit ES that is located in the focal point of a collimator lens KO. The radiation exits KO as a beam of parallel light. A prism system P breaks up the radiation more or less depending on the wavelengths. There is then a beam of parallel light for each wavelength that exits the prism system at a separate angle that depends on the wavelength. Finally, a camera lens KM focuses the beams of light on the focal curve. There sharp images of the entrance slit are displayed next to each other from red to blue. Each image corresponds to a wavelength and, therefore, one of the energy transitions described in Figure 3.14 that can be assigned to a particular element.

Recording of the spectra was done directly through observation with the eye or by exposing a photographic plate PL, the light-sensitive side of which was placed on the focal curve. From the brightness of suitable lines in the spectrum, it was possible to determine the concentration of the associated alloying element. If the spectrum was observed with the eye, the intensity of the analyte line was compared with that from a neighboring line of the matrix element. This method of homologous line pairs was introduced by Gerlach and Schweitzer [18]. It was their idea to compare line pairs not too far away from each other in the spectrum that made it possible to determine concentrations regardless of the luminosity of the spectral apparatus. Before that, attempts were made to base the analysis on absolute intensity levels, which was not very promising. Even today, it is necessary to keep the analytical results independent of the overall throughput of the optical system. Here, for a given wavelength, overall throughput should be understood as the relationship between the light that reaches the sensor divided by the radiation that enters the optical system. Spectrometer software algorithms take this into account; see also Section 3.9.4, Calculating Intensity Ratios, and Section 3.9.6, Recalibration.

Figure 3.50 shows a spectroscope for the visual assessment of spectra. Radiation enters the optic O, which contains several prisms placed in series, through the lens system L. By rotating the drum T, the desired section of the spectra can be selected for observation through the eyepiece OK.

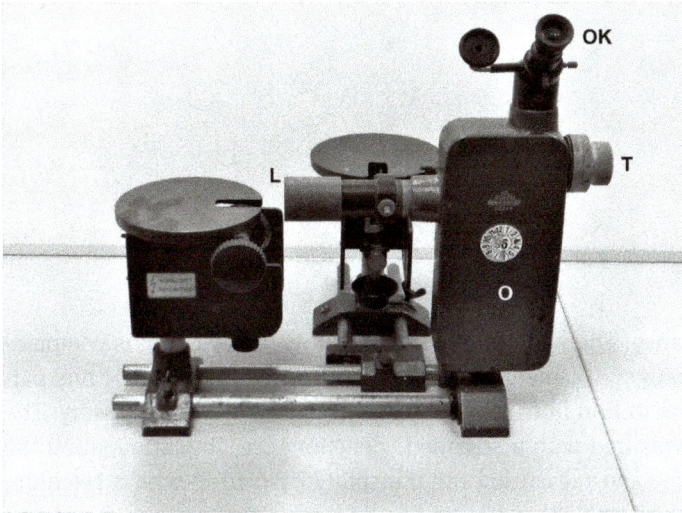

Figure 3.50: Spectroscope.

The spectral photographs were also evaluated by intensity comparisons of lines. First, the plate was developed, and then the density of the lines was photoelectrically determined. An example of such a photographic plate can be seen in Figure 3.51.

One drawback of photographic evaluation was the large amount of time required. But there were also benefits:

- When observing the spectrum with the eye, only one section of the spectrum is visible. While this can be shifted by rotating the prisms, examination of a larger number of elements is not practical. In contrast, the photo plate records the entire spectrum. Subsequent evaluation of any part of the spectrum is possible.
- The photo plate was the basis for the evaluation; in addition, the plate could be archived. So, it offered the possibility of documentation – and that long before computers, mass storage or printers were available.
- It is true that experienced users achieved astonishingly good results when comparing line intensities, but it important not to overlook the fact that visual observations are subject to variations and not all spectral examiners were capable of

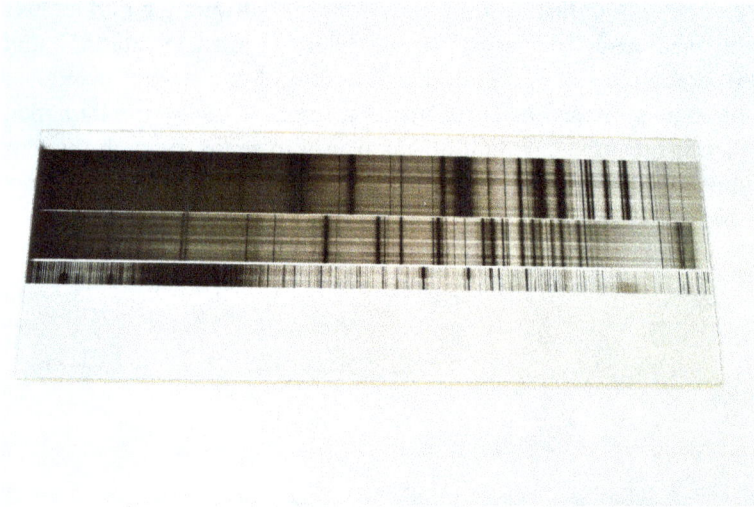

Figure 3.51: Exposed and developed photographic plate.

maximum performance. The photographic plate, on the other hand, is comparatively safe. The photographic plate could be aligned with characteristic line pat-
ters, making it easier to find lines and templates could be used if necessary. The
density could be measured with a so-called "densitometer," by illuminating the
spectrum from below and measuring the transmitted radiation with a selenium
photocell. This method was objective compared to visual assessment. It must be
noted that the transmitted light intensity could fluctuate and the exposure process was not always the same. But because it was the ratio between lines and an
internal standard that mattered, these fluctuations did not play a role.

Spectrometers with photographic plate recording were used in practice into the
1990s, e.g., for the testing of pure graphite. Spectrometers with photographic plates
became obsolete with the advent of systems that could record wide spectral ranges
with semiconductor multichannel detectors.

### 3.5.1 Rowland circle concave grating optics with photomultiplier tubes (PMT) as radiation detectors

The ability to simultaneously measure a large number of spectral lines and, at the
same time, to achieve rapid and objective results was first realized with the next generation of spectrometer optics. Optics of this design have slit-shaped apertures located on the focal curve, the so-called "exit slits." Each of these slits is adjusted so
that it isolates a single given wavelength from the total spectrum. A photomultiplier

tube (PMT) is mounted behind each exit slit so that the radiation from the spectral line falls onto the tube's photocathodes. The radiation is converted to a current in the PMT and can then be measured and evaluated. The way in which photomultiplier tubes and the associated readout electronics work are discussed in Sections 3.6.1 and 3.7.1. For decades, these modern optics were the preferred design in arc and spark spectrometers and are still manufactured in large quantities. The number of analyte lines available is limited due to cost and space constraints. For the same reasons, only a few internal standard lines can be installed. Frequently, only one internal standard is used per base element. The advantages resulting from the availability of the entire spectrum were sacrificed for the measuring speed and were only rediscovered in the subsequent optic generation.

The so-called "concave grating" is the central component for this type of optic. This is a spherical concave mirror on the surface of which are vertical grooves in very short, regular intervals. The resulting component has, on the one hand, the properties of a concave mirror, but on the other hand it can bend a portion of the radiation in a wavelength-dependent manner, i.e., diffract it.

Figure 3.52 shows the basic construction of a Rowland circle optic with the so-called "Paschen-Runge mount." For this type of assembly, the concave grating G, the entrance slit ES as well as several exit slits AS, of which only one is drawn in Figure 3.52, are mounted on a circle whose diameter corresponds to the radius of the grating curvature. The circle touches the center of the grating and forms a right angle to the grooves of the grating. The line N that is formed by the vertical groove at the center of the grating is called the "grating normal." The term "in the normal position" is often used for the intersection $S_N$ of this line with the focal curve. It becomes clear from the context whether "normal" means N or $S_N$. An optic with the form shown in Figure 3.52 is also known as a polychromator because the simultaneous measurement of multiple line intensities is possible with it.

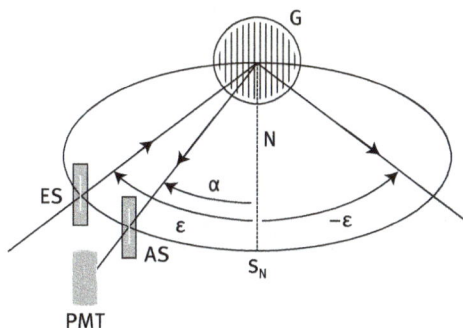

Figure 3.52: Rowland circle optics in Paschen-Runge mount.

Imagine a point-like source of light on the focal curve. The angle between the light source to the center of the grating and the grating normal is then ε. Due to the mirror properties of the concave grating, it follows that the image of the light source appears on the opposite side of the normal, i.e., with an angle of -ε.

However, the grating is not just a simple concave mirror. A portion of the radiation is bent. The locations on which the diffraction images appear are dependent on the wavelengths of the radiation source. The correlation between the wavelength falling at an angle of ε and the angle of the diffracted beam is calculated with the following equation:

$$\lambda = \frac{\sin(\alpha) + \sin(\varepsilon)}{N*G}$$
(3.8)

Where:
ε    the angle of incidence (angle between grating normal and the line center of grating-entrance slit)
α    the angle of diffraction (angle between grating normal and the line center of grating-line position)
λ    the wavelength (in mm) that is diffracted with the angle of diffraction
G    the number of grooves per mm on the grating (grating groove number)
N    the diffraction order

The diffraction order N is an integer number that can also have a negative sign. The radiation is generally distributed over several diffraction orders, whereby the intensity distribution depends mainly on the shape of the grating grooves. This subject is explained in more detail together with the so-called "blaze angle" in Section 3.5.5.

Angles of incidence between 40° and 45° are usually chosen; the exit slits are then mounted between the entrance slit and the grating normal (see Figure 3.52). Gratings with 1,200 to 3,600 grooves per mm are generally used in optics for arc/spark spectrometers. If an angle of incidence of 43° is assumed, the spectral regions listed in Table 3.3 can be used in the first and the second diffraction order.

Table 3.3: Wavelength covered between 0° and 43° with an angle of incidence of 43°.

| Grooves per mm | Wavelength range between 0° and 43°, first order of diffraction (nm) | Wavelength range between 0°and 43°, second order of diffraction (nm) |
|---|---|---|
| 1200 | 568.3–1136.7 | 284.2–568.3 |
| 1800 | 378.9–757.8 | 189.4–378.9 |
| 2400 | 284.2–568.3 | 142.1–284.2 |
| 2700 | 252.6–505.2 | 126.3–252.6 |
| 3600 | 189.4–378.9 | 94.7–189.4 |

The focus of Rowland-circle gratings approximately obeys the equations (see e.g., [52]):

Spectral focus

$$\frac{\cos^2(\alpha)}{la} - \frac{\cos(\alpha)}{F} + \frac{\cos^2(\varepsilon)}{le} - \frac{\cos(\varepsilon)}{F} = 0 \qquad (3.9)$$

Meridional focus

$$\frac{1}{la} - \frac{\cos(\alpha)}{F} + \frac{1}{le} - \frac{\cos(\varepsilon)}{F} + N\lambda C = 0 \qquad (3.10)$$

The terms correspond to those used in eq. (3.8) with the following additions:

$F$   Radius of spherical curvature of the blank
$le$   Distance between entrance slit and center of the grating
$la$   Distance between exit slit and center of the grating
$C$   Constant, determined by grating fabrication, for details see Section 3.5.3.

At larger angles, the spectral ("sagittal") focus and the vertical ("meridional") focus are increasingly divergent. A point-like entrance slit generates an image perpendicular to the optical bank with a height $h_s > 0$ for a slit width $b_s \sim 0$ in the sagittal focus and an image parallel to the optical bank with a width $b_m > 0$ for a slit height $h_m \sim 0$ in the meridional focus.

Since the exit slit must, of course, always be set in the spectral focus, the lines as diffraction images of the entrance aperture are longer than the illuminated height of the entrance aperture. It is easy to calculate the extension of the aperture images using the theorem of intersecting lines from geometry, the difference of the focal lengths and the illumination height of the grating. A point-like (stigmatic) entrance aperture leads to aperture images in line form. For this reason, the effect is called "astigmatism."

Long slits are unfavorable (adjustment problems, noncurved, long slits lead to defocusing either in the middle of the slit or on the ends of the slit) and are, therefore, not used. Short apertures lead to loss of light.

It has already been mentioned that grating groove numbers between 1,200 and 3,600 mm$^{-1}$ are common. In most cases, the gratings are round and have a diameter between 50 and 70 mm. Common Rowland circle diameters lie between 300 and 1,000 mm. The resolution capability of Rowland circle optics is limited by diffraction effects; however, optics with smaller Rowland circle diameters are still possible. They are, however, seldom found in combination with PMTs as detectors. One factor limiting miniaturization is the necessity to manually adjust the exit slit so that it is aligned with the spectral line. Slit widths from 15 to 50 μm are common; rarely found are wider slits up to 150 μm (e.g., for the internal standard). A detailed discussion about the dimensions of slit widths can be found in Clark [53] (p. 40 et seq.). The entrance slit should always be narrower than the exit slit width so that the lines are completely within the exit slit even with the presence of small drifts (changes in position of the lines due to changes in the environmental parameters).

Drift effects cannot be completely avoided; they are caused by different heat expansion in the grating area of the optical system as well as changes in dispersion of the air due to pressure fluctuations. This is especially critical when the shifts for the internal standard and the analyte are different.

Positioning must be performed with an accuracy of better than one-tenth of the slit width. This process is also known as profiling. Reducing the focal and slit widths leads to very high requirements on the absolute positioning accuracy, which is problematic in practice.

The space available on the focal curve for slits and photomultipliers represents a further problem. Common photomultiplier tubes have a diameter of $1^1/_8$ inches (28 mm) or $1/_2$ inch (13 mm) (see Section 3.6.1); they are then very large compared to the slit widths. This can lead to problems when important spectral lines are close together.

### Here a Numerical Example:

If the Rowland circle diameter is reduced to 150 mm and assuming a grating groove number of 2,400 mm$^{-1}$ and an angle of incidence $\varepsilon$ of 43°, then there is a mean reciprocal dispersion of about 2.5 nm/mm between 0 and 43°.

The wavelength range from 284 to 568 nm lies between normal and the entrance slit. It is only 112.5 mm long. There is room for four big or eight small PMTs, all of them mounted side by side (for common dimensions of PMTs see Section 3.6.1). If the mounts for the slits are 5 mm wide, the minimum possible distance between two lines (for a reciprocal dispersion of 2.5 nm/mm) is 12.5 nm. However, important lines that can hardly be replaced frequently lie even closer together. For example, the important lines Al 396.2 nm, W 400.8 nm, Mn 403.4 nm and Pb 405.7 nm are within a wavelength interval of only 9.5 nm, which corresponds to less than 4 mm on the focal curve.

A polychromator with a Rowland circle diameter of 750 mm, a grating groove number of 2,400 mm$^{-1}$ and an angle of incidence of 43° enables minimum line distances of 2.5 nm with slit mounts of the same width. Here, too, it may be necessary to use lines that are, from an analytical viewpoint, less suitable because the optimal lines are located in places where there are a large number of lines to be measured. Approximately twenty large or 40 small PMTs can be mounted side by side. Frequently, it is necessary to locate the photomultiplier tube where there is space and to use mirrors to transmit the radiation coming through the entrance slit to the PMT cathode. Figure 3.53 shows the principle. By using mirrors, a larger number of PMTs can be mounted. Nevertheless, the space available is limited.

Especially for mobile spectrometers, polychromators with long focal lengths are too heavy and impractical to be held directly on the workpiece to be tested. For this reason, fiber optics are used to connect the spark probe with the optical system. The properties of fiber optics are addressed in Section 3.5.5.2. There it is also described that the use of fiber optics is only possible for wavelengths above 185 nm.

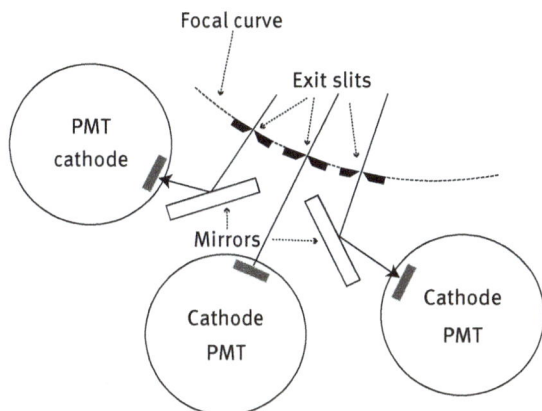

**Figure 3.53:** Mirroring of radiation passing through the slit onto the PMT cathodes.

However, for the analysis of steel, it is often necessary to determine the elements phosphorous and sulfur, for which the main detection lines lie below this limit with 178.3 nm and 180.7 nm, respectively. A polychromator that is compact and still achieves a high resolution can be designed, if only a limited range of short wavelengths has to be covered. The short wavelengths enable a high grating groove number in combination with reasonable angles as can be calculated using the grating equation, eq. (3.8).

A grating with a 150 mm focal width and 3,600 grooves $mm^{-1}$ achieves in the second order the same dispersion as an optic with 2,400 grooves $mm^{-1}$ and a focal length of 450 mm in the first order. For $\varepsilon = 40°$ and using eq. (3.8), calculate:

$$\alpha_{P\ 178.3} = 39.9°$$
$$\alpha_{S\ 180.7} = 41.2°$$

A narrow design, mainly limited by the diameter of the grating, is possible due to the small angle differences. By taking advantage of the properties of the concave mirror of the grating, the entrance slit can be located above or below the exit focal plane. Flushing this small optic with argon enables the light paths to be transparent for short wavelength radiation. With a suitable construction, an optic volume of only 200 ml to be flushed can be achieved. On the previous page, the average reciprocal dispersion was calculated for a diffraction angle of 43° for a Rowland circle optic with 2,400 grating grooves per mm and 150 mm focal length. For this purpose, the width of the wavelength range covered ($\lambda_H$ bis $\lambda_L$) was divided by the length of the accompanying Rowland circle section. The section length is determined as follows:

First, the angles of diffraction $\alpha_H$ and $\alpha_L$ associated with the boundaries of the wavelength range are calculated. The desired arc then covers an angle of $2 \times (\alpha_H - \alpha_L)$ with respect ot the center of the Rowland circle. With the circumference of the Rowland circle

$\pi \times D$, one then obtains $\pi \times D \times 2 \times (\alpha_H - \alpha_L) / 360$ as the circular arc length. When this is limited to the first order of diffraction, the following equation is obtained:

$$RDISP_{AVG} = \frac{360^*(\lambda_H - \lambda_L)}{D^*\pi^*2^*(\alpha_H - \alpha_L)} \tag{3.11}$$

In this equation, $RDISP_{AVG}$ is the mean reciprocal dispersion in nm/mm and D is the Rowland circle diameter in mm. The angles are to be used in degrees and $\lambda_H$ and $\lambda_L$ in nm. The factor two is included because according to the grating equation the diffraction angle is used to calculate the angle between the line positions on the focal curve and the center of the grating. The associated angle is exactly twice as large with respect to the center point of the Rowland circle. (see Figure 3.54).

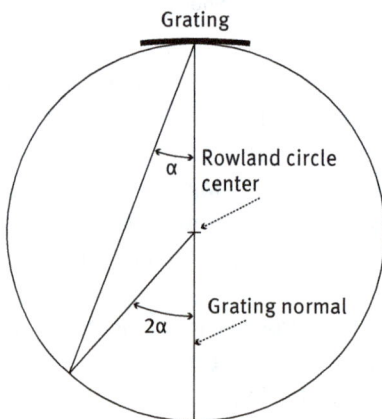

Figure 3.54: Diffraction angle and associated center angle.

If the angles $\alpha_H$ and $\alpha_L$ are known, then the wavelengths $\lambda_H$ and $\lambda_L$ are easy to determine using the grating eq. (3.8) and vice versa. Equation (3.12) shows the form of the equation when the wavelengths are expressed in the associated angles.

$$RDISP_{AVG} = \frac{360^*(\sin(\alpha_H) - \sin(\alpha_L))^*10^6}{D^*\pi^*2^*(\alpha_H - \alpha_L)^*G} \tag{3.12}$$

As in eq. (3.8), G denotes the grating groove number per mm. The multiplicator $10^6$ in the numerator ensures that the result still has the unit nm per mm. Note that the choice of the angle of incidence has no influence on the mean reciprocal dispersion.

In practice, the question often arises as to how the reciprocal dispersion is at a given angle $\alpha$ or for the associated wavelength $\lambda$. This information is easily obtained by differentiating eq. (3.8) with respect to $\alpha$:

$$\frac{d\lambda}{d\alpha_B} = \frac{\cos_B(\alpha_B)}{N^*G} \tag{3.13}$$

The term "$\alpha_B$" is the angle $\alpha$ expressed in radians. $\cos_B$ is the cosine function for arguments in radian measure. In our examples, we have until now always used eq. (3.8) so that the angles were specified in degrees and the cosine function, which demands arguments in degrees, was also used, although the grating equation is also valid for notation in radian measure.

If eq. (3.13) is converted back to the more manageable degree notation, the following is obtained:

$$\frac{d\lambda}{d\alpha} = \frac{\cos(\alpha)*2*\pi}{N*G*360} \tag{3.14}$$

The term on the right side of the equation is now the wavelength coverage per degree that applies for an angle of reflection $\alpha$. If this is divided by the term $D \times 2 \times \pi/360$, which corresponds to the length of the focal curve section per degree diffraction angle, then, after reducing, the desired equation for reciprocal dispersion $RDISP_\alpha$ is obtained. $RDISP_\alpha$ gives the reciprocal dispersion at the location on the Rowland circle where the angle between the grating normal and the diffracted beam has an angle of $\alpha$. Once again, the factor $10^6$ is used to obtain a result with the manageable unit nm/mm.

$$RDISP_\alpha = \frac{\cos(\alpha)}{N*G*D}*10^6 \tag{3.15}$$

Again, there is no dependence on the angle with which the incoming beam hits the grating. In our numerical example, we calculated the mean reciprocal dispersion for an optic with a 2,400 groove per mm grating and a 150 mm Rowland circle diameter for the range of angles between 0° and 43°. It was approximately 2.5 nm/mm.

If the extreme angles and the middle angle are used in eq. (3.15), the following is obtained:

$RDISP_{0°}$ = 2.7777 nm/mm
$RDISP_{21,5°}$ = 2.5844 nm/mm
$RDISP_{43°}$ = 2.0315 nm/mm

The reciprocal dispersion thus decreases with increasing angle of diffraction by the cosine of this angle.

## 3.5.2 Optics with Rowland circle gratings and detector arrays for detection

In the previous section, it was shown that it is the size of the large photomultiplier tubes, compared to the small slit width, that limits the construction of compact optical systems. Developments in the area of semiconductor technology have produced detectors that are significantly smaller than photomultiplier tubes and that offer the possibility of seamlessly detecting complete wavelength ranges.

Light-sensitive semiconductor components have been known for a long time. The phototransistor was already invented by John Northrup Shive in 1948 and has found commercial use since the 1950s, e.g., in punch card readers. However, the radiation sensitivity of the early components was far from sufficient for spectrometric applications. This changed with the invention of the so-called "CCD detectors" around 1970. Each of these detectors consists of many light-sensitive elements, the so-called "pixels" (derived from picture elements), that are arranged as rectangles or as rows. When arranged in rows, the pixels can be manufactured in dimensions similar to those for the slit openings. The pixels are connected with an analog shift register that, in a kind of bucket chain, leads the measured charges pixel by pixel to an amplifier. Its output is connected with an IC housing pin and can be read out and processed from there. Figure 3.55 shows a CCD chip and a photomultiplier tube. The CCD shown here can be understood to be a sequence of 2048 exit slits of 14 μm width and 200 μm in height. Recently, there are also detectors using the CMOS technique, which are designed with a different semiconductor technology, but can be integrated into the optic in a similar way as CCDs. Section 3.6.2 deals extensively with the CCD and CMOS technologies and other semiconductor detectors.

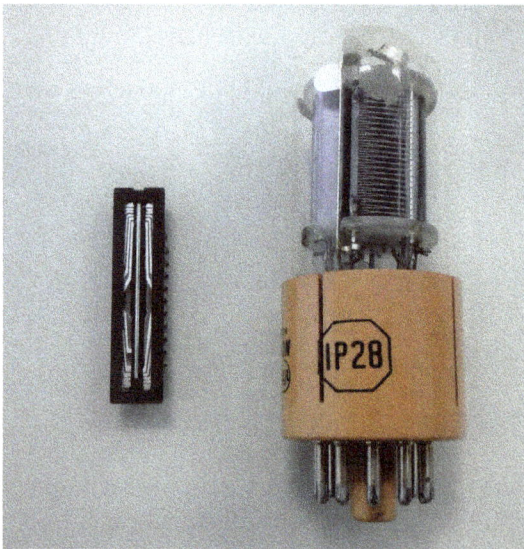

Figure 3.55: CCD chip (left) and photomultiplier (right). Printed with the friendly permission of SPECTRO Analytical Instruments GmbH, Boschstr. 10, 47533 Kleve, Germany.

Focal curve

Locations with minimal
defocussing

Locations of the pixel
surfaces

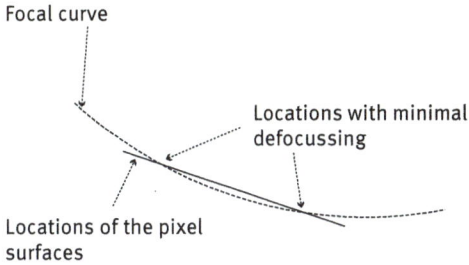

Figure 3.56: Adaptation of the linear sensor array to the focal curve.

The detector arrays can be used by lining up the detectors along the focal curve.

However, there are two problems:

- The focal curve is a circular arc; the detectors are, in contrast, linear. This leads to the fact that the detector lies exactly on the focal curve on only two points (see Figure 3.56). Pixels that are not on the intersections view the spectrum more or less heavily defocused; i.e., the lines are broader there.
  The line broadening $\delta_{LB}$ is calculated with the theorem on intersecting lines:

$$\delta_{LB} = \frac{B_G \cdot |la - la'|}{la} \qquad (3.16)$$

  $B_G$ denotes the illumination width of the grating, $la$ the distance between the center of the grating and the focal curve and $la'$ the distance between the center of the grating and the detector (all units in eq. (3.16) are meters).

- The housing of the detector chip extends far beyond the optically active width on both sides. If the detector chips were simply lined up chip by chip along the focal curve, there would be large gaps in the recorded spectrum due to these protrusions. Thus, mirrors are usually positioned in front of the focal curve, as seen from the grating. The mirrors are placed at an angle of 45° and reflect the radiation alternately upward and downward. Figure 3.57 shows the principle. The width of the mirrors is sized so that they lead radiation only toward the light-active area of the detector. When the mirrors are placed edge on edge, it is possible to record the entire spectrum without gaps. Since the spectrum is somewhat blurred on the edges of the mirrors, a spectral line located there may appear at the end of one detector as well as at the beginning of the next. The width of the radiation in the transition area can once again be calculated with eq. (3.16). Example: If the mirrors are located la − la' = 20 mm in front of the focal curve, that is, at a distance of la = 500 mm from a $B_G$ = 20 mm wide illuminated grating, then the radiation of a spectral line sharply depicted on the focal curve is dispersed to a width of 0.8 mm. The patent application "Spektrometeroptik mit nicht-sphärischen Spiegeln" (Spectrometer Optics with

Non-Spherical Mirrors) [54] describes the state of the technique and explains an additional advantage associated with the use of these 45° mirrors: If they are designed as parabolic cylinder mirrors, the height of the spectral line can be compressed onto the detector pixel. With typical pixel heights of 200 µm, there is an increased light density on the pixels and the ratio between usable signal and detector noise is improved. In principle, it is also possible to use spherical mirrors. However, in this case imaging errors that can lead to losses in resolution must be taken into account.

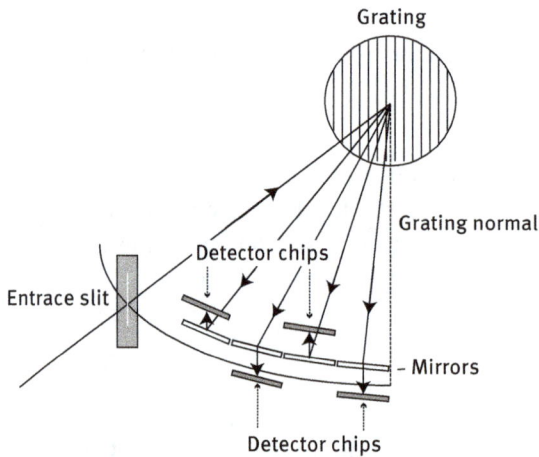

Figure 3.57: Mirroring for gap-free acquisition of the spectrum with linear detectors.

## 3.5.3 Optics with corrected concave gratings

The sizes of the slits and the photomultiplier tubes were given as the main obstacles for the miniaturization of the Rowland circle optics in Section 3.5.1. It has been discussed in Section 3.5.2 that there are now much smaller detectors, which do not require slits and are able to completely record the spectra. However, it was also noted there that placement of the linear detectors along the spherical spectral curve leads to defocusing. Of course, this becomes more serious, for a given detector length, when the Rowland circle diameter gets smaller. Only with large grating focal lengths F (F ≥ 300 mm) and short detector lengths $L_S$ ($L_S$ ≤ 30 mm) does the defocusing remain in the range of the depth of focus, which is dependent on diffraction effects. In Section 3.5.5, we will see, during the discussion of the appropriate slit width, how the depth of focus is estimated. To keep the distance between linear detector and the focal curve small, the so-called "flat field corrected concave grating" is used for small grating focal lengths. With this type of grating, the grooves are not parallel, but the distance between them changes with the distance from the center of the grating. As a consequence, the spectrum does not appear on a circular arc but is almost linear over wide spectral ranges.

Gratings of this type are exclusively holographically manufactured. The effect of the varying groove distances is achieved with the appropriate arrangement of interfering laser sources.

The focal terms differ from eq. (3.9) and eq. (3.10) (see e.g., [52]):
Spectral focus (corrected concave grating)

$$\frac{\cos^2(\alpha)}{la} - \frac{\cos(\alpha)}{F} + \frac{\cos^2(\varepsilon)}{le} - \frac{\cos(\varepsilon)}{F} - \frac{N\lambda S}{\lambda'} = 0 \qquad (3.17)$$

Meridional focus (corrected concave grating)

$$\frac{1}{la} - \frac{\cos(\alpha)}{F} + \frac{1}{le} - \frac{\cos(\varepsilon)}{F} - \frac{N\lambda A}{\lambda'} = 0 \qquad (3.18)$$

The symbols have the same meanings as in eqs. (3.8)–(3.10) with the following additions:

$\lambda'$   Wavelength of the laser used to make the grating
S, A Constants, set by the angles of incidence of the two interfering laser beams
    The constants S and A result from:

$$S = \frac{\cos^2(\gamma)}{Q} - \frac{\cos(\gamma)}{F} - \left(\frac{\cos^2(\gamma')}{Q'} - \frac{\cos(\gamma')}{F}\right) \qquad (3.19)$$

$$A = \frac{1}{Q} - \frac{\cos(\gamma)}{F} - \left(\frac{1}{Q'} - \frac{\cos(\gamma')}{F}\right) \qquad (3.20)$$

Whereby:
$\gamma$   Angle of the first laser light source to the center of the grating
$\gamma'$   Angle of the second laser light source to the center of the grating
Q   Distance of the first laser light source from the center of the grating
Q'   Distance of the second laser light source from the center of the grating

Figure 3.58 shows the arrangement of the lasers required to generate the interference pattern.

The classical Rowland circle grating is a special case with the coefficients A = 0 and S = 0. The curvature of the spectral focal curve can be straightened through suitable selection of the coefficient S.

Details about the properties, production and testing of gratings are described, e.g., by American Holographics [52], Zeiss [55–58], Agilent [59] and Dobschal/ Kröplin/Reichel/Rudolph/Steiner [60].

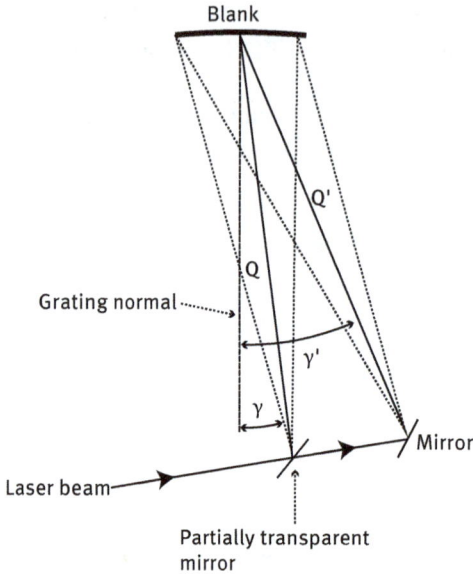

Figure 3.58: Laser setup for exposure of corrected gratings.

### 3.5.4 Echelle optics with two-dimensional detector arrays

Since the mid-1990s, commercial spectrometer systems equipped with two-dimensional CCD or CID arrays have been available on the market.

Here, Echelle gratings with sawtooth-shaped groove profiles that are operated in high diffraction orders are used (Figure 3.59). The radiation falls on the short sides of the sawtooth structure.

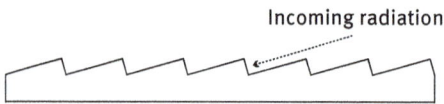

Figure 3.59: Groove profile of an Echelle grating.

Figure 3.60 shows the basic design of an Echelle optic. For such systems, the light is transmitted through an entrance slit and is parallelized by a collimator (concave mirror or lens). The Echelle grating diffracts the light. It is optimized for the higher-order range (e.g., 35th to 55th order). The so-called "order sorter," which consists of a prism or a second grating, fans out the different orders vertically.

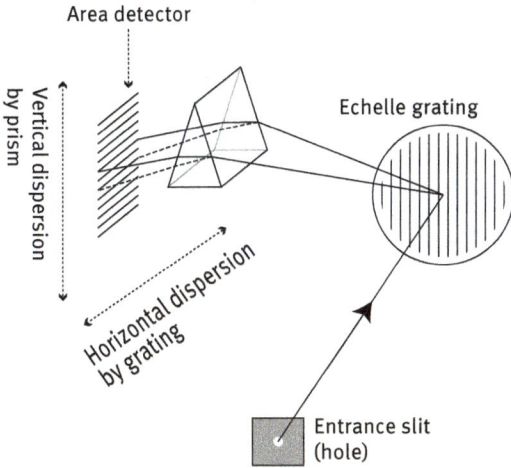

Figure 3.60: Structure of an echelle optic.

Finally, the camera mirror (or camera lens) sharply images the spectrum onto a two-dimensional detector. There, the different orders appear in rows due to the dispersion properties of the order sorter. Collimator and camera mirror are omitted from the illustration in Figure 3.60 for better clarity.

This type of optic is rarely used in today's arc and spark spectrometers. There are two reasons for this:

Area sensors generally cannot be read out as quickly as line detectors. If the area sensor has n pixel columns and m rows, then n × m pixels must be read for every readout process. This information is usually only connected to one or two readout circuits. This means that even with high readout rates, only a limited number of total images (frames) can be recorded. If the same pixel count is distributed on z line detectors, each line can be assigned to an analog/digital converter that reads out the lines simultaneously. It is possible to read out the complete spectrum after every single spark even for high sparking frequencies.

For area sensors, the entire spectrum is concentrated onto a comparatively small area. A very bright analytical line in line i can disturb a nearby detection line in line i−1 or line i + 1 with stray light, thus increasing the signal. For Rowland circle optics, the spectrally widely removed spectral lines are also spatially widely apart (at least when higher diffraction orders are suppressed as is usually the case when using line detectors). In polychromators with 750 mm focal lengths and diffraction angles between 0 and 43°, the spectrum covers about 563 mm. The largest possible distance for two-dimensional sensors is between two opposite corner points. Common two-dimensional sensors for imaging purposes usually have frame diagonals of less than an inch (25.4 mm). It is possible to make larger sensors specifically for use in Echelle spectrometers. The fundamental problem, however, remains; it can only be mitigated.

Echelle optics are occasionally used in systems in conjunction with lasers as the excitation source, where there are frequently less stringent requirements for accuracy and detection sensitivity.

Detailed descriptions of such optics are found, for example, in Grobenski/ Radziuk/Schlemmer [61], Okruss [62] and Skoog/Leary [26].

### 3.5.5 Optical components

This section addresses important aspects of individual optical components. Especially, the possible errors that result from the differences between ideal and actually existing components are discussed.

#### 3.5.5.1 Gratings

Calculation of the angle of diffraction and the focus of concave gratings was already discussed in Sections 3.5.1 to 3.5.3.

This section will discuss the diffraction efficiencies that can be achieved with gratings as well as typical errors.

#### Production of original gratings

Gratings can be produced in two ways:
- The traditional method for production is to mechanically score the grooves using a so-called "grating ruling machine." With this method, the grooves are engraved with a diamond stylus. In addition to plane gratings, concave gratings can also be engraved, provided that the curvature is not too large. The grating blanks can be solid metal, but usually they consist of a glass ceramic that has a low coefficient of expansion. For ceramic carrier materials, the surface is coated with a sufficiently heavy layer of metal. After being engraved in the ruling machine, the grating is often passivated with a protective layer. $MgF_2$ passivation, which can be used from 115 nm, is common.
- The idea of making gratings photo-technically is quite old. Gratings with low groove density were already generated by photographing an interference pattern around 1900. However, photographic techniques gained in practical significance only after short-wave lasers were available with which it was possible to generate line patterns with sufficient density. Production is as follows: A grating blank is first treated with photoresist. Then it is exposed with the interference pattern from two suitably positioned lasers. Such a positioning was shown in Figure 3.58. A development process follows by which the photoresist – depending on the process – is stripped either on the exposed or on the nonexposed surfaces.

A surface structure that is suited to diffraction has already been formed. There can still be a posttreatment, e.g., ion etching. After the desired surface structure has been created, the grating is coated with a metal film. Finally, as with engraved gratings, a surface passivation can be conducted.

### Replicating gratings

Both engraved and holographic, i.e., produced with lasers, gratings can be replicated with an impression technique. Especially the mechanical engraving of gratings is a very time-consuming process, so that creating a mold contributes to the cost-effectiveness of grating production.

The impression process consists of several steps:
- First, the surface of the original is covered with a thin layer of a release agent.
- Then a metal layer is applied by vapor deposition onto the surface.
- An epoxy resin layer is applied and the copy-blank is placed on this.
- After hardening of the resin, the copy is removed from the original.
- Finally, the copy surface can be post treated by vapor deposition and, if necessary, the surface passivated with, e.g., $MgF_2$.

The impression is often conducted in two steps: First a daughter copy is made from the original. Concave grating originals generate convex daughter copies. Grand copies, which are once again concave, are generated by making an impression of the daughter copy. The primary gratings are referred to as originals and the copies as replicas.

Replicas made with the impression process have a high quality. Only when a large number of impressions have been made from a blank is there noticeable deterioration in the quality, for example in terms of the scattered light component. The grating manufacturers, however, ensure with individual testing that only the gratings corresponding to the specifications are delivered. Zeiss [56] explains that, among other things, the dimensions of the grating area, groove density, groove profile, surface layer and efficiency are tested. The grating manufacturers usually provide a test certificate, in which key data for the given grating is stated, for every grating on request. Table 3.4 shows the reflectivity for gratings of one type. This is a random sampling from spectrometer optic production collected over several years. The information is taken from the manufacturers' test certificates and was checked for plausibility as a part of quality assurance.

A certain variation in the reflectivity can be observed. This is unavoidable and can be considered rather insignificant. Even a series of original gratings has variations. In the context of emission spectrometry, always achieving the same absolute optical throughput is in any case a hopeless undertaking. The importance of absolute intensities in arc/spark spectrometry is discussed in Section 3.5.6.

Table 3.4: Reflection of different gratings of a series.

| Specimen | Reflection at 200 nm | Reflection at 240 nm | Reflection at 500 nm |
|---|---|---|---|
| 1 | 44% | 37% | 25% |
| 2 | 46% | 40% | 25% |
| 3 | 43% | 38% | 26% |
| 4 | 41% | 36% | 25% |
| 5 | 41% | 37% | 26% |
| 6 | 46% | 40% | 26% |

## Grating errors

Real gratings have errors that cause the diffracted radiation not to be diffracted in exclusively the direction determined with eq. (3.8).

The following errors should be distinguished:
- *Grating phantoms* are found especially on mechanically ruled gratings. They are caused by periodic irregularities during grating engraving. The so-called "Rowland phantoms" appear right next to the line position calculated with eq. (3.8). Lyman phantoms are generated by overlapping of several ruling errors. They can be found throughout the entire spectrum and are, according to the *Lexikon der Optik* (Lexicon of Optics) [63], a factor of $10^3$ to $10^4$ weaker than the spectral line.
- Errors, referred to as "*grass*," are caused by the fact that the groove positions vary slightly around their target positions. These variations are random in contrast to the effects described in point 1. The result is an increase in the signal originating from the radiation of neighboring wavelengths.
- *Stray light*, on the other hand, is caused by a grating groove surface that is not completely smooth. According to the *Lexikon der Optik* (Lexicon of Optics) [63], the scattered light component of holographically produced gratings is an order of magnitude of a decimal place smaller than for mechanically produced rulings. This is easy to see, because irregular roughness can be created on the edges of the grooves during engraving. So-called low-scatter laminar gratings, having an extremely low stray light level, can be made. Such gratings are produced by exposure to an interference pattern, development of the photoresist, ion etching and, finally, complete stripping of the photo-sensitive layer. This creates a rectangular groove profile. The low stray light level of such gratings is easily understandable. On the one hand, the elevated areas have their original smooth structure after stripping of the photoresist; on the other hand, ion etching also creates areas of low roughness in the pits. Details about the production of such gratings are described by Zeiss [58].

Like grass, stray light increases the signal level. In contrast to grass, this interference consists of broadband radiation. On a given point on the focal curve, there is a stray light component with wavelengths that deviate greatly from what is expected at that point.

## Blaze

Table 3.4 showed that the diffraction efficiency is not the same for all wavelengths. It depends more on the form of the grating grooves. Figure 3.61 shows a sawtooth-shaped grating profile. For mechanically ruled gratings, such shapes with widely differing angles of inclination θ can be produced simply by selecting a suitable angle for the diamond stylus. Optimal efficiency is provided when the light falls in and out perpendicularly to the long side of the triangular profile. This direction of radiation is labeled s in Figure 3.61. However, the sketched beam path is obtained at most for one wavelength in the polychromator arrangements described in Sections 3.5.1 to 3.5.3, because the radiation from other wavelengths results in other angles of diffraction in accordance with eq. (3.8). Compromises can be found for the wavelength range to be measured. Mechanically engraved gratings of the type shown in Figure 3.61 are referred to as echelette gratings. Gratings with a similar groove profile (see Figure 3.59) are used for Echelle optics like those described in Section 3.5.4. However, the grating groove numbers with values below 100 are relatively small and reflection occurs on the short sides of the triangles. Such gratings are called "Echelle gratings."

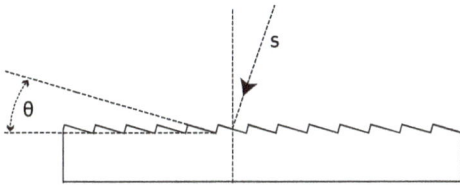

Figure 3.61: Echellette-Grating.

Originally, gratings made with laser exposure had a sinusoidal groove profile. Today, gratings with a sawtooth-shaped groove profile can also be produced using the holographic method. Lands and pits are rounded, but the flanks have the same shape and slope as those for echelette gratings. This makes it possible to optimize holographic gratings for different wavelength ranges.

## Handling

Grating surfaces are very sensitive. They must not be touched. If a grating has been inadvertently touched, no attempt should be made to clean it. This would result in destruction of the surface.

### 3.5.5.2 Fiber optics

Figure 3.62 shows the way in which an optical fiber works. Such fiber optics consist of an elongated cylinder Z made of a radiation transparent material with a refractive index of $n_z$. This body is encased by a layer S of an optically less dense material with an index of refraction of $n_s$. The surrounding material is called the "cladding." If the angle $\varepsilon$ of the incoming radiation to the fiber axis is smaller than the critical angle $\varepsilon_0$, there is repeated total internal reflection inside the fiber, which is almost free of loss. Equation (3.21) shows how $\varepsilon_0$ depends on the indices of refraction $n_z$ and $n_s$. The radiation exits again at the end of the fiber-optic cable.

$$\varepsilon_0 = \arcsin\left(\sqrt{n_z^2 - n_s^2}\right) \tag{3.21}$$

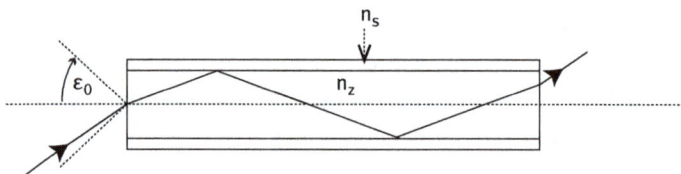

**Figure 3.62:** Function of a fiber optic.

The properties described here apply to straight fibers. The conditions become more complicated with fiber curvature. Details of the way in which fiber optics work can be found, e.g., in Schröder [64].

Using fiber optics is not without its problems:
1. Fiber optics cannot be used for the entire wavelength range of interest. According to current state-of-the-art technology, measurements below 185 nm with quartz/quartz fiber optics (core and cladding are both made of quartz glass with different indices of refraction) are hardly feasible in practice.
2. A second problem is the "progressive blindness" of the fiber optic for wavelengths below 250 nm due to harsh UV radiation. One possible explanation of this effect is as follows: The fiber optic consists of a pure quartz core encased with a fluorine-doped quartz glass (cladding). Total internal reflection is possible because the fluorine-doped quartz has a smaller refraction index than the pure quartz core. Harsh UV radiation supports diffusion of fluorine from the cladding into the core. The sharp "step" in the refraction indices is blurred. This leads to hindrance of total reflection particularly for short wavelengths. Figure 3.63 shows the transmission before and after aging the fiber optic with irradiation (60 min of continuous sparking with high-energy parameters). The attenuation increases from 250 nm and reaches a local maximum at about 215 nm. Below this wavelength the attenuation

Transmission of a fiber after aging (blue),
Normalized to intensities before aging (black)

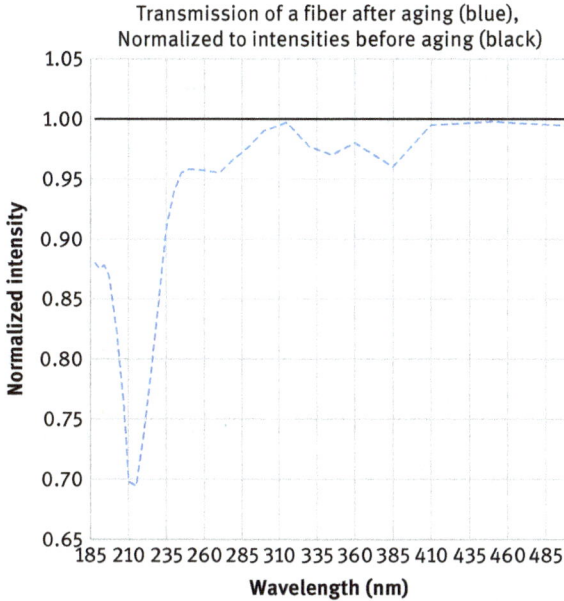

Figure 3.63: Transmission curve of an optical fiber optic.

is somewhat reduced. Measurement of the important carbon line at 193 nm re-mains possible even with aged fiber-optic cables.

Constructive measures can be taken against the effect, also known as "solarization," described in point 2 above:
- Blocking input into the fiber optic during the high-energy pre-spark with a mechanical shutter;
- Avoiding high peak light intensities during the measurement with a properly designed excitation generator;
- Damping the harsh UV radiation by filtering out wavelength ranges that are not required.

Combining these measures can almost completely prevent blindness of the fiber optic. The wavelength range between 185 nm and 250 nm becomes usable, whereby the range around 215 nm usually remains problematic.

There are significant differences in the transparency of short wavelengths be-tween batches of fiber optics. The reason for this is found in the varying purity of the quartz blocks ("preforms") used to draw out the optical fibers.

### 3.5.5.3 Slits and the smallest optimum slit width

The openings in spectrometer slits are either cut with a laser or made galvanically. The inner slit edge should be as straight and even as possible. A "cutting edge" shape of the slit is advantageous in terms of light throughput.

The slits are always mounted in an upright position, i.e., in the direction of the grating grooves. If this parallelism is not perfect, then the slit appears to be widened. If the slit has an illuminated height of h and an inclination of v degrees in respect to the grating grooves, the widening $b_{add}$ is:

$$b_{add} = h*sin(v) \tag{3.22}$$

Constructive measures must ensure that this widening remains small. If this cannot be guaranteed, the slit inclination must be adjustable.

Repeatedly, attention has been drawn to the existence of a smallest optimum slit width. Not always do narrower slits lead to an improvement in resolution.

It is necessary to consider the diffraction effects at the slit to determine the appropriate slit width. If the light bundle of perfectly parallel radiation passes through the slit at a right angle, the radiation is partially diffracted. The intensity of the diffracted ray at an angle $\varphi$ (see Figure 3.64) obeys eq. (3.23).

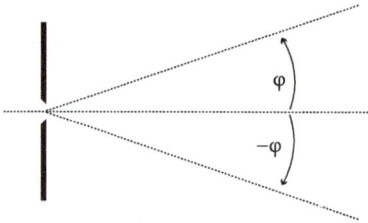

Figure 3.64: Angle $\varphi$ of a diffracted beam from the slit to the direction of the incident beam.

$$I_\varphi \sim b^2 * \frac{sin^2\left(\frac{\pi b}{\lambda}\sin\varphi\right)}{\left(\frac{\pi b}{\lambda}\sin\varphi\right)^2} \tag{3.23}$$

Here b refers to the slit width and $\lambda$ the wavelength. Derivation of this equation can be found in Bergmann and Schaefer [65].

The following special cases result from eq. (3.23):

$sin\,\varphi = 0$     : The intensity is maximal in a straight line, it grows proportionally to the square of the slit width

$sin\,\varphi = \frac{\lambda}{B}$     : The intensity is 0 at this point

Figure 3.65 shows the intensity distribution in the range of angles between −0.24 and 0.24 (angles recorded in radians). The wavelength of 500 nm and a slit width of 10 µm were chosen for this example.

Equation (3.23) shows that a perfectly parallel bundle of light diverges as it passes through the slit.

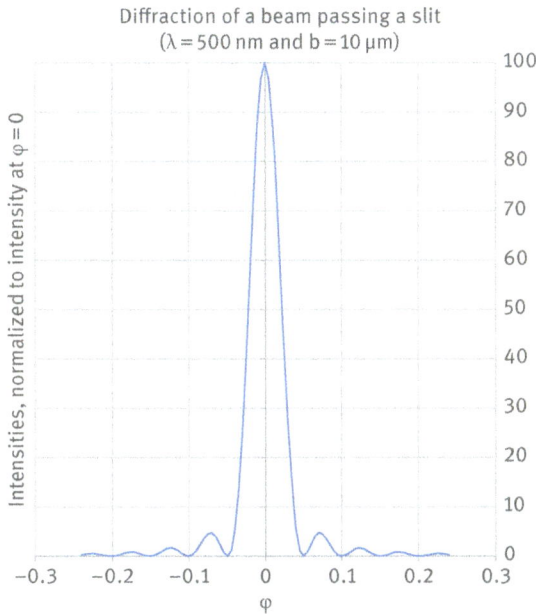

Figure 3.65: Intensity distribution of the radiation after diffraction through a slit.

An example shows how wide the divergence can be. The following boundaries are assumed (note that the wavelength deviates from the example displayed in Figure 3.65):

- Only radiation between the first minima, the so-called "central image," is considered.
- The slit has a width of 10 μm.
- The wavelength of the radiation that passes through the slit is 250 nm.
- The distance between the slit and the grating is 500 mm.

With these parameters, the radiation of the central image is distributed over a width of $\frac{0,00025}{0,01} *2*500$ mm = 25 mm on the grating.

The radiation divergence is not disturbing as long as the grating has a usable width of at least 25 mm. The concave mirror characteristics of the grating ensure that the spectrum is sharply imaged on the focal curve. Conversely, however, it is precisely the limits of the light spot on the grating that limit the resolution capacity of the optic. If, as in the example above, this has a width of 25 mm, there is a diffraction effect that affects the radiation passing from the grating toward the focal curve. The illuminated zone of the grating takes over the role of the slit. The sizes from above result in a broadening of the lines on the focal curve of $\frac{0,25}{25,000} *2*500,000$ μm = 10 μm.

The line images on the focal curve are, then, (for radiation around 250 nm) at least 10 μm wide, even when the entrance slit width approaches zero. Selection of a slit width less than 10 μm only costs intensity without improving the resolution. From eq. 3.23, it is clear that the influence of the slit width on the intensities is even quadratic. If the illumination width of the grating, the wavelength range and the distance between the grating center and the focal curve are known, a suitable slit width can be determined with the help of eq. (3.23). The slit width, from which further narrowing no longer leads to a gain in resolution, is known as the smallest optimum slit width.

### 3.5.5.4 Mirrors, lenses, windows and refractors

*Optical windows* and *lenses* must consist of a material that is transmissive for the wavelengths to be detected. The window materials normally used for arc/spark spectrometry are not critical in terms of the transmission on the long wave end of the spectral range. Table 3.5 gives an overview from which lower wavelength the common materials should be used. It is based on the availability of alternatives and attempts to remain at a minimum transmissivity of 50% for a material thickness of 2 mm.

Table 3.5: Recommended use of window materials.

| Material | Usable from (nm) |
|---|---|
| Borkron glass BK7 | 330 |
| Common quartz glass | 220 |
| Quartz glass, UV grades | 180 |
| Calcium fluoride | 125 |
| Magnesium fluoride | 120 |
| Lithium fluoride | 116 |

Two further criteria must be considered in addition to transmissivity when selecting window or lens materials:
- Some materials form color centers, i.e., they develop filter properties after they have been exposed to radiation. This condition is undesirable.
- Some materials are hydroscopic. Sodium chloride is, from its transmissivity, suitable as a window material for wavelengths from 180 nm. It is, however, only rarely used because it is water soluble and attracts moisture.

*Refractors* are plane parallel plates of transparent material. They have the purpose of causing a parallel offset of the radiation. If the refractor is set perpendicularly to the path of the radiation, the light passes without changing its direction (see Figure 3.66, above). However, the refractors consist of a material with a higher optical density than the environment. If the refractor has the thickness D and is made of a material with a refractive index n, then the light path D through the refractor corresponds a light path D × n through vacuum. This light path extension must be considered when using refractors. If the refractor is slightly turned, the radiation that enters the plate is refracted toward the perpendicular of its front side, passes through it diagonally and exits on the rear side (see Figure 3.66, below). In the process, the exiting beam undergoes a refraction, which negates that from entering the plate. The following law applies at an interface between two optical media:

$$n^*\sin(\varepsilon) = n'^*\sin(\varepsilon') \qquad (3.24)$$

Where:

n    Diffraction index of the medium in front of the interface
$\varepsilon$    Angle between incoming beam and the normal to the interface
n'   Diffraction index of the medium behind the interface
$\varepsilon'$    Angle between the beam behind the interface and the normal to the interface

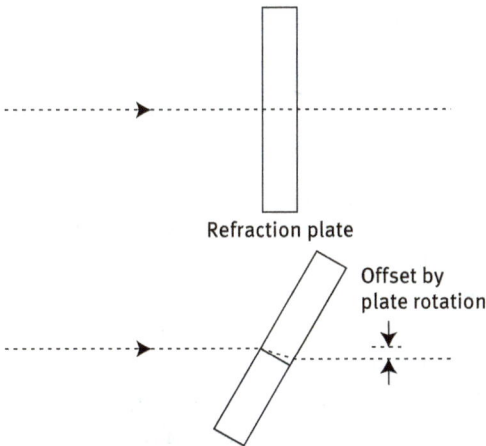

Figure 3.66: Beam path through a refractor.

Thus, rotating the refractor can lead to a parallel shift of the spectrum passing through it. The entire spectrum can be shifted in general with refractors behind the entrance slit. Refractors in front of exit slits offer a simple way of shifting spectral lines into exactly the slits intended for them. The material requirements for refractors are similar to those for windows.

In optical systems based on gratings, radiation of different diffraction orders is present on the focal curve. With selection of the appropriate material for the refractors, it is frequently possible to suppress higher diffraction orders.

Example: If a line with a wavelength of 400 nm is to be measured, the wavelength 200 nm can appear in the same place in the spectrum in the second order of diffraction. If BK7 is chosen as the material for the associated exit slit refractor, the 200 nm radiation does not pass through the refractor and, thus, does not pass through the exit slit.

One drawback of refractors is that the radiation is reflected on both inner sides and then exits the refractor – after passing through it three times. As a result, the usable spectrum is overlapped by a second strongly defocused spectrum, which is weak, but mimics line overlaps and can increase the background spectrum. This effect is similar in its consequences to the grating error discussed as "grating grass" in Section 3.5.5.1.

Aluminum is the most popular material for *mirror* surfaces. Glass ceramics with low expansion coefficients usually serve as mirror-carrier materials. Mirror surfaces are generally passivated with a thin magnesium fluoride layer. The thickness of the $MgF_2$ layer is critical for the reflectivity for short wavelengths and radiation input with high angles of incidence. It must be optimized depending on the application.

### 3.5.5.5 Optic housing

The container in which the optic is installed is a frequently underestimated component. Usually, there are also wavelengths to be measured that are so short that oxygen and water vapor from the ambient air lead to an unacceptable radiation absorption. To make these measurements possible, there are three different approaches:

1. The optic housing is flushed with a gas that is transparent for ultraviolet radiation. This is a viable solution for smaller optical systems. Either nitrogen or argon can be used as the flushing gas. Argon is usually preferred, as it is already needed as the operating gas for the sparking process. For reasons of cost, the flushing rate is kept as small as possible. The required flow rate depends on the air-tightness of the housing, the size of the internal surface of the housing and the materials used in the optic. The argon flow must remove oxygen and water vapor that has entered the system. It is especially time consuming to remove water vapor, as it is only slowly released from the inner surfaces of the housing.

   Organic compounds in the gas phase in the optic can be broken down by harsh UV radiation resulting in deposits on mirrors, windows, lenses or the grating. Table 3.6 shows that radiation present in the optic has a higher energy than the binding energies of many organic materials. There is, then, a risk that these bonds may be broken. Lubricants should be avoided. If they must be used, care must be taken that they are used sparingly and that the variety selected has the lowest possible vapor pressure.

Table 3.6: Binding energies of some radicals, expressed in wavelengths.

| Radical | Wavelength of a light quantum whose binding energy corresponds to the binding energy of the radical (nm) |
|---|---|
| C–H | 288 |
| C–C | 342 |
| C = C | 193 |
| C–O | 337 |
| C = O | 167 |
| O–H | 257 |

2. The flushing flow rate required for spectrometer optics with long focal lengths and covering a wide spectral range would be too high for economical operation. In this case, the optic housing can be constructed as a vacuum chamber. The vacuum should be generated and maintained with a dry, i.e., oil-free pump combination, e.g., a combination of a membrane pump with a turbomolecular pump. The oil used in a rotary vane pump can be problematic. The harsh UV radiation can lead to the decomposition processes described in point 1. Deposits that decrease the light transparency may develop particularly on the inner side of the window or lens through which the radiation enters the optical system. It is advantageous to take constructive measures to enable quick and easy replacement of the optical inlet components.

3. The third possibility for ensuring transparency of the light paths for short wavelength radiation is to fill the optical system with a transparent inert gas and to continuously circulate the fill gas through a membrane pump. Here, the argon stream is led through a purifying cartridge as discussed in Section 3.4.3. All in all, this solution seems to have advantages in terms of the contamination of the entrance to the optic. Personal experience has shown this. An astronautic experiment reaches results supporting these observations. Demets et al. [66] report about magnesium fluoride windows that were exposed to VUV radiation in space. If there was an argon atmosphere in the chamber behind the windows, they remained clean. However, if the chamber was evacuated, a typical brownish deposit was formed.

The circulation-flushing solution has the disadvantage that leaks are not immediately detected. In this case, rapid exhaustion of the cartridge's cleaning capacity is to be feared.

### 3.5.6 The importance of absolute intensity levels

Until the 1930s, the use of arc and spark spectrometers for the quantitative determination of elements was considered to be hopeless. The reason for this was that it was, and still is, very difficult to reproduce absolute intensity levels and to keep the system in such a state. As has been frequently mentioned, it was to the credit of Gerlach and Schweitzer [18] that spark spectrometry was freed from the shackles of absolute measurements and evaluations based on a comparison of line intensities was established.

An example shall explain why it is difficult to maintain a constant radiation level. It is based on a typical light path through a PMT-equipped optical system. Let us assume that the radiation is generated by a perfectly stable source and then takes the following route:

- First, it passes through the window that separates the spark stand atmosphere from the optic's atmosphere.
- Then it passes through a lens and is parallelized.
- It passes through the entrance slit.
- After that, an entrance slit refractor that is required for global profiling is transilluminated.
- Now it reaches the grating where the radiation is diffracted.
- Then it passes through an exit slit refractor, which is used to direct the spectral line exactly into the exit slit.
- There is usually another deflection mirror behind the exit slit, because the photomultiplier tube is rarely mounted directly behind the slit due to space limitations.
- Finally, the radiation reaches the glass housing of the photomultiplier tube.

In this example, the radiation strikes 13 optical surfaces between the spark stand window and the PMT housing. There is harsh radiation in the optic, which can lead to radiation absorbing deposits. This circumstance was explained in Section 3.5.5.4. If it is assumed that over time only 4% of the radiation is lost on every slit, every window, lens or mirror surface, then the radiation that reaches the photomultiplier tube is reduced to $0.96^{13}$, i.e., 51%. The signal is cut in half. A 4% reduction is not much: In connection with the slits, it was shown (eq. (3.23)) that narrowing the slit by 2% is sufficient to reduce the intensities that pass through by 4%.

Certain reductions in the absolute intensities have to be tolerated, especially because some components, such as the grating, can hardly be cleaned. More important than the absolute size of the signal is the requirement that the standard deviation of the spectral background must dominate over that of the readout noise. The spectral noise must be separately determined for every base, e.g., through repeat measurements on a pure sample of the given base. For example, an electrolytic copper sample can be used for such measurements for copper base.

Readout noise is determined by completely blocking the light path and then again conducting the repeat measurements. If the standard deviation of the background is significantly greater (e.g., by a factor of 5) than that for the blind measurements, then, in principle, even at a reduced intensity level, measurement without major reduction in the analytical performance is possible. The spectrometer software ensures that concentrations can still be determined correctly by using the formation of ratios and recalibration algorithms (see Sections 3.9.4 and 3.9.6). Normally, the software sends out a warning when there is a state of error. However, this does not relieve the user of the necessity checking the analytical performance of the system after a loss in intensity by measuring control samples.

## 3.6 Optoelectronic detectors for atomic emission spectrometers

Various optical concepts were presented in the previous section. There it was explained that radiation can be detected with either a combination of exit slit and photomultiplier tube or alternatively by using multichannel semiconductor detectors complete spectral ranges can be recorded. Both detector technologies are described in Section 3.6.

### 3.6.1 Photomultiplier tubes

Photomultiplier tubes work based on the photoelectric effect, by which radiation removes electrons from a metal surface. Electron emission into the surrounding atmosphere was first observed by Heinrich Hertz in 1886. With the so-called "gold-leaf electroscope," his assistant Hallwachs constructed the first measuring device that made use of the external photoelectric effect. A metal electrode charges under the influence of radiation. A thin piece of gold leaf that is attached on one side to the electrode rises away on the unfixed end because charges with the same polarity repel each other.

Julius Elster and Hans Geitel, both former students of Kirchhoff and Bunsen, invented the vacuum photocell around 1890 [67] and published the results of their work in the *Annalen der Physik und Chemie* (Annals of Physics and Chemistry) between 1890 and 1894 [68–71]. The physicist Philipp Lenard systematically examined the photo effect in vacuum for the first time around 1900. He recognized that the wavelength of the radiation falling onto the cathode is crucial to whether the effect occurs, whereas for sufficiently energy-rich radiation the radiation intensity only determines the number of electrons ejected. In 1905, the work of Albert Einstein provided the theoretical framework for complete explanation of the effect. He was awarded the Nobel Prize in Physics in 1921 for this. In 1929, L. Koller and N. R. Campbell developed a cathode material, still known today as S-1, that is made of silver, oxygen and cesium [72]. A thin layer of cesium directly on the surface is optoelectronically active. This type of photocathode had a two-decimal order of magnitude higher sensitivity than all until then known cathode surfaces. In the following years, various cathode materials were developed, most of them made of alkali elements and those of the boron and nitrogen groups, such as the combination of gallium and arsenic or of In, Ga and As.

However, the sensitivity of the photocells produced in this way was still not competitive with the then practiced photographic detection, although there had been attempts in this direction. It was only by coupling the photocells with secondary electron multiplication that the detector sensitivity could be increased to such an extent that they could compete with photographic detection.

This combination of photocell and electron multiplier is called "photomultiplier tube" (PMT). Figure 3.67 shows the principle. The radiation passes through a transparent window into the interior of the photomultiplier tube and strikes the photocathode. The radiation ejects electrons from the cathode due to the photoelectric effect. Between the cathode and the first dynode, a flat electrode with a magnesium oxide or beryllium oxide coating, is a difference in potential on the order of 100 V. The electrons released from the cathode are accelerated in the direction of the dynode and strike there. The kinetic energy of each of these electrodes releases several more electrodes from the dynode material. These electrodes are accelerated in the direction of the second dynode, which again has a difference in potential of about 100 V from the first dynode. Additional dynodes follow the second. There are typically ten dynodes in modern photomultiplier tubes. The number of electrons, and with them the current, multiplies from dynode to dynode. The dynodes are simultaneously anode and cathode, because they are both target for and source of electrons. The electrons reach the last electrode, which is just an anode. There, the output current can be measured against ground. It is nearly proportional to the radiation that originally reached the

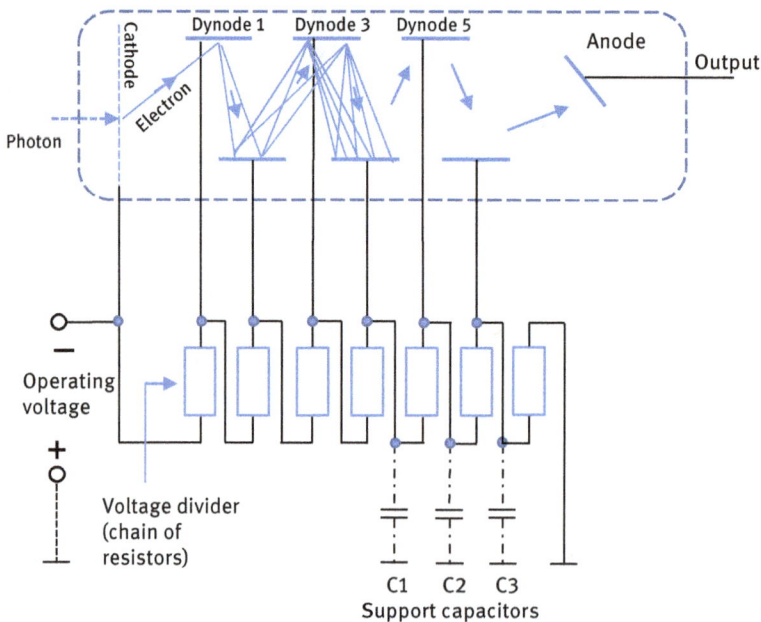

Figure 3.67: Basic structure of a photomultiplier tube.

cathode. The resistors in Figure 3.67 form a voltage divider. They usually all have the same value, so that between two adjacent dynodes, there is a constant voltage. The current between the anode and the last dynode as well as between the last dynode pairs can be so large that the voltage between the associated resistors drops slightly, as the distances between the dynodes act like high-impedance resistors. This is especially problematic for radiation sources that deliver strong signals over a short time.

Figure 3.68 shows the effect if there is an additional high-impedance resistor between every two dynodes. The voltages change (dashed line). In reality, only the resistance between the last dynodes is noticeably reduced, and that is only the case when the photomultiplier sees a strong radiation. The capacitors C1 to C3, drawn with dashed lines in Figure 3.67, are added to prevent this effect. Their charging supports the voltage applied to the last dynode and the anode and prevents distortion of the kind sketched in Figure 3.68.

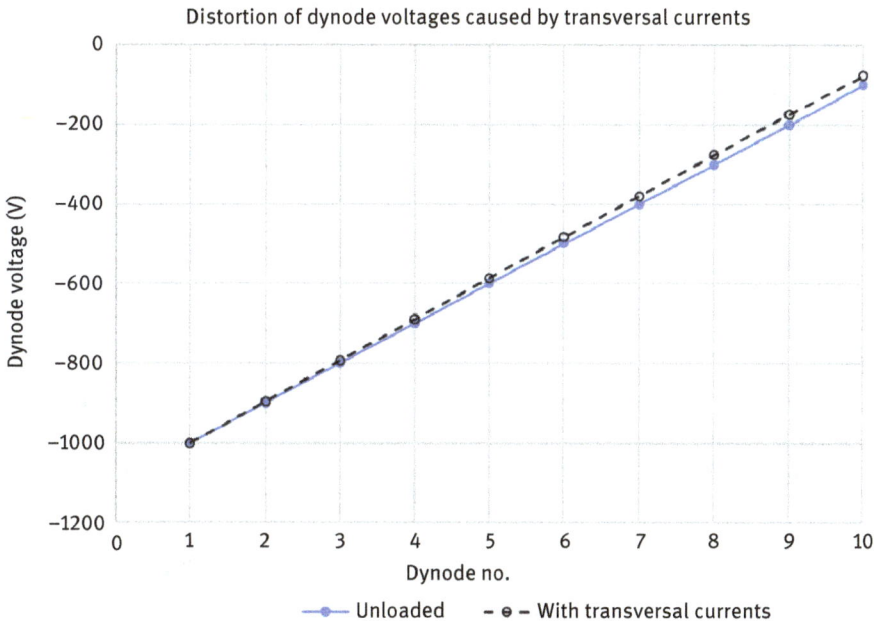

**Figure 3.68:** Distortion of dynode voltage drops due to crosscurrents.

According to Ohls [5], PMTs were used for the first time by Iams and Salzberg in 1935. In modern PMT designs, the light enters from the top (head on) or from the side (side on). The head on design is in principle still the same as shown in Figure 3.67. In side on PMTs, the dynodes are arranged a little differently. Figure 3.69 shows the setup. Here, the cathode is not a wire mesh, but a solid plate. This form is also called a "reflection mode cathode," because the photons strike the cathode with an angle of incidence and the electrons are ejected from the cathode

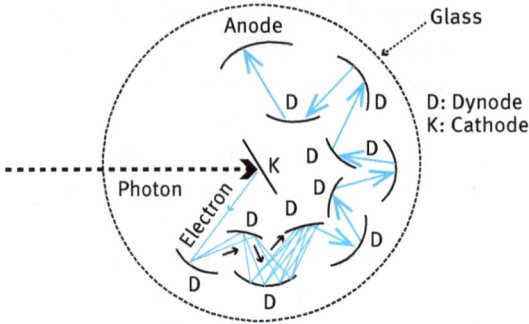

Figure 3.69: Construction of a side on photomultiplier.

with an angle of reflection. This process is similar to the reflection of radiation on a mirror.

Photomultiplier tubes are available in a wide variety of designs. The main differentiation criteria are as follows:

– *Component size and position of the window*

In its product line overview [73], the Hamamatsu company, a renowned Japanese manufacturer of high-quality photomultiplier tubes, recommends multipliers with side on light entry and diameters of 13 mm (1/2″) or 28 mm (1 1/8″) for optical emission spectrometry. The smaller diameters are often preferred for the construction of optics because they enable a larger number of PMTs to be placed behind the exit slits on the focal curve.

– *Cathode design*

The cathode can be semitransparent as shown in Figure 3.67. Alternatively, the compact cathode design can be used (see Figure 3.69). In most cases, compact cathodes operating with the reflection mode are used in the PMTs for arc/spark OES.

– *Cathode material*

The cathode material is the deciding factor as to for which radiation the photomultiplier tube is sensitive. Hamamatsu's data book [74] mentions, among others, the following cathode materials:

*Sb-Cs* cathodes are usable for the wavelength range between 185 nm and 650 nm and are mainly used for compact cathodes. Higher cathode currents are possible than those for the bialkali cathodes discussed below.

*Cs-Te* cathodes are only sensitive for wavelengths between 160 and 320 nm. Since no signal is generated for visible light, PMTs that are equipped with this type of cathode are called "solar blind." This characteristic is very practical when designing optics. As was explained in Section 3.5, the radiation from multiple diffraction orders may appear at one point on the focal curve. For example, radiation with a wavelength of 200 nm can simultaneously pass through a slit through which radiation from the first diffraction order with a wavelength of 400 nm passes. If the measurement is conducted with a PMT with a Cs-Te cathode, only the short-wave radiation is measured, because it is not sensitive for the long-wave radiation.

*Cs-I* cathodes are also not sensitive for visible light. Their spectral sensitivity lies between 115 nm and 195 nm. Thus, they are suitable for suppressing first order radiation above 200 nm.

*Sb-Na-K-Cs* "multialkali" cathodes are sensitive over a wide range from 185 nm to 850 nm.

*Sb-Rb-Cs, Sb-K-Cs* "bialkali" cathodes are usable between 185 nm and approximately 700 nm.

*Sb-Na-K* "low noise bialkali" are designed for low noise in a similar wavelength range.

In addition, the cathode material is responsible for the achievable quantum efficiency. This is normally given in percent and provides information as to how many electrons are released per photon from the cathode. Diagrams of quantum efficiency versus wavelength for the different types of cathode can be found in Hamamatsu's product catalog [73].

– *Window material*

*Borosilicate glass* allows only radiation with wavelengths of 300 nm and longer to pass through. PMTs with this window material are rarely used for arc and spark spectrometers. Filters, with which absorption edges for other wavelengths can be realized, are usually used to block shorter wavelengths of below 300 nm.

In contrast, widespread window materials are *UV glass* with an absorption edge at 185 nm and *quartz glass*, which is transparent over a wide range of UV radiation starting at about 160 nm.

If shorter wavelengths are to be measured, multipliers with glued-on windows made of *magnesium-fluoride* are used. These windows are transparent from 115 nm and translucent into the infrared region.

Photomultiplier tubes are designed to measure low levels of light. High-intensity radiation can damage or even destroy them. Therefore, the high-voltage supply to the optic must be turned off before opening an optic equipped with PMTs. Operation or storage of photomultiplier tubes in an atmosphere containing helium should also be avoided. Helium permeation through glass is well known and can reduce the quality of the vacuum in the tubes, promoting effects such as the "afterpulsing" effect, which is still to be discussed.

A small current flows out of the PMT anode even in the absence of lighting. Skoog and Leary [26] state that the main cause for this so-called "dark current" is the thermal emission of electrons and report that this effect can be completely eliminated by cooling to −30 °C. Additional sources for anode current not caused by incoming radiation are described in Hamamatsu's PMT manual [74]:

- Photocurrents caused by scintillation of the glass housing
- Emission of electrons from the electrodes caused by strong electric fields
- Currents caused by residual gas ionization
- Currents due to imperfect insulation of the socket and glass housing: Since high voltages are applied and very low currents are measured, even high impedance resistors in the giga-ohm range between the negative voltages and the anode can result in measurable currents.
- Noise due to cosmic radiation, gamma radiation from the environment and radioactive isotopes contained in the glass housing

The so-called "afterpulsing" is a further effect. It is expressed as follows: If the cathode is irradiated with a short pulse, the associated signal appears at first at the anode. With the time delay of a few microseconds, however, a second smaller output pulse follows. In the context of time-resolved spark spectroscopy, this is a disturbing effect, because the afterglow of the plasma is detected at a time when the thermal background has subsided. The signal intensities are much lower compared to those during the current conducting phase of the spark.

Hamamatsu [74] names the following causes for afterpulsing:
- Electrons that wander back to the cathode from the first dynode
- Emission of radiation that arises in the PMT and whose photons return to the cathode

Hamamatsu's manual recommends artificial aging of the PMT as a countermeasure and recommends that the PMT not be subjected to mechanical or thermal impacts in order to prevent afterpulsing. Hamamatsu's manual [74] is an excellent source of information on photomultiplier tubes. Jennewein [75] has an easily readable discussion about photomultiplier tubes with practical advice about usage, which is sufficient for a general overview.

Special demands are placed on the voltage constancy of the PMT power supply. A proven rule of thumb states that a relative increase in the high voltage of one percent results in a photocurrent increase of 15%. The high-voltage must also be free of superimposed AC. For example, if a ripple component is present, it can be disturbing for the detection of single spark intensities. The level of the photocurrent can be adjusted by changing the PMT power supply within certain limits. However, the high voltage should not be chosen higher than necessary. Experience has shown that the stability of the dark current worsens with high supply voltages.

## 3.6.2 Semiconductor-based sensor arrays

The description of semiconductor sensors in this chapter is an updated version of information based on an older publication by one of the authors [76].

Photosensitive detectors in the form of CMOS and charge-coupled device (CCD) arrays are used in large numbers in camcorders, scanners and photocopiers. Such arrays have also been used in spectrometry for a long time.

The first works (e.g., from Cox [77]) date back to the second half of the 1970s. Semiconductors have a crystalline structure. The outer shell electrons of an atom (valence electrons) are shared together with those from the neighboring atoms so that the outer shell is fully occupied. For example, Germanium has four electrons in the outer shell and shares one electron with each of four neighboring atoms forming a lattice.

One electron can be removed from this structure by introducing energy. A minimum energy $W_g$, the so-called "band gap" is required. The band gap depends on the material and the temperature. If a voltage has been applied to the semiconductor, the electron moves in the direction of the anode. Not just the electron moves, but also the hole that it leaves behind. The reason is that one of the neighboring electrons that is located in the direction of the cathode is attracted by the anode and fills the hole. That is, the hole moves toward the cathode.

To a very small extent, electrons are raised into the conduction band by thermal excitation. The resulting conductivity is called "intrinsic." If foreign atoms with a different number of valence electrons are added to the semiconductor crystal ("doped"), there is either an electron surplus or electron deficiency at this place depending on whether the foreign atoms have more or less valence electrons than the semiconductor. These built-in disturbances increase the conductivity. If there is a surplus of electrons, it is called an "n-type semiconductor." If there is an electron deficiency, it is a p-type semiconductor. Foreign atoms with an electron surplus are called "donors;" atoms with electron deficiency "acceptors."

When photons strike a semiconductor material, they release bound electrons from the valence band and raise them to the conduction band. However, this can only happen if the energy of the photon is larger than the energy difference between the conduction band and the valence band.

$$Wp = hv = \frac{h^*c}{\lambda} \geq Wg \tag{3.25}$$

$W_p$   Energy of the photon (J)
$W_g$   Band gap for silicon (in J, 1.12 eV or $1{,}79 \times 10^{-19}$ J at room temperature)
$v$   Frequency ($s^{-1}$)
$h$   Planck's quantum of action ($6.626 \times 10^{-34}$ J × s)
$c$   Vacuum speed of light ($3 \times 10^8$ m $s^{-1}$)
$\lambda$   Cutoff wavelength (m)

The result from eq. (3.25) is that wavelengths <1.1 µm can be detected with silicon semiconductors. The usable range is limited by absorption effects for short wavelengths.

The so-called "photoresistor" is a semiconductor detector with a very simple structure (see Figure 3.70). It consists of an undoped semiconductor to which two electrodes are attached. When illuminated, the electrons are raised into the conduction band and a current begins to flow. Photoresistors are rarely used for spectrometric purposes. Even high-purity semiconductor crystals contain low-residual levels of foreign atoms that act as donors and acceptors. This causes a so-called "dark current," which superimposes the current induced by irradiation, to flow when power is applied to the photoresistor. In addition, some of the charge carrier pairs recombine again before the charges reach the electrodes, as the field strength in the crystal is quite low. Other components are better suited to the detection of small amounts of light.

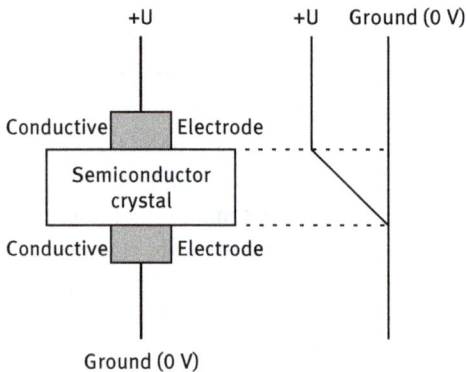

Figure 3.70: Structure of a photoresistor with associated potential curve.

### 3.6.2.1 Photodiode arrays

Figure 3.71 shows the structure of a photodiode and the field strengths present in it. A photodiode is nothing more than a reverse-biased diode that has been optimized for the detection of radiation.

The doped p-type and n-type regions are relatively low resistance. At the junction of the two regions, the electrons are transported toward the n-type region and the holes toward the p-type region. A small region that is free of charge carriers is formed. Because this region surrounds the complete volume of the p-silicon, the circuit is interrupted. Current cannot flow. The potential changes from 0 to +U; the field strength is highest in the depletion zone. If a photon ejects an electron there, the electron is immediately accelerated in the direction of the n-silicon by the high field strength. The recombination probability is small. As noted above, thermal effects can also create electron/hole pairs. While in the case of the photoresistor, the entire semiconductor volume contributes to the dark current, for the photodiode, only the volume of the depletion zone contributes.

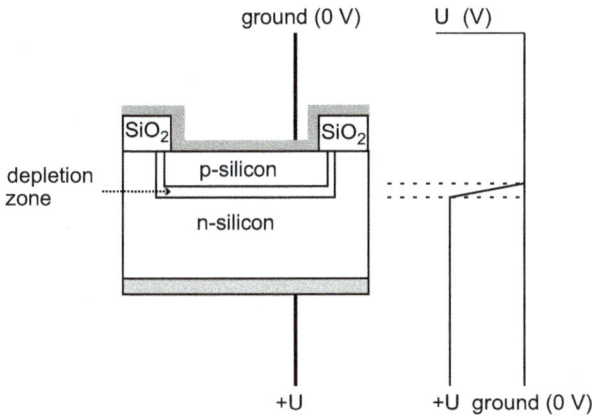

**Figure 3.71:** Structure of a photodiode with associated potential curve.

The thickness N of the depletion zone in the n-type doped region can be calculated using the following equation (according to Krüger [78]):

$$N = \sqrt{\frac{2*\varepsilon*(U*V)}{e*D*(1+\frac{D}{A})}} \tag{3.26}$$

The following applies for the thickness P of the depletion zone in the p-type doped region [78]:

$$P = \sqrt{\frac{2*\varepsilon*(U*V)}{e*A*(1+\frac{A}{D})}} \tag{3.27}$$

Whereby:

e     the elementary charge ($1.602 \times 10^{-19}$ C)
$\varepsilon$     the dielectric constant of the semiconductor material ($\frac{C}{V*m}$)
V     the amount of the forward voltage of the diode (about 0.7 V for Si diodes)
A     the concentration of the acceptors ($m^{-3}$)
D     the concentration of the donors ($m^{-3}$)
U     the applied reverse biased voltage (V)

Production is done by heavily p-doping the surface of a weakly n-doped crystal. A is, then, much larger than D. It follows that P is small compared to N.

The absorption increases exponentially with the penetration depth into the p-layer. For this reason, the p-doped layer must be as thin as possible.

A photodiode array is made up of a linear arrangement of structures as shown in Figure 3.71. A light-insensitive "gap" between two photodiodes (corresponds to the $SiO_2$ structures in Figure 3.71) is unavoidable; it is required to insulate the anodes from each other. Each diode is connected to an electronic switch that transmits the charge stored in the diode to an integrated operational amplifier.

The output of the operational amplifier leads to a detector chip contact. The analog switch also has a second function: After the readout process for a pixel, it switches through a reset voltage, which resets the photodiode to the zero state. Common diode arrays have 128 to 2048 individual diodes. Therefore, not every analog switch belonging to the photodiodes can be switched via a dedicated pin.

There are two methods to control the analog switch:
- *Control via (digital) multiplexer*
  Digital multiplexers are logical combinatorial circuits with n address inputs $A_1 \ldots A_n$, $2^n$ outputs and an "enable" line. Address lines and enable lines are connected to housing contacts, the outputs of the multiplexer lead to the control lines of the analog switch. The address A of the analog switch to be selected is encoded as an n-bit binary word A and connected to $A_1 \ldots A_n$. Activation of the enable line causes the analog switch to open. A description of the digital multiplexer can be found in Tietze and Schenk [15].
  An array with 2048 photodiodes requires $\log_2 (2048) = 11$ address lines. The advantage to this type of addressing is that there is direct access to every pixel.
- *Control using a digital shift register*
  Shift registers consist of a chain of 1-bit data storage elements, the so-called "flip-flops." Flip-flops are switched so that with a clock impulse, flip-flop n + 1 takes over the content of the *n*th flip-flop (see Tietze/Schenk [15] and Koestner [79], p. 359 ff, for information about shift registers). To read out a pixel k, the first flip-flop is set. The logical high-signal of the first flip-flop is then transported by the k-1 clock impulse to the *k*th pixel, where it (together with an enable signal) causes the *k*th analog switch to close, thus connecting the photodiode of the *k*th pixel to the data line. Any number of photodiodes can be addressed with a data line, a clock line and the enable signal; however, the *n*th photodiode only after the output of n-1 clock signals. This delay can be a hindrance when immediate access to a photodiode is required, for example, for time-resolved spectroscopy.

One disadvantage to diode arrays is the fact that the charge of the photodiode discharges through the analog switch into the structure that connects the analog switch outputs to each other. Unfortunately, it is not possible to make this as small as may be desired, because all the outputs over the entire width of the array must be connected. This is why it has a capacity that is usually larger than that of the photodiode. This worsens the signal/noise ratio, which is a disadvantage compared to the CCDs described in the next section.

### 3.6.2.2 Charge transfer devices

charge-transfer device (CTD) is the umbrella term for charge coupled device (CCD) and charge injection device (CID).

### Charge coupled devices (CCDs)

For historical reasons (because of patents for CIDs, see Sweedler/Ratzlaff/Denton [80], p. 50]), there are more providers of CCDs than of CIDs. CCDs are currently the most widely distributed type of detector arrays for spectroscopic purposes. Figure 3.72 shows the structure of a CCD pixel.

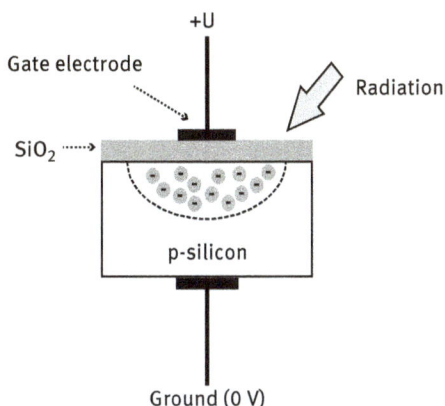

Figure 3.72: Structure of a CCD pixel.

The so-called "substrate" is the base for the functional unit. It consists of p-doped silicon and usually carries ground potential. A lightly p-doped epitaxial layer is built on this; this is the photoactive region and has a thickness of several μm. An insulating $SiO_2$ layer of only approximately 0.02 μm separates it from the gate electrode. This consists of highly doped, i.e., low-resistance, silicon.

There is a positive voltage on the gate. If radiation with wavelengths below 1,000 nm falls on the surface of the structure, it passes the gate and $SiO_2$ insulator and results in the formation of a free electron in the epitaxial layer due to the inner photoelectric effect. The inner photoelectric effect relevant here is qualitatively different from the junction photo effect of the diode arrays (see e.g., Perkampus [81], pp. 49, 447). The free electrons collect below the gate electrode because the gate is positively charged. However, they cannot penetrate the nonconductive $SiO_2$ layer.

After the end of the integration time, the charge collected under the gate electrode is, ideally, proportional to the amount of radiation acquired. "Ideally" means here that there are so few electrons produced that the repulsion of additional new electrons by those that are already stored does not yet play a role. If it does play a role, then an effect called "blooming" occurs; it is explained below.

The charge of the pixel must now be read out. In the simplest case, the pixels in the CCD form an analog shift register. This makes it possible to sequentially clock out the charge of each pixel at an output.

Figure 3.73 shows how the charge is shifted. There are three auxiliary electrodes to the left of the gate electrode for each pixel. In the first step, only the left electrode is positively charged. Thus, the electrons collect under this electrode. If a positive voltage is applied to the electrode to the right of this, the electrons are distributed under both electrodes (step 2). If the control voltage on the first electrode is switched off, the charge is shifted by a quarter of a pixel to the right (step 3). Imagine that the other pixels in the array are aligned to the left and right and that the gate electrodes 1 to 4 for each pixel are connected to one another, the charges can be shifted from left to right over the chip. A two-phase control can be realized by overlapping the gate electrodes. Single-phase control is also possible by implanting charge barriers (see, e.g., Hynecek [82]).

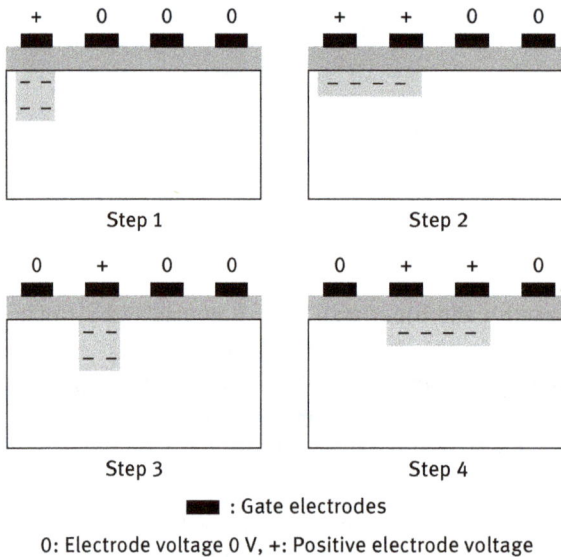

Figure 3.73: Charge transport in the CCD.

It is, however, disadvantageous to use the pixels themselves as analog shift registers. The light that enters during the shifting process produces a charge that cannot be assigned to the correct pixel. Imagine that a photon generates an electron at pixel 1,000 in a 2,000-pixel CCD. Continue to assume that this happened during the readout process after 500 charge shifts and a shift direction towards lower pixel numbers. The electron would then be added to a charge that was originally collected (before shifting began) below pixel 1,500. For this reason, the pixel charges are first transferred into a separate shift register covered with a radiation-impermeable layer. In this way, it is possible to integrate and read out simultaneously.

Figure 3.74 shows the principle structure of a CCD component.

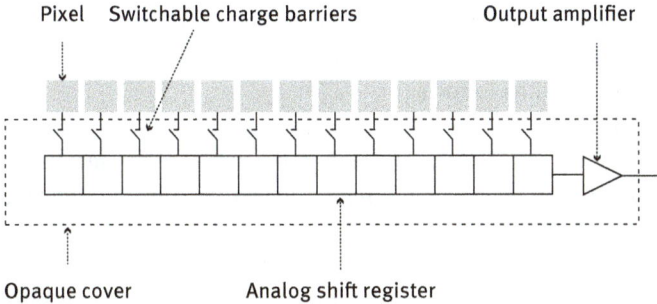

Figure 3.74: Structure of a CCD line detector.

The entire chip surface, with exception of the pixels, is covered by an opaque aluminum layer (area inside the dotted line).

After the integration time, the charge is transferred to the analog readout register. This is done by deactivating a charge barrier. The charge barrier can be thought of as a gate electrode to which 0 V is applied during the integration. Transfer into the readout register is performed as shown in Figure 3.73 (the barrier gate corresponds to the second gate connection).

Note that the voltage drop through discharge of a small capacity into a larger one as described for photodiodes does not occur. On the contrary: The pixel capacities can be chosen to be large compared to the transfer bin capacities; in this way, even voltage amplification is possible.

The charge generated by photons is completely transmitted. An explicit deletion process ("reset") is not necessary. This results in a significantly reduced signal to noise ratio for the CCD compared to photodiode arrays.

One drawback of commercially available CCDs is their limited wavelength range. The gate structures form a filter that absorbs short wavelengths. The absorption edge is between 140 nm and 400 nm depending on the construction of the chip. Sweedler, Ratzlaff and Denton [80], p. 28, speak, in general, of absorption by front-illuminated CCDs (CCDs where the radiation enters the light sensitive area after passing the gate electrodes) below 400 nm, which is certainly correct for detectors mounted in housings with glass windows. If this window is removed, it is possible to find types of CCDs for which absorption starts to play a role below 160 nm. This is due to the fact, that modern CCD sensors often use photodiodes for charge generation and CCD structures to transport the charges.

If a fluorescence layer is applied to the detector surface by vapor deposition or painting, the detectors can be used starting at the shortest wavelengths of interest (approximately 115 nm). CCDs developed specifically for spectrometry are often designed so that the light enters from the back, i.e., through the substrate. The substrate must then form a particularly thin layer to prevent recombination. These CCDs are also called "back thinned CCDs" or "back illuminated CCDs." Such detectors can also be used for all wavelengths that have to be measured by spark instruments.

The blooming effect, already mentioned above, is an additional problem with using CCDs. This effect comes about as follows:

At first, light quanta produce electrons that accumulate in the space charge region below the gate electrode (see Figure 3.75a). Through repulsion of the negatively charged electrons, the volume of the space charge region increases with the growing number of electrons (see Figure 3.75b). If the electrons are pushed to the outer side to such an extent that the attraction forces of the neighboring gate electrodes dominate, they "jump" over to them (see Figure 3.75c). Note that for simplification of the illustration only one electrode per pixel has been drawn in Figure 3.75.

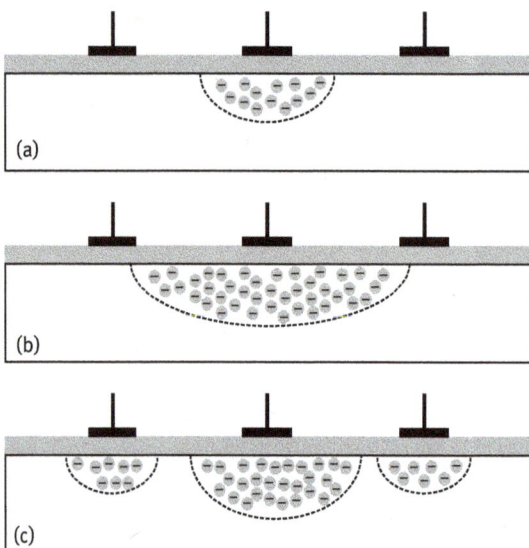

Figure 3.75: Blooming.

Two dimensional CCD arrays are available in addition to CCD line sensors. These arrays can, in principle, also be used for spectroscopic purposes, e.g., in combination with Echelle spectrometers. Here, however, the use of commercially available detectors creates difficulties:

To achieve high resolution for acceptable chip dimensions, distances ("pitches") between pixels of $7 \times 7$ μm are common. Part of this area is lost, as space is also required for the connecting structure. Therefore, the pixels usually have micro lenses etched onto them to concentrate all the light onto the effective pixel surface. As explained above, a fluorescent coating is required for wavelengths below 300 nm. This is, however, located directly on the lenses' surfaces, so that the lenses can no longer fulfill their purpose – focusing the radiation onto the optically active zones of the pixels. Even worse: The fluorescent layer acts as a diffuse source of light and irradiates the surrounding pixels.

An additional problem arises from the number of pixels. High-resolution CCDs with up to 63 megapixels have been available for several years [83]. However, the charges cannot be shifted with unlimited speed. For a pixel clock rate of 2 MHz, the readout process for such a large CCD requires almost 32 seconds. A 2,000-pixel linear detector with the same pixel clock rate can be read out a thousand times per second. This has two consequences for megapixel CCDs:

–   Blooming can hardly be avoided.
–   The dynamic range is strongly limited because the total number of readout processes for a given integration time is relatively small.

As mentioned above, modern CCDs frequently have structures that deviate from those described in the beginning of this section. The so-called "photodiode CCDs," which use diodes for photon detection, are very popular. Transmission of the charges is realized with CCD structures. Photodiode CCDs are also usually simply called CCDs, as the charge transport influences the properties of the detector more substantially than the method of the charge generation. An example for such photodiode CCDs is the surface sensor from Sony manufactured with their hole accumulation diode (HAD) technology. Details about this can be found by Sony [84].

## Charge injection devices (CIDs)

Like CCDs, CIDs are based on MOS structures. Figure 3.76 shows the structure of a pixel.

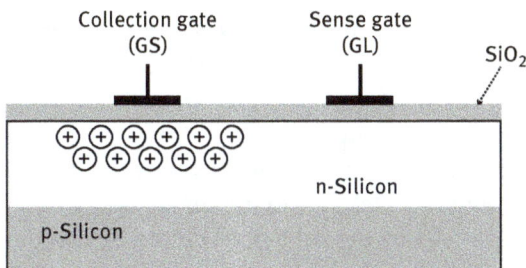

Figure 3.76: Structure of a CID pixel.

For historical reasons, CIDs use reverse polarities compared to CCDs: The epitaxial layer is n-doped and the gate is negatively charged, i.e., holes, instead of electrons, are collected beneath the $SiO_2$ layer. Each pixel has its own collecting electrode, the readout electrodes of all pixels are connected with each other. During the integration phase, both the collecting electrode GS and the readout electrode GL are negatively charged. (Figure 3.77a). However, since $-U_{GS} = 2 \times -U_{GL}$, the charge carriers only accumulate under GS. The readout process proceeds as follows:

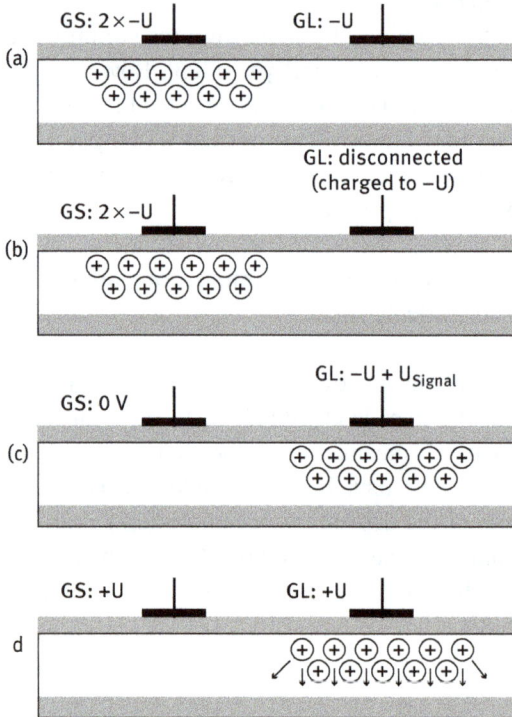

**Figure 3.77:** Reading a CID pixel.

First, the readout electrode is separated from the negative voltage by opening an electronic switch (Figure 3.77b). The readout electrode GL now forms a capacitor plate that is charged with a voltage -U. The n-silicon forms the opposite capacitor plate, the $SiO_2$ layer the dielectric layer of the capacitor.

If the $U_{GS}$ is reduced to 0 V, the charge carriers flow under the readout electrode, which is still negatively charged (Figure 3.77c). Due to the charge $Q_{Signal}$ flowing in, the voltage on the capacitor $C_{GL}$ formed by the readout electrode GL and the n-silicon changes by a voltage $U_{Signal}$. $U_{Signal}$ is calculated using the following equation:

$$U_{Signal} = \frac{Q_{Signal}}{C_{GL}} \tag{3.28}$$

The voltage on the readout electrode GL is now $U_{Signal} - U$. $U_{Signal}$ is obtained by measuring the voltage on GL and adding U.

By once again raising the collecting electrode GS potential to $-2U$ (return via Figure 3.77b to Figure 3.77a), the charge can be returned to the collecting electrode and the integration can be continued. Destruction-free readout is, thus, possible. Opposing this advantage is the disadvantage of the large capacity of the readout electrode structure that leads to a voltage drop of about two orders of magnitude (see Sweedler/Ratzlaff/Denton [80], p. 53]). This effect is already known from the discussion about diode arrays.

If a positive voltage is applied to GL, the holes are repelled in the direction of the pn-junction and recombine there (Figure 3.75d). A comprehensive description about CID construction details, for two-dimensional cases too, can be found in Ninkov [83, 85].

### 3.6.2.3 CMOS detectors

The CMOS process is currently the most common standard procedure for the production of integrated circuits. CMOS stands for complementary metal oxide semiconductor: n or p-doped structures are applied to the surface of silicon substrates. The surface can be partially free or covered with a $SiO_2$ layer. Aluminum layers can in turn be vapor deposited above that as electrodes. The C in CMOS stands for complementary, which refers to the fundamental logic structure of the circuits produced using this technique, namely a pair of complementary field effect transistors connected in series.

In this context, complementary means that one of the transistors is an n-channel type and the other a p-channel MOSFET. Figure 3.78a shows the circuit. The gate electrodes for the two transistors are connected so that either the p-channel transistor conducts and the MOSFET with n-channel blocks or vice versa, depending on the polarity on the gate electrodes. No current flows after switching because one of the transistors always blocks. Only in the moment when switching takes place do the parasitic capacities of the transistors have to be charged. Figure 3.78b shows an equivalent circuit. Microprocessors and memory devices, but also analog circuits such as operational amplifiers and analog switches can be produced using the very large-scale integration (VLSI) possibilities of the CMOS process. Photodiodes can also be implemented. In contrast, analog shift registers cannot be easily realized. As a consequence of this, the charge of the photodiode cannot be transported to the output amplifier easily and without loss, as is possible with the analog shift register of the CCD. There remains the possibility of implementing a collecting line, to which the pixel signals are connected one after another. This solution and its drawbacks have already been introduced together with diode arrays: The collecting line must extend over the full length of the chip and, therefore, has quite a high capacity. If the individual pixels are simply connected to this structure using analog switches, then a charged small capacity

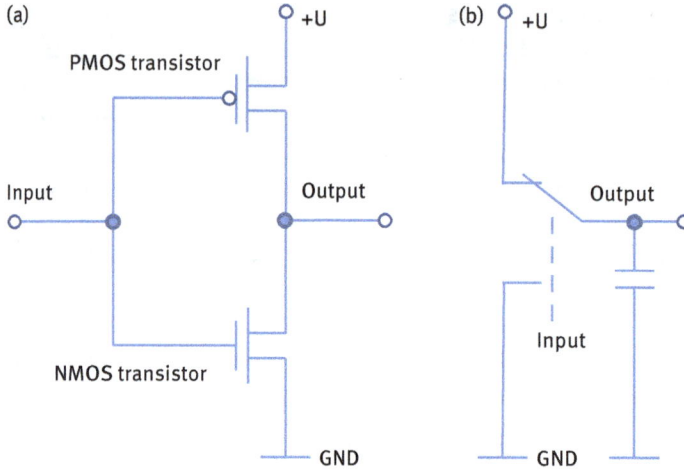

Figure 3.78: CMOS inverter with equivalent circuit diagram.

(that of the photodiode) is combined with an uncharged larger capacitor (the capacity of the collecting line). The charge balances, and the voltage on the collecting line is low compared to the original pixel voltage. For this reason, a buffer amplifier, which raises the voltage on the collection line to the level of the pixel photodiode, must be provided in the output for every pixel.

Figure 3.79a shows a pixel realized with the CMOS technique and Figure 3.79b the associated equivalent circuit. Since every pixel has its own amplifier, the type of detector described here is also called an "active pixel sensor" (APS).

In the case of CCDs, the charge is transmitted virtually without loss to a single amplifier via analog shift register. CMOS components, on the other hand, have one amplifier per pixel. This harbors the risk that the individual pixels of CMOS detectors may have large deviations in dark current and transmission function. The transmission function is the function that is obtained when the amount of radiation and the measured value resulting from it are plotted against each other. A further drawback that is frequently mentioned about the CMOS technique is that the active detector technology takes up space on the sensor surface. This space that is, then, no longer available for radiation detection. The so-called "fill factor" is less than 100%. In the case of two-dimensional detectors, this is a serious objection. In the context of linear sensors, however, this is irrelevant: The additional circuits can be attached above or below the pixels. The pixels here have typical lengths of 200 μm for widths of 7 μm or 14 μm. This results in pixel areas that are large compared to the space required by the auxiliary electronics.

(a)

(b)

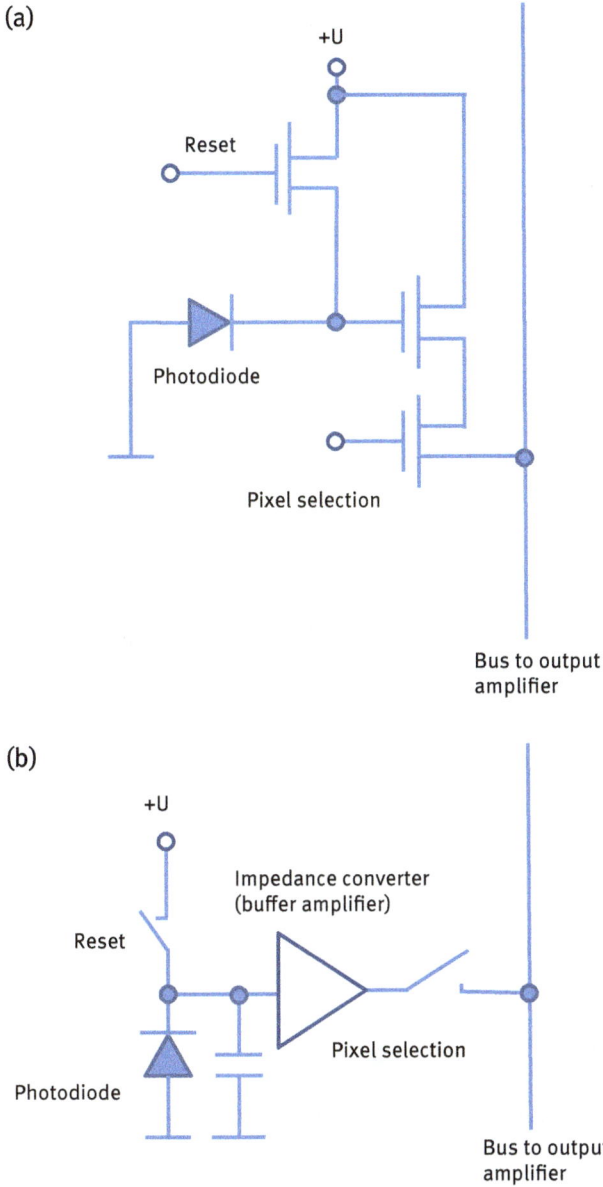

Figure 3.79: CMOS pixel with equivalent circuit diagram.

## 3.6.2.4 Comparison of CCD, CID and CMOS arrays

In the following, the advantages and disadvantages of the line sensors discussed in Section 3.6.2 will be presented in a catchword-like manner.

## CMOS detectors
Advantages:
- CMOS detectors can be produced with the most advanced and cost-effective IC standard processes. Therefore, they can be less expensive than CTDs. This applies particularly for CIDs that, in contrast to CCDs, are only produced in small quantities. Here are, additionally, the largest advances in development, as the CMOS process is the most common semiconductor technology.
- Electronics for control and signal processing can be integrated onto the detector chip.
- Blooming effects do not occur.
- By using the highly integrated standard process, it is, in principle, possible to build chips that enable single pixels to be read out; allowing dynamic problems to be solved.
- CMOS chips have clearly caught up in terms of sensitivity and are almost comparable to CCD components.
- The readout speed is high.

Disadvantages:
- Higher fixed pattern noise.
- CMOS chips use active pixels, which can lead to different transmission functions between radiation and measured values from pixel to pixel.

## CCDs
Advantages:
- Low noise
- Very high sensitivity

Disadvantages:
- Blooming effects may occur.
- Quick readout of individual pixels is not possible.
- Limited dynamic range

## CIDs
Advantages:
- Nondestructive readout of individual pixels is possible.
- The noise level is low.
- No blooming

Disadvantages:
- Usually higher price
- Limited opportunities for the implementation of special functions compared to CMOS chips.

CCD and CMOS detectors are compared in Göhring [86] The key messages made there still apply today. It appears that the future belongs to CMOS detectors. The advantage in sensitivity that CTDs have enjoyed for years has shrunk. The various opportunities for innovative chip details, the security of no blooming, the possibility of a higher readout rate as well as the somewhat lower price are all factors in favour of CMOS detectors. Even the drawbacks of higher fixed pattern noise and the lower uniformity of the pixel sensitivities are not as serious today (2018) as they were 16 years ago. In addition, microprocessors are available with ever-increasing processing power. They enable compensation for these drawback effects by using the appropriate algorithms. McCormick [87] compares linear CCD detectors commonly used for spectroscopic purposes with a modern CMOS detector from the company Hamamatsu [88] and concludes that the decision as to whether CMOS or CCD detectors should be used depends on the application. The CMOS detector is superior or equal to CCD detectors in several important respects. According to [87], the CMOS detector does have a slight disadvantage concerning the dark current. This characteristic is likely to be insignificant for many applications.

## 3.7 Measuring electronics

The measuring electronics, also frequently called the "readout electronics," form the interface between the optoelectronic detectors and the superordinate computer. The requirements on the electronics for readout of the photomultiplier tubes differ from those for the circuits intended to control multichannel semiconductor detectors.

### 3.7.1 Measuring electronics for photomultiplier tubes

In the early direct reading spectrometer systems, the charge that flowed from the anodes of the photomultiplier tubes was collected in capacitors. This type of circuit is illustrated in Figure 3.80. Before the beginning of the measurement, the switches S1 to Sn were closed for a short time and the tube voltmeter bypassed to discharge the capacitors C1 to Cn. After this, S1 to Sn were opened again. The photo-current of each PMT charged the capacitor assigned to it. After the end of the measurement, the capacitors were connected to the tube voltmeter sequentially and the absolute intensities were read. Such readout electronics were found frequently in publications from the 1950s, e.g., in Pfundt [89].

The type of integration described here had a major drawback: The dynode bias was realized with a voltage divider (see Section 3.6.1). The voltage difference between

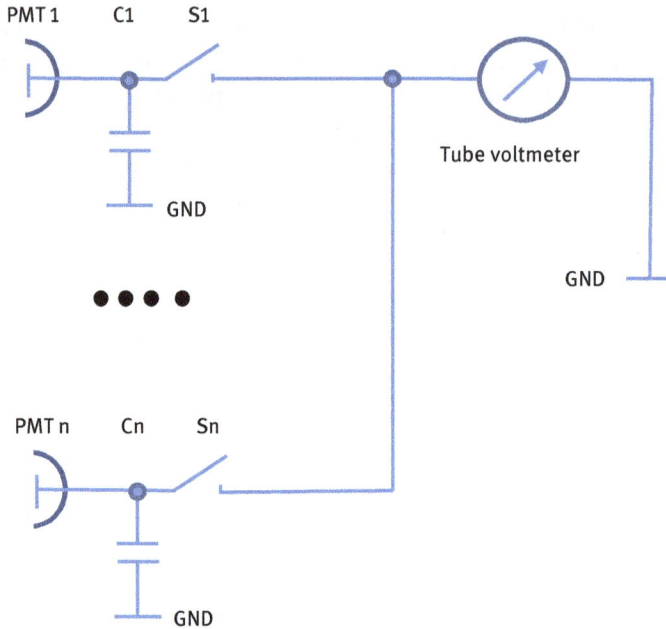

Figure 3.80: Charge integration via single capacitors.

the last dynode and the PMT anode decreases constantly with increasing charge on the capacitors C1 to Cn. If $U_{C\_End}$ is the voltage that drops across the capacitor C (here C stands for any of the capacitors C1 to Cn) at the end of the measurement and $U_D$ is the voltage difference between the last dynode and ground, then there is a voltage difference $U_D$ between the last dynode and the anode at the beginning of the measurement. However, at the end of the measurement this voltage is only $U_D - U_{C\_End}$. Photomultiplier tubes react strongly to changes in the acceleration voltages as has been explained in Section 3.6.1. As a general rule of thumb, for common multipliers that are operated with voltages around −800 V, a reduction of the voltage to −850 V results in a doubling and an increase to −750 V to a halving of the output current. If the output current depends on the state of charge of the capacitor, then the charge of the capacitor no longer grows proportionally with the strength of the radiation.

The circuit according to Figure 3.81 does not have this disadvantage. The anode of the photomultiplier tube PMT is connected to the inverting input of the operational amplifier, OP. If there is a current flowing out of the PMT, the inverting input of OP becomes slightly more negative than the non-inverting input and the voltage on the output increases a little, which results in a positive current across the capacitor C. Exactly the same current flows through C as the one from the PMT, but with the opposite sign, thus holding the voltage on the inverting input of the operational amplifier at zero. The minus input of the OP is always at

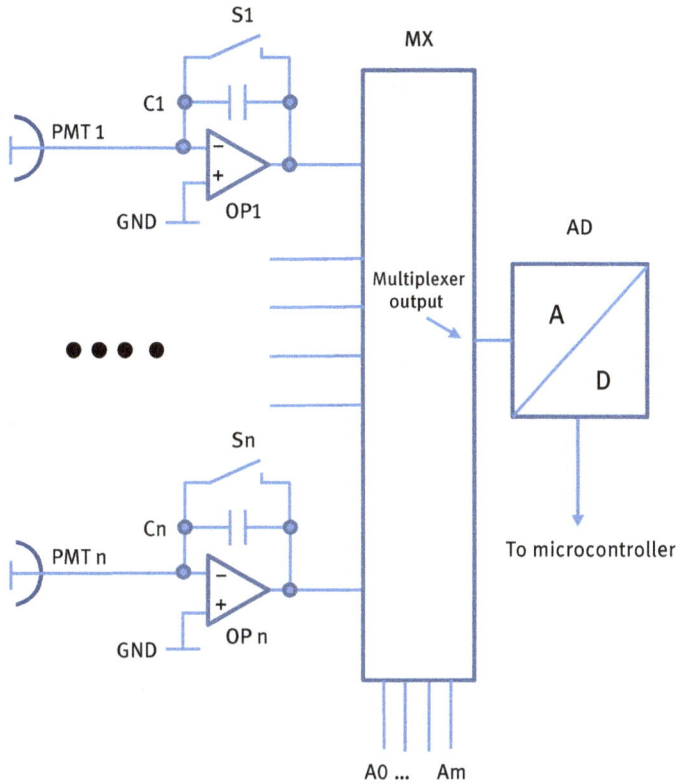

Figure 3.81: Readout system with inverting integrators.

ground potential. The capacitor C is charged in the process and in the end contains exactly the charge that flowed from the PMT. PMT, C and OP stand here, once again, for any of the photomultiplier tubes, capacitors and operational amplifiers grouped in the integrator illustration in Figure 3.81. A detailed description of the reverse-integrator presented here, also known as a Miller integrator, can be found in Tietze and Schenk [15] (p. 756 ff). The voltage $U_C$ that drops on the integration capacitor C can be measured on the output of the OP, as the other side of the capacitor is always at ground potential. It has, in contrast to the capacitor voltage in Figure 3.80, a positive sign, which enables it to be used directly behind the operational amplifier component that can only process unipolar, positive voltages. In Figure 3.81, the positive output voltage is processed by the analog multiplexer MX with which multiple integrator outputs can be connected to the analog-digital converter AD by selecting a digital address information A0 to Am. Analog multiplexers and analog-digital converters are usually operated by a dedicated microcontroller.

This proceeds as follows:

1. First, the integration capacitors are discharged by briefly closing the switch S that can be found parallel to each capacitor C. A memory location Int[i] in the microcontroller is reserved for each integrator i. These are initialized with zero.
2. Then the measurement is started.
3. During the measurement, the integrators are sequentially connected to the A/D converter by selecting their address. An A/D conversion is initiated and a voltage U is measured. If U approaches the maximum possible voltage, which is usually slightly less than the positive supply voltage of the operational amplifier, then the memory location Int[i] assigned to the integrator is increased by U and the associated capacitor C is reset by actuating the electronic switch parallel to it. This method makes it possible to process a capacitor charge many times per measurement. During spark operation it is necessary to ensure that the readout and especially the resetting of the integrator takes place between two sparks; i.e., when no usable signal is generated.
4. When the measuring time is over, the measurement is stopped.
5. The residual charges are read out exactly as described in point 3 and the intensities stored in the Int array increased accordingly.
6. Finally, the measured raw intensities now available in the Int array are transmitted to the superordinate computer.

In modern instruments, photomultiplier tubes are preferred when the measurement of time-resolved signals is required.

The ability to record the signal of the photomultiplier tube in a time-resolved manner has various advantages:
-  No usable signal is generated in the first microseconds of the spark. The signal to background ratio can be improved by switching off the PMT output during this time.
- If the signal is blocked for the complete current conducting phase of each spark, then the ion lines disappear but the atom lines are still excited. If an atom line is interfered by an ion line, then this interference can be reduced using time resolved measurement. At this time, the spectral background caused by electron recombination (bremsstrahlung) is absent, too.
- If only a time slice at the beginning of the spark is integrated, the cold atom cloud has not yet formed around the plasma. Therefore, self-absorption effects are less pronounced and the concentration limits, at which lines that are prone to such effects can still be used, are raised.

The circuit according to Figure 3.81 can be enabled for time-resolved measurements by introducing additional switches (SZ1 to SZn in Figure 3.82). The signal is only integrated when SZ is connected to the integrator. In the other switch position, the current from the anode of the photomultiplier tube PMT is led directly to ground.

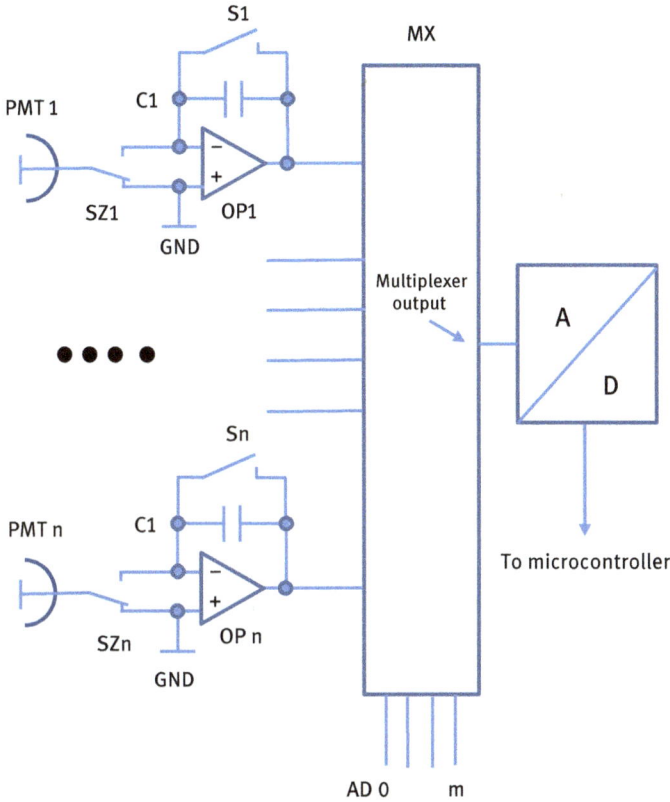

Figure 3.82: Readout system with the possibility of time-resolved acquisition.

Interruption of the connection between the PMT and the operational amplifier input would not be effective: The photomultiplier tube is a current source. When the switch is open, the charges on the parasitic capacitance formed by the anode and the input line connected to it would charge and this charge would reach the integrator when the switch is closed the next time.

The disadvantage to the circuit in Figure 3.82 is that only a single spark phase can be integrated per integrator. It could, however, make sense to use multiple integration windows of an analyte line. For example: The entire signal from a trace line is used once without the first microseconds for a calibration curve (curve 1). At the same time, however, a second calibration function (curve 2) is to be created that takes only the first part of the discharge into account. If a sample contains the analyte in trace concentrations, curve 1 is used. If, however, the analyte content is so large that self-absorption begins, curve 2 is taken. Algorithms for line switching between different curves of one element are described in Section 3.9.7.

The flexibility required here is offered by the circuit illustrated in Figure 3.83. It looks like a simplified version of the circuit from Figure 3.81. Since the integration

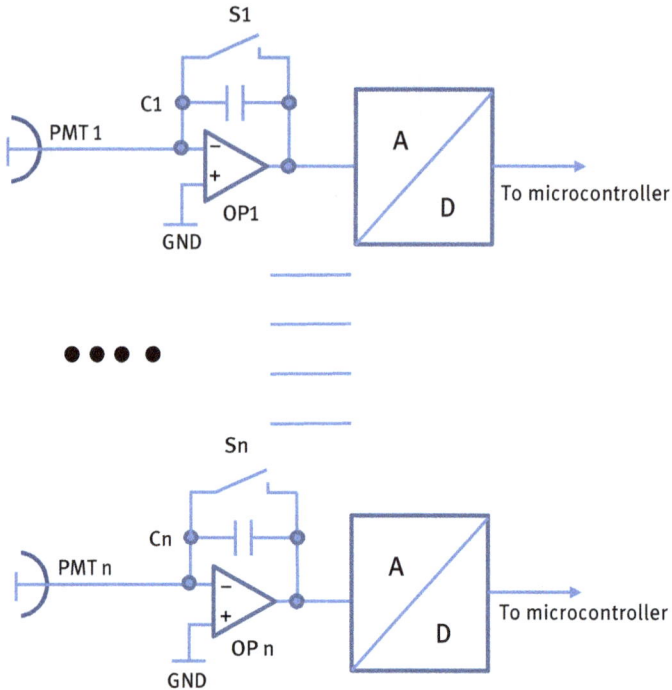

**Figure 3.83:** Readout system suitable for the time-resolved acquisition of several time windows per PMT.

capacitors are addressed in very small time intervals, it makes sense to assign an A/D converter and a microcontroller to every integrator. The integration capacitor is small compared to those in Figure 3.81. If capacities from 10 to 100 nF are common for the circuit according to Figure 3.81, the capacitors C1 to Cn in Figure 3.83 only have capacities on the order of magnitude of ten to several hundred picofarad.

The measurement is carried out as follows:

1. A memory location is reserved in the microcontroller for each time interval. With intervals with a length of 5 μs, memory locations are reserved for the signals between 1 and 5 μs, 6 and 10 μs, 11 and 15 μs, etc. All memory locations are initialized with zero.
2. The measurement is started.
3. As soon as the excitation generator indicates the beginning of a spark, C is scanned with the specified time grid and the associated memory location (i.e., at 5 μs grid, the first memory location, the second after 10 μs, etc.) is increased by the measured intensities. After each measurement, the switch S is briefly closed to discharge the capacitor C.

4.  When the end of the measuring time has been reached, the measurement is stopped.
5.  Finally, the measured raw intensities are transmitted to the superordinate computer. An array of raw intensities is sent for each line. A single raw intensity from one such array contains the sum of the intensities from all the sparks that were integrated from a time t µs after the time of ignition (e.g., t = 5, 10, 15 and 20) over a fixed duration of d µs (e.g., d = 5).

The circuit shown in Figure 3.83 is, in principle, simple, but the smallest of charges have to be measured. An intensity range is quickly reached where the readout noise dominates the source noise, thus reducing the analytical performance (see Section 3.5.6). Therefore, very careful implementation of the electronics is required to achieve the necessary performance level.

### 3.7.2 Measuring electronics for multichannel semiconductor detectors

Figure 3.84 shows a typical example for the control of linear semiconductor detectors. Here, these detectors can be CID, CCD or CMOS detectors. Such detectors are described in Section 3.6.2.

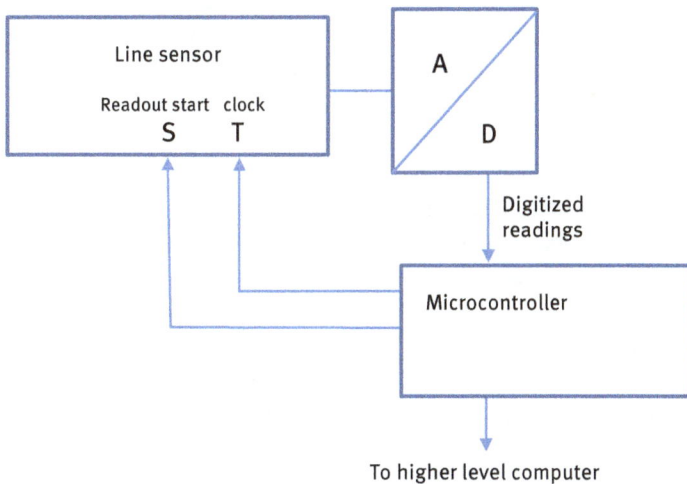

**Figure 3.84:** Readout system for controlling multichannel semiconductor sensors.

The outputs of the detectors are connected to the inputs of analog/digital converters. These are usually connected to a microcontroller with a serial interface, for example a "Serial Peripheral Interface" (SPI). The microcontroller also sends control signals to the detectors. A start impulse (line S in Figure 3.84) transfers the signal integrated in the pixels to an output register, which may be, for example, an analog shift register. Then impulses are applied to a clock line T. The clock pulses ensure that the signals from the pixels are transmitted one after the other to the output line connected to the A/D converter. While a microcontroller can generally process more than one linear detector, the number is limited due to the required processing speed and the limited number of serial interfaces available in the controller. A microcontroller can rarely capture the data from more than eight detector lines. Therefore, it is frequently necessary to use multiple microcontrollers to control the detector lines.

Data collection proceeds as follows:

1. In the microcontroller, a two-dimensional integer array Int is created in which a memory location for the recorded intensities is available for every pixel with the number P from every sensor S. Every memory location of this array is initialized with zero.
2. The detectors are reset. This is usually done by issuing first a start pulse and then a number of pixel clock pulses corresponding to the number of pixels.
3. Then the measurement is started.
4. First, a short, specified amount of time is waited. Times between 2 and 100 ms are common. During this so-called "micro-integration time," charge accumulates in the pixels. After the micro integration time has elapsed, a start pulse is sent through the line S to the line sensors and causes the transfer of charges from the pixels to the output shift registers. Then right away, the timer is started again in order to determine the correct time for the next readout. Now the signals for the current integration, present in the output shift registers, can be processed. The signal from the first pixel is currently located at the detector's analog output. It is converted and the memory location assigned to the pixel is increased by the measured value. Then a pixel clock signal is applied to the line T. The next signal from the second pixel is converted and processed by increasing the assigned memory location in Int. This process is repeated until all pixels have been read out. After the end of the next micro-integration phase, the process described in step 4 is repeated.
5. When the end of the measuring time is reached, the measurement is stopped.
6. The residual charges are read out and the intensities stored in the Int array are increased accordingly.
7. Finally, the measured raw intensities now present in the Int array are communicated to the superordinate computer.

Modern spectrometer systems often use both parallel multichannel line detectors and photomultiplier tubes. In this way, the advantages of complete spectral capture can be combined with those from recording the signal progression of the intensities during and after the spark event. It is possible to build optical systems that combine both detector types while enabling them the same view of the spark. The German patent DE19853754B4 [90] describes such a system.

## 3.8 Superordinate computer

Computers with $80 \times 86$ compatible processors and Microsoft Windows or Linux operating systems are usually used as the superordinate computer. Handheld devices often have other hardware and software platforms, such as ARM processors with Android operation systems.

If non real-time operating systems are used, separate microprocessors or "Digital Signal Processors" (DSPs) usually take over control of the integrators or semiconductor detectors. This is necessary because particularly CCD arrays require reproducible, exact timing. Interruption of the readout process can lead to malfunction. The communication between slave processor and master must take place over a fast interface. Usually, Ethernet or USB interfaces are used.

## 3.9 Spectrometer software

The spectrometer software is becoming increasingly important. Many functions that were implemented by hardware in the last decades have been transferred into the software. The most important algorithms, which are used for the following purposes, are explained in Section 3.9:
1. Creation of the calibration functions
2. Calculation of the element concentrations from the measured values of unknown samples using the calibration functions determined in point 1
3. Recalibration (standardization) of individual line pairs or entire spectra

Other algorithms, e.g., those for sorting, for control of compliance with predetermined material specifications or for finding material grades fitting to an analysis, are described in Section 7.7. These calculations are of particular interest in the context of mobile spectrometers.

### 3.9.1 Calculation of the line intensities from the spectra

Optics equipped with exit slits provide intensities for the wavelength ranges that pass through the slits. The situation is much more complicated for systems equipped with multichannel detectors. The pixel widths do not exactly fit to the lines to be measured. Computational steps are required to make the intensities of the desired lines available.

#### 3.9.1.1 Interpolation of spectra
The measured spectra consist of a two-dimensional array Int of integers in which an intensity i is stored after the measurement for every detector number s and every pixel p, i.e., in symbols: $Int[s,p] = i$, whereby s is an integer between 1 and the

number of detectors found in the system $S_{max}$; p is an integer between 1 and the number of pixels per detector $P_{max}$. Every pixel has a defined width on the order of 10 micrometers. For the purpose of simplification, it is assumed that every pixel of the detector s captures a wavelength interval with the constant width $\varepsilon$ and the first pixel of s captures the wavelength interval $[\lambda_s, \lambda_s + \varepsilon]$ (more precise would be the notation as a half open interval, $[\lambda_s, \lambda_s + \varepsilon[$. In favor of a better readability closed intervals are written). Then the $p$th pixel captures the interval $[\lambda_s + (p - 1) \times \varepsilon, \lambda_s + p \times \varepsilon]$. However, only information about the sum of the intensities from spectral areas that are bordered by wavelengths that conform to $\lambda_s + p \times \varepsilon$ with $1 < p < P_{max}$ and $1 < s < S_{max}$ are available. Unfortunately, the wavelengths of the pixel boundaries $\lambda_s + p \times \varepsilon$ are not constant. They can easily change due to aging or pressure or temperature drifts, so that the spectral ranges found on s pixels are shifted. Figure 3.85 shows the problem using a section of a spectrum. Above is the "real" spectrum as recorded by a detector with many, very narrow pixels. A detector, whose pixel width is approximately half the width of the spectral line, is drawn in the second and third rows. The second row shows the intensity distribution of the original spectrum, and the third row shows the distribution that would be recorded if the spectrum is shifted 5/7 pixel widths. Hereby, the pixel boundaries coincide with the vertical grid lines. It can be seen that the detector position plays a large role in recording the spectrum in the upper row. Small shifts of the detector or the spectrum have a large impact on the distribution of the intensity distributions on the pixels.

The so-called "spline interpolation" is used to reconstruct the spectrum as accurately as possible independently from such shifts. With the spline interpolation, a cubic polynomial $F_P$ is assigned to every pixel number p. The function formed by the concatenation of spline polynomials is so constructed that they can be differentiated on the borders between the individual polynomials, i.e., they merge into each other "without kinks." Spline interpolation algorithms can be found, e.g., in Engeln-Muellges/Reutter [91] (p. 145 et seq.). Using curve sketching of the spline polynomials, it is possible to determine the position and height of the spectral lines with a resolution of a fraction of a pixel.

However, it should not be concealed that the spline interpolation is not a universal remedy. Artifacts can appear in the spline-interpolated spectrum. If the pixel width is not much smaller than the full width at half maximum of the lines and if there are, at the same time, strong lines with low background, the spline-interpolated spectrum can have "undershoots," i.e., negative intensities in front of and behind the lines. The use of chips with narrower pixels can help here.

### 3.9.1.2 Virtual exit slits

Any given wavelength range can be integrated if the spline functions are available for all pixels. Translated to the conditions in an optic equipped with exit slits; this means that slits that are as wide as desired can be placed on any position.

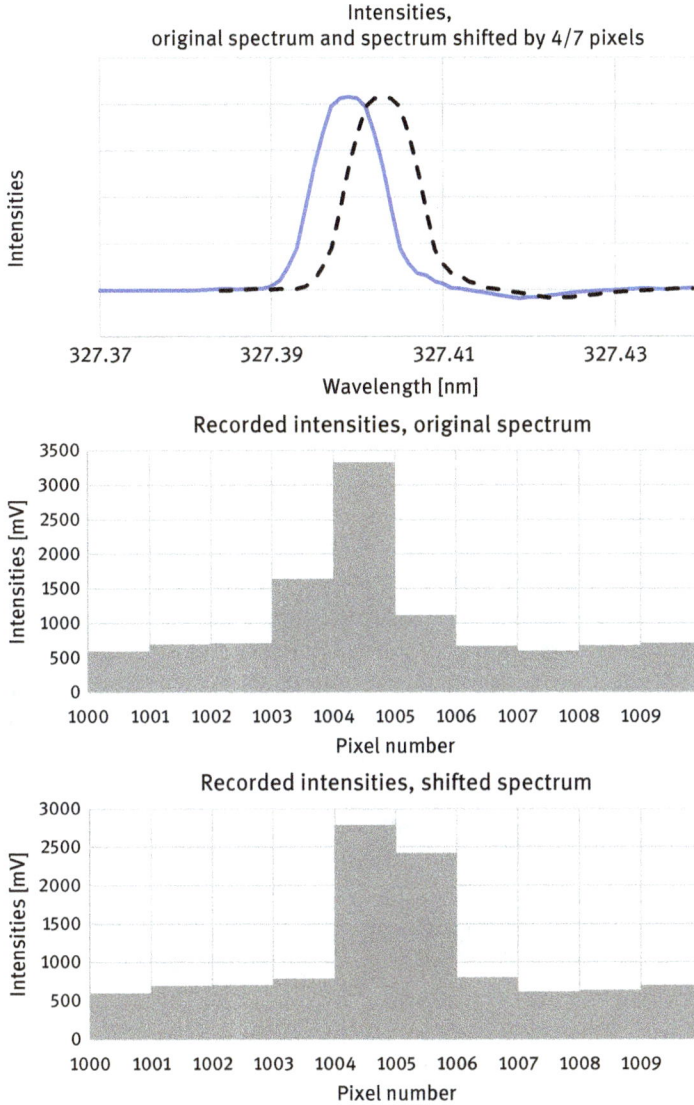

**Figure 3.85:** Real spectrum and detected intensities with slight spectral shift.

Pixel ranges can be defined for any spectral line used as an analytical line or as an internal standard. The limits of these pixel ranges do not need to be integers – pixel fractions are also possible.

## 3.9.2 Wavelength calibrations and profiling

In the case of optics equipped with slits, the exit slits are mounted in the correct positions on the focal curve and then fine adjustments are made during production. The spectrum can, however, shift slightly in the course of the instrument's lifetime. Therefore, it is necessary to ensure in regular intervals that the spectral lines for the analytes and standards are as exactly aligned with the exit slits as possible. This process is called "profiling."

For optical systems that capture the entire spectrum with semiconductor detectors, the spectral lines do not always have to be captured in the same place on the detector. However, it is necessary to be able to predict where and on which detector a spectral line can be found. A function must be determined that calculates a sensor number and pixel number for a given wavelength.

### 3.9.2.1 Profiling an optic equipped with exit slits

This kind of optic is equipped with a mechanism to shift the spectrum so that it is possible to profile it. Profiling can be done by moving the entrance slit using a motor in small steps along the focal curve. Alternatively, a refractor positioned behind the entrance slit can be rotated, again preferably using a motor. The way in which refractors work is described in Section 3.5.5. In both cases, the angel of incidence ε is slightly modified and the whole spectrum on the focal curve is shifted according to eq. (3.8). For profiling, it is also necessary to have a sample in which as many analytes as possible are present in sufficiently high concentrations. Profiling is conducted by driving the entrance slit in steps (or rotating the refractor in small angular increments) and briefly measuring the sample after every change in position. One measured value is obtained for every position for each spectral line. Finally, the entrance slit is driven to the position for which the maximum intensity was measured for as many lines as possible. Only the lines for the analytes contained in the profiling sample are considered.

### 3.9.2.2 Wavelength calibration of optics equipped with semiconductor detectors

The wavelength calibration of such optics can be conducted by measuring samples that provide spectra having few lines with uniquely identifiable spectral lines. The approximate association between pixel and the wavelength measured with it is determined using the angle of incidence and the angle at which the detector is mounted. This enables an inaccurate, approximate wavelength calibration. It is sufficient to predict a pixel range where a line is supposed to appear. If a line in this range is identified, the pixel position at which it appears is recorded. Then the exact wavelength is taken from a table such as the wavelength atlas from Saidel, Prokofjew and Raiski [92]. These exact wavelengths are plotted against the pixel positions where the spectral lines were actually found. It is not sufficient to determine the integral number of the pixel as the pixel position. Rather, the position of the line on the detector must be precisely determined to a fraction of a pixel. Either the line maximum is determined from the spline function mentioned in Section 3.9.1 or a so-called "Gaussian fit" is calculated. Using the method of the smallest square fit, a Gaussian bell curve is matched as closely as possible to the line profile. The advantage to this type of calculation is that, in addition to the exact position in the subpixel range, it provides information about the line width that can be checked for plausibility. If several lines with the associated wavelengths and pixel numbers are identified per detector, a polynomial can be determined with a regression calculation to enable the conversion of wavelengths into pixel numbers. With the same input data, the reverse function can also be determined as a polynomial. When the pixel number is entered, the associated wavelength results.

This procedure is easily automated. However, it is not necessary to do this for each instrument. If the complete spectra recalibration is used (see Section 3.9.6.2), then, for every instrument in a series, the spectrum from any sample can be converted so that it matches the spectrum that was measured for that sample on the master instrument for the series. Therefore, only a single wavelength calibration is required for each type of instrument even when thousands of instruments of this type are produced.

## 3.9.3 Background correction

If a calibration covers a wide range between non-alloyed and high-alloyed materials, the spectral background may vary. In a nickel screening program, the background for pure nickel or Ni-Cu alloys is generally much lower than that for, e.g., Ni-base superalloys. This can lead to an increase in the standard error of the calibration curve for small concentrations. To compensate for such background fluctuations, one or two background positions are defined for each analytical line at fixed intervals from the line positions. The background intensities are determined in the same way as the line intensities.

The background intensity is subtracted from the line intensity for lines for which only one background position is defined. If background positions are defined to the left and the right of the analytical line, then the background at the line position is interpolated and this interpolated value deleted from the line intensity. Details about background correction can be found in Slickers [24].

## 3.9.4 Calculating intensity ratios

Calibrations are generally based on ratios of analyte intensities to the intensities of so-called "internal standards," sometimes also referred to as "reference lines" [24, 93]. An internal standard line is a line for the main element.

During development of their method of homologous line pairs, Walther Gerlach and Eugen Schweitzer [94] established that it is more advantageous to work with "well-matched" line pairs to determine element concentrations than to use only the intensity of an analyte line to calculate the concentration. They observed that line pairs can be found that react with the same changes in relative intensity to fluctuations in the plasma temperature. Such temperature fluctuations are always present, even with perfectly reproducible electrical parameters. This compensation is important as small changes in the plasma temperature can lead to large changes in intensity.

The formation of ratios has another advantage. The light transparency of the optical components can be negatively influenced by contamination effects. The ratio, however, remains almost the same. This is even the case for wavelength-dependent contamination, provided that the wavelengths of the analyte and the internal standard are not too far from one another.

### 3.9.5 Calculating calibration functions

Standard sample intensity ratios are measured as an integral part of instrument production. Every standard is depicted as a point in the first quadrant of a Cartesian coordinate system, whereby the intensity ratio is located on the x-axis. The concentration ratios, additively and multiplicatively corrected if necessary, of the standard are displayed on the y-axis. Occasionally, the functions of the axes are reversed. The corrected concentration ratios are then on the abscissa, the intensity ratios on the ordinate axis. A calibration function K is calculated between the intensity ratios and the corrected concentration ratios with a regression calculation (see, e.g., Slickers [24] and Thomsen [93]). K is a polynomial that best approaches the points. It is calculated in such a way that the sum of the squares of the deviations between the ordinates of the points representing the samples and the curve is at a minimum (least squares method). Figure 3.86 shows intensity and concentration ratios for several standards with the associated best-fit polynomial; here, a linear equation. The best-fit line is such that the sum of the areas of all the illustrated discrepancy squares is minimal.

As already stated, concentration ratios are plotted against intensity ratios. Concentration ratios are obtained by dividing the concentration of the analyte by that of the base metal.

Example: If a chrome-nickel steel contains 20% chromium and 80% iron, then the chromium concentration ratio is 25%. It is not beneficial to plot intensity ratios against concentrations instead of concentration ratios, as the following consideration shows:

Assume there are three samples. Sample 1 consists of pure iron, sample 2 of 20% chromium and 80% iron and, finally, sample 3 of 20% Cr, 20% Ni, 20% Co and 40% Fe. The chromium line delivers 10 V per percent chromium, the iron line 1 V per percent iron. If these points were then drawn into the Cr coordinate system, the point for sample 1 would be at (0,0), for sample 2 the point (2.5,20) would be obtained and for sample 3 a point at (5,20). The intensity ratio is twice as large for the third

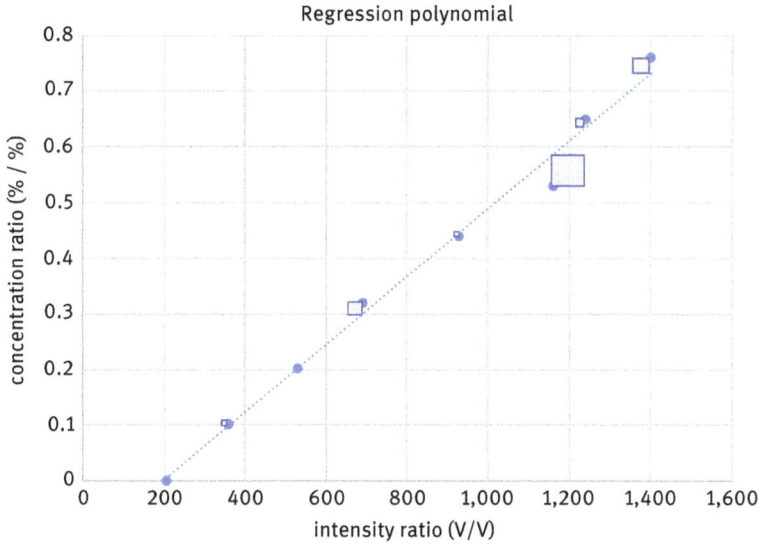

Figure 3.86: Samples in the Cartesian coordinate system with corresponding regression line.

sample as that for sample 2, without the ordinate value having changed. A regression line can only be drawn if very high deviations are accepted (see Figure 3.87a) and that, although an ideal, linear behavior of the spectral lines was assumed.

The situation is different when concentration ratios are used (see Figure 3.87b). The points (0,0) for sample 1 (2.5,25) for sample 2 and (5,50) for sample 3 are obtained. All the samples lie on a perfectly straight line.

In the beginning, we had mentioned that a correction is applied to the concentration ratios of the standards before the best-fit polynomial is calculated. Here, one distinguishes between interferences due to line overlaps and interferences by which individual elements influence the plasma, the so-called "interelement interferences."

### 3.9.5.1 Line interferences

If lines of other elements have either the same wavelength or are so closely neighboring to the analyte line that they partially fall through the real or virtual exit slit for the analyte line, the elemental content of the associated third-party element influences the measured values. Potentially interfering lines can be found by examining wavelength atlases for lines neighboring the analytical line. If a potentially interfering element is identified, then the interference, usually expressed as percent concentration increase of the analyte per percent interfering element, is calculated on a trial basis. However, the correction should only be used if it is plausible:

– The sign must be correct; if there is a line overlap, a positive value must be subtracted for every percent of the interfering element.

(a)

Intensity ratios plotted against concentrations

(b)

Intensity ratios plotted against concentration ratios

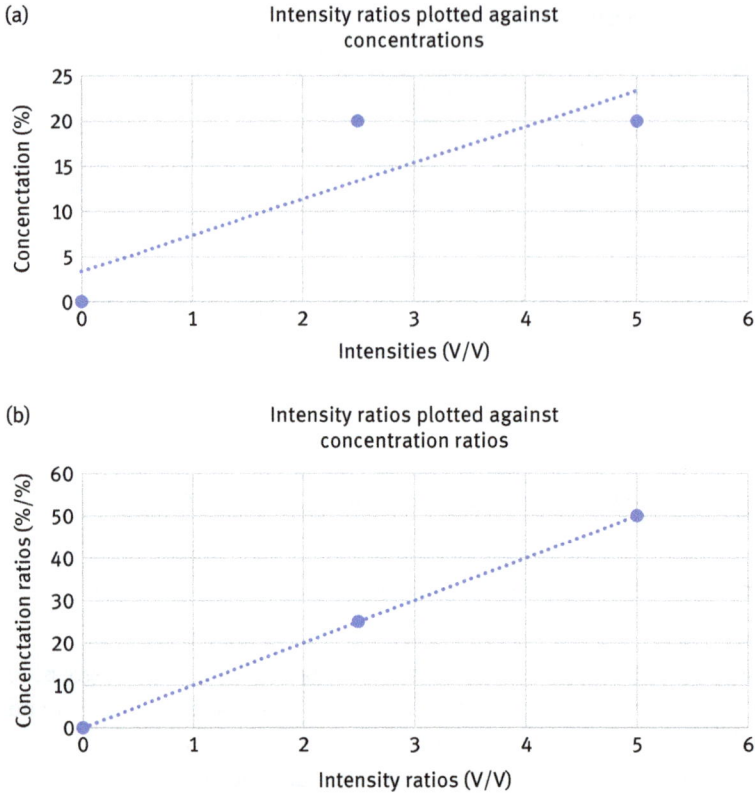

Figure 3.87: Necessity of using concentration ratios.

- At least some of the standards used for the calibration must contain the disturbing element with high contents (usually at least several percent). In addition, standards with low contents or completely without the interfering element should also be considered.
- If the interference is very large, e.g., when 1% must be subtracted from the analyte per percent disturbing element, then the use of an alternative analyte line should be considered.
- If the interference is very small, e.g., a few ppm per percent disturbing element, then it may be better not to use the correction. This is the case when the differences between the concentration ratios of the sample and the course of the best-fit polynomial are not significantly reduced by the correction.
- If the interfering element lies within the expected range, the lower section of the calibration curve should be considered, e.g., a range between 0 and twice the background equivalent. If in this lower section of the curve the dispersion of the calibration function improves, i.e., the points representing the samples approach the best-fit polynomial, then the line correction should be used.

### 3.9.5.2 Interelement interferences

Interelement effects are caused by the fact that individual elements can influence the position and temperature of the plasma. It is usually assumed that these elements increase or decrease the intensities of the analyte. A corrective value c that is multiplied by the interfering element concentration k is determined. The number one is added to this term. The result is multiplied with the concentration ratio that already has an additive correction thus obtaining the additively and multiplicatively corrected concentration ratio. The addition of one to the product of the corrective value and the concentration of the interfering element has the following background: If the interfering element is either not present or the correction is not needed, then the corrective value would be 0, then it is simply multiplied by one, i.e., the concentration ratio remains unchanged. The larger the corrective value c and the higher the concentration of the interfering element k, then the further the multiplier used is from 1: When c is negative, it is smaller than 1; when positive, then it is larger.

Of course, it is also necessary to check for plausibility when using multiplicative corrections.

### 3.9.5.3 Calculating the calibration function

The entire calibration function K is composed of:
- a $n$th degree regression polynomial
- m additive correction terms with interfering elements percentage $k_s$ and factors $b_s$ $(1 \leq s \leq m)$
- l multiplicative correction terms with interfering elements contents $k_t$ and factors $c_t$ $(1 \leq t \leq l)$

K is:

$$K(x) = \left( \prod_{t=1}^{l} 1 + c_t k_t \right) \left( \sum_{i=0}^{n} a_i x^i + \sum_{s=1}^{m} b_s k_s \right) \tag{3.29}$$

Polynomial coefficients, line interferences and interelement corrections are simultaneously calculated by solving a system of equations. The given coefficient tuple $(a_0, \ldots a_n, b_1, \ldots b_m, c_1, \ldots c_l)$ is determined, which leads to a function K. The function K approaches the standards' corrected concentration ratios as close as possible. "As close as possible" means in this context that the sum of the deviation squares is minimal (deviation squares are shown in Figure 3.86).

The degree of the calibration polynomials is chosen by the calibrator. They can be linear, square or cubic functions. In rare cases, the use of polynomials of higher degrees may be useful. Again, it is necessary to consider the plausibility. If only as many calibration standards are used as the coefficient tuple has components, then the polynomial inevitably goes through all the points representing the standards and the sum of the square errors is zero. This is also the case when one of the points is completely incorrect, e.g., due to incorrect measurement or incorrect input. If there are only a few more standards than are present in the sum of the polynomial degree and the number of disturbing elements, a polynomial can be calculated that approaches the points without reflecting physical reality. The analysis of unknown samples with such a calibration curve can lead to incorrect analyses. It is not plausible if the polynomial has points of inflection, i.e., takes the form of a stretched "S". If the polynomial has such a form, it is better to re-evaluate using a lower degree polynomial.

For this reason, many standards, often several hundred, are used to calibrate for common material grades, e.g., for the calibration of low alloy steels. Modern spectrometer software provides the possibility of giving individual standards, e.g., a blank sample made of a pure material, a higher statistical weight than the others. Generally, the software also enables rapid and easy elimination of standards or to display them without any weighting on the graph for informational purposes without including them in the calculation.

### 3.9.5.4 Recording high- and low-sample expected values for recalibration purposes

A set of so-called "recalibration samples" are measured together with the standards used for the calibration for instruments that do not use complete spectra recalibration. These samples are used to bring the instrument channel by channel back to the state when the calibration was conducted. Here channel means the line pair consisting of analytical line and associated internal standard. The samples are chosen so that there is a sample with a low concentration (low sample) and a sample with a high concentration (high sample) for each channel. The analyte concentration of the low sample should not exceed the background equivalent; usually, a pure or ultra pure material of the base metal is used. For the concentration of the high sample, it is advantageous when its elemental content is in the upper third of the working range of the calibration curve. The exact elemental concentrations in the recalibration sample do not need to be known. However, the material must be homogeneous, so that for every recalibration, material with the same elemental composition is ablated as that which was used to determine the expected values (one low-sample expected value and one high-sample expected value for each channel) at the time of the calibration.

Spectrometers are recalibrated in periodic intervals. Before recalibration, the recalibration samples must be ground. This is associated with sample consumption. If replacement of such a sample is required, it can be replaced with a sample from the same heat. However, after many years, samples of the same heat can often no longer be obtained. In this case, the instrument is carefully recalibrated for a last time with the original samples that have been almost used up. Then the spectrometer system is switched to the display mode "recalibrated intensity ratios" and the samples for the new set of recalibration samples are measured. The original low and high-sample expected values are replaced with the newly obtained values. The calculations conducted for the recalibration are explained in Section 3.9.6.

It is usual and necessary to use a set of recalibration samples for each metal base to be measured. These sets generally consist of two to ten samples. The recalibration of a multibase instrument can be a very time-consuming affair, as each of these samples must be measured several times before taking an average.

For instruments that work with complete spectrum recalibration, it is necessary to capture the state of the instrument that existed during the first calibration. The spectrum of a suitable adjustment sample is utilized for this. Unlike a conventional calibration, a single sample with a line-rich spectrum is usually sufficient. For the second and every further calibration (for example the calibration of Al-base methods after completion of the Fe-base), a complete spectrum recalibration is conducted before measuring the reference materials to convert the instrument back to the hardware state of the first calibration. The algorithms used for the complete spectrum recalibration are also explained in Section 3.9.6. Of course, the adjustment sample is also consumed over time. Here a new data set is usually supplied together with the new sample. The instrument software modifies the existing reference scan so that is matched the new sample heat.

### 3.9.5.5 Combining calibration functions into methods

If an analysis is to be conducted with a modern spectrometer system, a suitable method must first be chosen. Such a method includes calibration functions for all the elements relevant to the analysis of the samples to be measured. The calibration functions are determined with reference materials similar to the samples that are to be later analyzed. In other words, standards and unknown samples come from the same alloy group.

At least, general methods for the metal bases (such as iron, aluminum, nickel and copper base) to be measured with the spectrometer are provided. However, there are usually submethods within the metal bases to increase the accuracy.

Such submethods cover in, for example, iron base, the following alloy groups:
- Nonalloyed and low alloyed steels,
- Chromium and chrome-nickel steels,
- Nonalloyed and low alloyed free-cutting steels,
- Manganese steels,
- Nonalloyed and low alloyed cast iron
- High-speed steels
- High chromium cast irons
- High-nickel cast irons

Excitation parameters and measuring times can be specifically optimized for each of these alloy groups. The composition when considering the main elements is often similar within the alloy group submethods. Then the ablation behavior, the level of the spectral background and the magnitude of the line interferences caused by the main elements are also alike. The dispersion of the calibration functions in a submethod are, thus, generally better than for the functions in the general overview methods.

However, it can also happen that the main elements occur in a wide variety of combinations within the above-named groups. This is the case for, e.g., the high-speed steels: They contain tungsten at between 0 and 19%, molybdenum and cobalt from 0 to 9% and vanadium between 0 and 5%. The elements tungsten, molybdenum and vanadium have, as carbide forming elements, similar metallurgical functions. In addition, high-speed steels usually contain about 4% chromium, 1% carbon and only traces of nickel and copper. However, the similarities mentioned ensure that an efficient submethod for high-speed steels can be easily created.

Conversely, it is advisable to use two different methods for nonalloyed and low alloy steels and nonalloyed and low alloy free-cutting steels, although the free-cutting steels are only distinguished from the nonalloyed and low alloy qualities by the addition of small quantities of sulfur and/or lead. Together with manganese, sulfur forms inclusions that are preferably attacked by the sparks. The signals from manganese and sulfur are, thus, very high at the beginning of the spark phase and need some time to approach a stable level. This is why methods for free-cutting steels usually work with longer pre-spark times.

### 3.9.6 Recalibration

When dealing with recalibration, it is necessary to distinguish between recalibration of channels, complete spectrum recalibration and type recalibration.

As already stated, a channel consists of a combination of an analytical line with an internal standard. The algorithms for channel recalibration use scalar values to determine a factor and an offset for each channel that make it possible to recalculate the channel back to its state at the time of the calibration.

For a complete spectrum recalibration, the current spectrum is recorded for the adjustment sample. The spectrum for this sample was also measured at the time of calibration (see Section 3.9.5.4). A set of parameters is then determined from the original and the current spectrum of this adjustment sample. These parameters enable the current sample spectrum to be transposed to the original spectrum measured during the calibration of the master instrument. Any spectra that would have been measured on the master instrument can be transposed back with this set of parameters.

The type recalibration is used to improve the accuracy for the samples for a particular type of alloy.

### 3.9.6.1 One and two-point recalibration

If an instrument is operated for a longer period of time, the intensities obtained for the analyte and internal standard lines change. This is due to changes in the light transparency of the optical elements caused by soiling as well as aging effects such as those for fiber-optic cables. It is possible to partially compensate for such effects by dividing the analyte intensities by the intensities of the associated internal standard. However, it is not possible to compensate for every effect in this way. For example, aging of the photomultiplier tubes can be quite different for the analyte and internal standard. Therefore, a method that recalculates the channel intensities back to the state at the time of calibration is required.

The so-called "two-point recalibration" is the customary method of compensating for these instrument drifts with exit slits and photomultiplier tubes. This is described in detail in, for example, Luehrs and Kudermann [95]. As mentioned in Section 3.9.5.4, a set of recalibration samples, which provide an intensity ratio for a low and a high concentration for each channel, was measured at the time of calibration. These measurements provide an expected value for a low sample and a high sample (TS and HS) for each channel. During routine operation of the instrument, the measurement of these samples is repeated and the current values for the low and high samples are obtained – the actual values for the low and high samples (TI and HI). TI and HI are required to calculate the recalibration factor and the recalibration offset.

A factor and an offset can be determined as follows:

$$\text{Factor:} = (\text{HS} - \text{TS})/(\text{HI} - \text{TI}) \tag{3.30}$$

$$\text{Offset:} = (\text{HS*TI} - \text{HI*TS})/\text{TI} - \text{HI}) \tag{3.31}$$

Recalculation of an intensity ratio $\text{IV}_{akt}$ to the intensity ratio measured during the calibration $\text{IV}_{Kal}$ can be done simply by applying the factor and offset:

$$\text{IV}_{Kal}: = \text{Factor} * \text{IV}_{akt} + \text{Offset} \tag{3.32}$$

Frequently, the practitioner is faced with the problem that there are no recalibration samples for individual elements. In this case, a so-called "one-point recalibration" may help. The pure material serves as the *high sample* in this case. The recalibration is then conducted using the signal originating from the spectral background. There is only one expected value (S) for such channels. Together with the current value (I), the signal measured at the time of the recalibration, it can be used to determine only one factor:

$$\text{Factor:} = \text{S}/\text{I} \tag{3.33}$$

Recalculation to the state at the time of the calibration is done simply by multiplying the current value by this factor:

$$\text{IV}_{Kal}: = \text{Factor} * \text{IV}_{akt} \tag{3.34}$$

The terms "one-point standardization" and "two-point standardization" are frequently used as synonyms for the expressions "one-point recalibration" and "two-point recalibration," respectively.

### 3.9.6.2 Complete spectrum recalibration

Multichannel detectors enable the capture of complete spectral ranges. This makes the spectral environment available for every analyte and internal standard line. This fact opens possibilities that do not exist for systems that only measure individual lines.

The development of spectrometer systems usually proceeds in the following manner: First, the hardware components (excitation generator, optical systems, sample stand, etc.) and the instrument software are created. Then, a close-to-production prototype is calibrated and the calibration is checked. This can be an iterative process: If there are shortcomings in the hardware or software, these must be corrected, after which it may be necessary to partially or completely repeat the calibration. At the end of development, there is an instrument with all the calibrations intended for the instrument series. This will be referred to below as the master instrument of the series.

If there is a possibility to convert back the measured spectra from any instrument in a series to those from the master instrument, it is possible to immediately use the master instrument's calibration functions.

This would save the time required for individual calibrations and prevent mistakes that can be made during these work steps. Of course, there would be other advantages as can be seen in the following.

In the conventional approach, calibration work leads to the construction of a calibration function K, which calculates the concentrations from the raw intensities. Somewhat formalized, it can be written:

$$\text{Concentration} = K(\text{Raw\_spectrum}) \tag{3.35}$$

When the complete spectrum recalibration approach is used, the calibration function breaks down into an instrument-specific part $K_G$ and an instruments-independent function $K_U$, which are executed in succession. The complete spectrum recalibration assumes the role of the instrument-dependent calibration function $K_G$. $K_U$ are the calibration functions that were created in a conventional way using the master instrument as described in the previous sections. $K_U$ is the same for all instruments in a series.

$$\text{Concentration} = K_U(K_G(\text{Raw\_spectrum})) \tag{3.36}$$

To be able to construct the function $K_G$ for a recalculation of the spectra from the instruments produced in series to those that were measured for the same sample on the master instrument, it is necessary to have at least one sample whose spectrum was measured on the master instrument. From the same sample such a spectrum measured on the instrument produced in series is required. As a rule, these two spectra from a single sample actually suffice to determine the function $K_G$. This sample shall be referred to as the adjustment sample.

To construct a suitable function $K_G$, it must first be clarified what distinguishes the individual instruments in a series or which changes may occur during the lifetime of the instrument:

The pixels of the multichannel detectors may, even when detectors of the same type are always used, have a slightly different radiation sensitivity from one another and/or from the corresponding detectors in the master instrument. The causes for this may include dust particles on individual pixels, varying thicknesses of a possible fluorescence coating or simply the production spread of the detector chip.

Pixels have a width on the order of 10 µm. Within a series of spectrometer optics, however, the variation of the detector chip positions is typically larger. Within the series it is entirely possible that they fluctuate in a range of ±0.2 mm. This corresponds to about ±20 pixel widths. A spectral line that appears on pixel i in the master instrument, can, when the entire series is considered, be found anywhere in the range i–20. to i+20. It is also possible that the detector can be slightly tilted, which manifests itself in a slight compression or expansion of the spectrum.

The full widths at half maximums of the lines may vary a little from instrument to instrument. As was explained in Section 3.5.2, the detectors cannot be mounted so that they lie perfectly on the curved focal curve (see Fig. 3.56). The optimal focus of the pixels can also differ from instrument to instrument. Variations in the entrance slit widths also influence the full widths at half maximums.

The transparency from windows and lenses, the reflectivity of mirrors and the transmission of fiber optics are subject to small production variations and result in intensities with different heights. The differences between the master instrument and the systems produced in series caused by these factors can lead to considerable deviations in the light transparency because there are many optical components between the radiation source and the detectors (see Section 3.5.6). Within the series, fluctuations in transmission values from half to twice the serial average are to be expected. It is typical for light transmission that it does not change abruptly with wavelength changes. When a narrow wavelength interval is considered, e.g., a nanometer, then the transparency is almost the same everywhere and only superimposed by a small increase or decrease, which, of course, must be taken into account.

In summary, the following partial functions of $K_G$ must be constructed:
- Correction of the pixel sensitivity $K_P$
- Correction of the pixel offset $K_V$
- Correction of the resolution $K_R$
- Correction of the wavelength-specific sensitivity $K_E$

These partial functions all have a spectrum as the argument and result in a spectrum as a functional value. It is advisable that they be executed in the order in which they are listed above, so that the following can be written:

$$K_G(\text{Raw spectrum}) = K_E(K_R(K_V(K_P(\text{Raw spectrum})))) \tag{3.37}$$

### Correction of the pixel sensitivity, partial function $K_P$

The pixel-specific sensitivity function $K_P$ is determined for a pixel n by exposing the detector to a source of light that irradiates all the pixels with as much the same brightness as possible. It is, however, hardly possible to achieve homogenous illumination from the first to the last pixel. The illumination looks approximately as drawn in Figure 3.88. Neighboring pixels measure a very similar amount of light. The illumination of the more distant pixels can differ. To correct the $n$th pixel in a multichannel detector, a best-fit line is formed over the values measured for a range from ± b pixels around the pixel n without the measured value for the pixel n itself. Whereby, the requirement on the uniformity of the illumination increases with a higher interval width b. It may be useful to weight pixels closer to n more than those further away. If $IM_{S,n}$ is the intensity that pixel n on detector S should have according to the best-fit line and $I_{S,n}$ is the intensity

### Calculation of the pixel specific sensitivity function

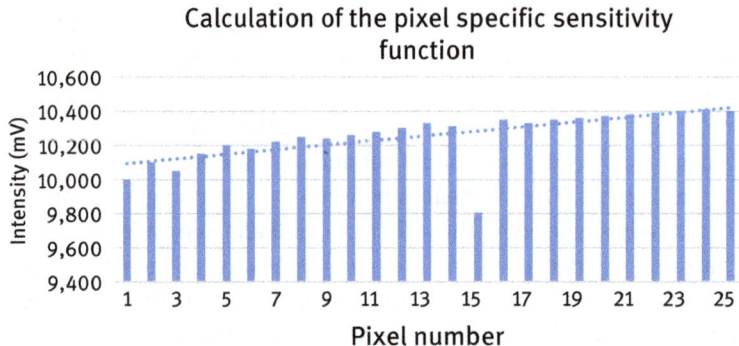

Intensity of pixel 15
$I_{S,15}$:  9805

Intensity acording to best-fit line, calculated from intensities of surrounding pixels
$IM_{S,15}$:  10284

Correction factor for pixel 15
$F_{S,15}$:  1.049

**Figure 3.88:** Continuously illuminated sensor section for pixel sensitivity correction.

that the pixel n on detector S delivers, the following is obtained for each pixel n on the detector S through division of $IM_{S,n}$ by $I_{S,n}$:

$$F_{S,n} := IM_{S,n} / I_{S,n} \qquad (3.38)$$

The function $K_P$ is applied by multiplying every pixel with the factor F assigned to it. Occasionally, pixels that change their sensitivity over time ("hot pixels") are found. This kind of malfunction can only be recognized by a repeated calculation of $K_P$ or by a pretest in a separate test device.

Detectors with such defects must be discarded. It makes sense to conduct testing for hot pixels or complete pixel failure ("bad pixels") with a separate device and to install only detectors that have been previously tested in the spectrometer system.

### Correction of the pixel offset, partial function $K_V$

As already stated, the spectrum can be offset by several pixels from instrument to instrument due to the inevitable imprecision in adjustments. Over the lifetime of the instrument, temperature changes and aging effects can lead to a slight shift of the spectrum, which are usually only in the range of fractions of pixels in pressure and temperature-stabilized optical systems.

This shift is not necessarily constant across the spectrum but can have different values on opposite ends of the detector and can vary from detector to detector. However, there are no abrupt changes in the pixel offsets for one sensor. Usually there are only continual changes. If the offset between pixel n on the detector S of an instrument produced in series has the value $\delta_{S,n}$, then the offsets $\delta_{S,n-1}$ and $\delta_{S,n+1}$ for the pixels neighboring pixel n are very similar.

A shift of the spectrum by d pixel to the right is simple. The measured values are just copied by d pixels: If the spectrum is located in an array Int with m pixels, the assignment $Int_{shifted}[i + d] := Int[i]$ is carried out for all pixel numbers i between 1 and m-d for a shift of d pixels to the right. The shifted spectrum is then in $Int_{shifted}$. A shift of the spectrum to the left is conducted analogously.

The physical pixels are, in reality, so wide that it is not sufficient to conduct the shift by an integer number. In most cases, a shift on the order of a hundredths of a pixel is desirable. However, the problem is easily reduced to a shift by whole pixels, as can easily be imagined. In Section 3.9.1, it was described how a continuous spectral profile can be replicated using the spline interpolation. Using the spline polynomial, it is possible to integrate over any fraction of a pixel and to also obtain the proportional intensity. If the exact determination of the pixel offset to the hundredth of a pixel is desired and if the spectrum has p physical pixels, then the spectrum can be distributed over an array of virtual pixels having $p \times 100$ components. The intensities of the virtual pixels are calculated using the spline polynomial. Of course, this operation must also be carried out for the spectrum from the master instrument's adjustment sample, in order to be able to conduct meaningful calculations.

Determination of the pixel offset between master and production instrument is carried out as follows:

First, a factor $F_{Norm}$ is determined for every detector S, with which the sum of the currently measured on S are matched to those of the master instrument. The intensities of both spectra are not simply summed up. Rather, the border areas are weighted less to minimize the influence of strong lines that appearing randomly on the borders due to shifts. For the first r pixels, the weight is linearly increased from 0 to 1; for the last r pixels, it drops from 1 to 0. The first and last r pixels are treated differently because instruments in a series may or may not see spectral lines due to minor deviations in the sensor positions. For inner pixels, the unchanged pixel intensities are added. The way in which the intensities are considered is shown in Figure 3.89. If the sum of the spectra determined in this way for the detectors S of the master instrument is $Sum_{first}$ and for the current instrument $Sum_{akt}$, the standard factor is calculated as follows

$$F_{Norm} := Sum_{first}/Sum_{akt} \tag{3.39}$$

Every pixel of the detector S is now multiplied with $F_{Norm}$, to adjust the intensities of the current spectrum to the level of that of the master instrument. Of course, a separate standard factor must be determined for every detector.

After that, a partial spectrum is defined in the center of the detector. The width is not critical, but it should be chosen so that it contains several spectral lines. These spectral lines have a characteristic pattern of distances and heights comparable with those from the master instrument's adjustment spectrum. Earlier, it was mentioned that shifts of ±20 physical pixels can be expected. In the simplest case, the spectral section is slid pixel for pixel over the area in which the pattern is expected to be. This sliding is conducted over the virtual pixels, i.e., fractions of physical pixels. The correct offset is found at the point where the sum of the deviation amounts between the spectrum of the master instrument and the spectral section

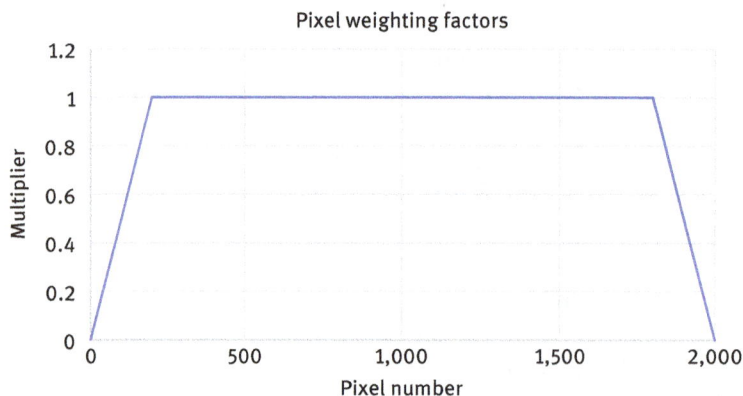

Figure 3.89: Multipliers for pixel intensities for the determination of the standard factor.

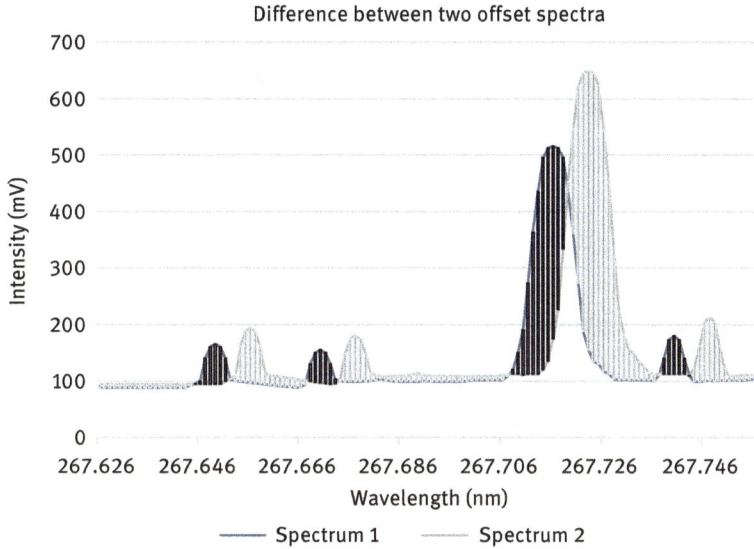

**Figure 3.90:** Difference spectrum as a criterion for finding the pixel offset.

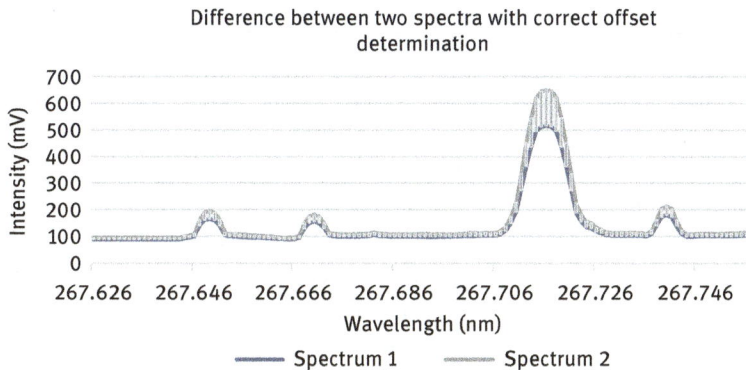

**Figure 3.91:** Difference spectrum with correctly determined pixel offset.

that is being slid around is at a minimum. Figure 3.90 shows the differences in two spectra as hatched lines for an incorrectly determined pixel offset; Figure 3.91 shows the differences for a correctly determined offset.

After these steps, the pixel offset in the center of the detector is known. It can now be tested as to whether a stretching or compression of the spectrum is present. Compression (to compensate for stretching) is realized by removing virtual pixels from the array at regular intervals by copying the following virtual pixel up one location. Conversely, broadening can be achieved by doubling a virtual pixel at fixed intervals. Compression or broadening is successful when the sum of the deviations is reduced by such a procedure.

Bit by bit, the offsets can be calculated for smaller and larger pixel numbers in the same way that the variations for the center of the spectra were defined. It is only necessary to determine stretching and compression as every area has to connect directly to the previous one.

### Correction of the resolution, partial function $K_R$

Differences in the resolution, just like the pixel offset, are caused by differences in the adjustment of detector arrays and the entrance slit. Again, like the pixel offset, resolution changes are not abrupt; the spectral resolution for neighboring wavelengths is similar.

Assuming a radially curved focal curve and a straight detector, the position of the detector intersects the focal curve at two points (see Figure 3.56). The resolution deteriorates steadily with increasing distance from the intersection points; it is best at the intersection points.

Production tolerances for the entrance slit (widening, slanting) can also worsen the resolution. These influences are constant across all the pixels. The pixel offset must already have been determined before the resolution of a production instrument can be compared with those of the master instrument.

When developing an algorithm for the correction of the resolution, it helps that the larger the variance (the term "variance" as is known from statistics) in the individual measured values, the better is the resolution in a pixel area (for the same sum of pixel intensities). A simple example: For the worst possible resolution, all pixels deliver the same signal and the variance is zero. The algorithm works according to the principle that to improve the resolution, intensities are proportionally subtracted from the flanks of a line and added to the intensities at the peak. To reduce the resolution, the reverse is done. The process can be done iteratively; in the end, the variance of the spectrum corrected using $K_R$ must correspond to the variance of the spectrum of the adjustment sample on the master instrument.

### Calculation of the wavelength specific sensitivity function, partial function $K_E$

After conduction of the previously described calculation steps, this correction is the last to be carried out, as the calculation can first be done after the pixel-specific influences, profile shifts and resolution changes have been taken into account.

It has proven to be advantageous to identify the areas of the line peaks in the spectrum of the adjustment sample for correction and to compare the peak areas $F_E$ from the adjustment sample spectrum measured on the master instrument with those measured on the current instrument $F_{act}$. To be able to carry out a correction for a pixel number n on a detector S, the quotients $F_E/F_{act}$ for the line peaks in a range around n are formed and averaged. It makes sense to consider peaks that are further away less strongly. If the adjustment sample is suitably selected, areas with a spectral width from 1 to 5 nm are sufficient to have a large enough number of line peaks available for correction purposes. In principle, it is possible to calculate a polynomial instead of a factor for wavelength-specific sensitivity correction. However, this is not necessary for a flawlessly constructed optic.

To carry out the correction, every pixel must be multiplied with the associated factor $F_E/F_{act}$.

The method described for recalculating the current spectra to those that would have been obtained with the master instrument for the instrument series assumes that the excitation always takes place in the same manner. Changing the excitation parameters can result in atomic lines being strengthened but ion lines being weakened, or vice versa. Lines with different excitation energies can react in completely different ways to such changes. Since atomic and ion lines as well as lines with different excitation energies can be present in every combination in the spectrum, the assumption that neighboring lines in the spectrum behave similarly no longer applies when the excitation conditions change. Therefore, the complete spectrum recalibration can no longer compensate for such effects. However, this fact also applies to calibration and conventional recalibration. They only deliver accurate values as long as the excitation conditions are the same. It is very easy to judge if the error described here occurs. If the function $K_G$ is applied to the spectrum of the adjustment sample, in the ideal case, the adjustment sample spectrum reproduces perfectly the spectrum from the master instrument. The differential area obtained when the spectra are subtracted from one another is practically zero. If there were changes in the excitation, there are errors and the differential area is larger. If the differential area is observed in relation to the total area of the spectrum, a measure of how well the adjustment has succeeded is obtained. An error message can be displayed when a preset limit is exceeded.

At the outset, it was mentioned that the complete spectrum recalibration offers further advantages in addition to the automatic transferability of calibrations.

These are clearly manifested after construction of $K_G$:

1. Recalibration samples (with the exception of a single adjustment sample) become obsolete. For this reason, extension of the set of calibrations to include additional metal bases does not lead to higher costs.
2. Operation is simplified. Instead of many recalibration samples, there is only a single sample to measure. This means a time savings for the user.
3. The source of error for a spoiling of the recalibration by "memory effects" in arc instruments is excluded, as mainly lines from the base metal and the main alloying element deliver the signal for the recalibration. Recalibration samples with high contents of trace elements are not required.
4. There are no recalibration samples available for many exotic elements. As already described, in such cases a one-point recalibration with the background signal

from the pure sample is the remedy, which often leads to considerable errors. This source of error is omitted.

5. Gradually progressing hardware changes are detected and can be compensated for within limits. It is possible to recognize beforehand when these limits are approached and service measures can be initiated. Troubleshooting is simplified. The parameters of the adjustment function $K_G$ combined with a reproducibility test make it possible to get a fairly accurate picture of the spectrometer system. Diagrams of the changes in state can be created if the data are stored regularly.

6. Under 5, data are determined, which helps customers to comply with documentation requirements as part of the ISO 9001.

7. Spectrometer systems become more flexible: Retrofitting of elements and postcalibration no longer mean transport of the instrument back to the factory; calibrations can be downloaded or received via email; waiting times are eliminated.

8. Frequently, the customer has a temporary testing application. A temporary transfer of calibrations becomes possible. Update services can also be offered on a subscription basis.

The procedure is described in detail in [76] and in the German patent DE10152679B4 [96]. In principle, it is also possible to use a similar algorithm for instruments equipped with exit slits and photomultiplier tubes. The information concerning the spectral environment is then obtained by moving a section of the spectrum sequentially over the exit slits by shifting the entrance slit. However, in order to be fully utilized, a motor-driven profiling must be possible for every single exit slit, which entails a considerable outlay for hardware. For this reason, only parts of this process are used in serial production at the moment. This procedure is disclosed in the European patent EP1825234B8 [97].

### 3.9.6.3 Type recalibration

As has already been said at the beginning of Section 3.9.6, the type recalibration serves to improve the accuracy. The instrument was previously recalibrated with one of the two other methods described in the beginning of Section 3.9.6. If a sample must be analyzed with high accuracy, after first calling up the respective software function, a so-called type standard is measured. It has a very similar composition to the unknown sample. After this measurement, every calibration curve is shifted along the concentration ratio axis so that the point for the type standard lies exactly on the curve (see Figure 3.92). Then the unknown samples and their contents are determined on the shifted curve. With this approach, the calibration functions are only used to interpolate the concentrations of the unknown samples based on the type standard, whereby the deviations of the type standard from the calibration functions are not random but are due to the composition of the standard. Similar deviations are to be expected by the

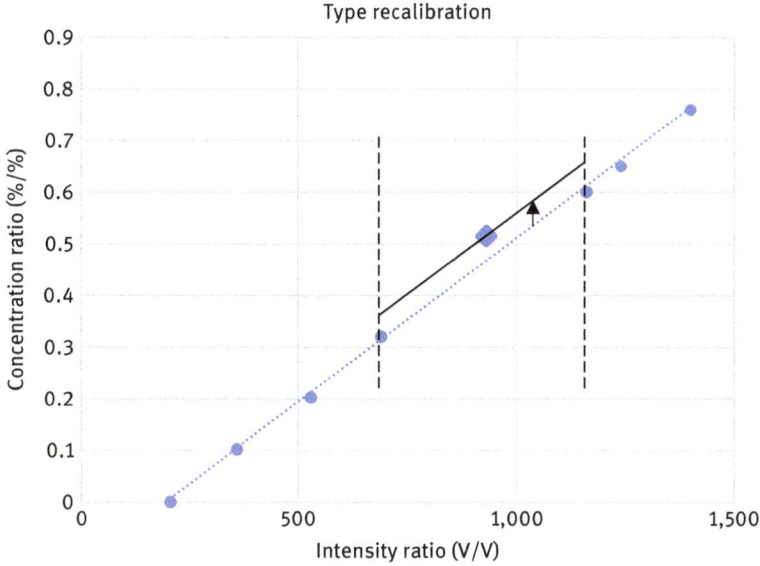

**Figure 3.92:** Displacement of the calibration curves when using the type recalibration.

measurement of unknown samples with similar compositions. In addition, the type recalibration compensates for small instrument drifts. The shifted calibration curves are only used in a narrow range around the concentrations of the type standard. This area is marked with upright dotted lines in Figure 3.92. Outside of this area, the calibration function that has not been shifted is used. The type recalibration requires only the measurement of a single standard, and it is, thus, quicker to carry out than a complete recalibration. This procedure is depicted in Lührs and Kudermann [95].

### 3.9.7 Line switches

The self-absorption effect was described in Section 3.2.1.3. It results in a calibration function where increases in concentration only lead to low increases in intensity and then, as the concentrations continue to increase, the intensities may even decrease (see Fig. 3.15). This effect is typical for particularly sensitive lines.

For this reason, it is customary to use more than one spectral line (or, more precisely, more than one channel, consisting of analyte line and internal standard) for elements that are analyzed in both trace ranges and at high concentrations. For each channel, a range of recalibrated intensity ratios is defined. A line may only be used when the recalibrated intensity ratio lies within this range. However, this is a necessary but insufficient criterion, because with a line in self-reversal, two concentrations can be assigned to one recalibrated intensity ratio.

This problem can be avoided by first testing the lines for the high concentrations for usability. These lines are usually less sensitive and show no self-absorption. The switch to the low line only occurs if the concentrations are so low that self-reversal cannot occur. It is advisable to set the line switch point at concentrations that are not present in common material grades. In this way, it is possible to avoid the problem that the concentrations displayed are taken from different analytical channels during multiple measurements of a sample. The transition from one spectral line to the other is frequently associated with deviations larger than the reproducibility expected from a multiple measurement. This is because the systematic deviation of a sample can vary for different calibration curves. For many elements, the repeatability in spark mode is well below 0.5% relative. If a larger deviation than expected is detected during a double measurement, which originates from the use of an analyte line switch, a lack off precision is suspected, although this is not the case.

### 3.9.8 100% calculation

Finally, the concentration ratios must be converted to concentrations. This is done by adding all the concentration ratios plus 100 for the reference element and then dividing the result by 100 [24]. The concentration ratios of all elements (except the base element) must be divided by this term to obtain the associated concentration. Finally, the concentration of the base element is obtained by subtracting the sum of all elements from 100.

### 3.9.9 Calculating the concentrations from PMT raw intensities or from complete spectra

Calculating the concentration for an unknown sample is carried out according to the following scheme for instruments equipped with exit slits and PMTs:
1. Measurement of the raw intensities
2. Performing background correction if necessary (see Section 3.9.3)
3. Calculating the intensity ratios as described in Section 3.9.4
4. Application of the recalibration factors and offsets (see 3.9.6.1)
5. Application of the calibration function (see eq. (3.29) in Section 3.9.5.3)
6. Performing a type recalibration if necessary (see Section 3.9.6.3)
7. Selection of the lines in the working range (see Section 3.9.7)
8. Performing the 100% calculation (see Section 3.9.8)
9. Output of the results

The process is slightly different for instruments with complete spectrum calibration:
1. Measurement of the raw spectra
2. Complete spectrum recalibration (see Section 3.9.6.2)

3. Background correction (see Section 3.9.3)
4. Interpolation of the spectrum (see Section 3.9.1)
5. Generation of the line intensities using integration of the spline function over the width of the virtual exit slit
6. Calculating the intensity ratios as described in Section 3.9.4
7. Application of the calibration function (see eq. (3.29) in Section 3.9.5.3)
8. If a type recalibration is turned on, it follows the description in Section 3.9.6.3
9. Selection of the lines in the working range (see Section 3.9.7)
10. Performing the 100% calculation (see Section 3.9.8)
11. Output of the results

After processing, further calculations can follow, such as determination of the material grade matching an analysis or testing for compliance with target concentrations. This type of post processing is important for mobile spectrometers. The most common algorithms are described in Section 7.7.

# Bibliography

[1]  Seidel T. Gleitfunkenspektrometrie – Eine neue atomspektrometrische Methode zur Untersuchung von Kunststoffen und anderen nichtleitenden Materialien. Duisburg, Dissertation, 1993.
[2]  Golloch A, Siegmund D. Sliding spark spectroscopy – rapid survey analysis of flame retardants and other additives in polymers. Fresenius J Anal Chem 358: 804–811, 1997.
[3]  DIN EN 61010-1/A1:2015-04;VDE 0411-1/A1: 2015-04– Entwurf Sicherheitsbestimmungen für elektrische Mess-, Steuer-, Regel- und Laborgeräte- Teil 1: Allgemeine Anforderungen (IEC 66/540/CD:2014). Berlin, Beuth Verlag, 2015.
[4]  Wikipedia article "Kohlebogenlampe". https://de.wikipedia.org/wiki/Kohlebogenlampe, downloaded from the internet August 24th, 2017.
[5]  Ohls K. Analytische Chemie – Entwicklung und Zukunft. Weinheim, Wiley-VCH Verlag GmbH & Co. KGaA, 2010.
[6]  Kirchhoff GR, Bunsen RW. Poggendorf Annalen der Physik, 1860, 110.
[7]  Zech P. Das Spektrum und die Spektralanalyse. München, Verlag Rudolf Oldenbourg, 1875.
[8]  Landauer J. Die Spectralanalyse. Braunschweig, Friedrich Vieweg und Sohn, 1896.
[9]  Liveing GD, Dewar J. Proceedings of the Royal Society, 1879.
[10] Liveing GD, Dewar J. Proceedings of the Cambridge Philosophical Society, 4, 882.
[11] Kayser H, Runge C. Über die Spektren der Elemente. Abhandlungen der Berliner Akademie, 1890.
[12] Demtröder W. Experimentalphysik 2, Elektrizität und Optik. Berlin, Heidelberg, New York, Springer Verlag, 2006, 64.
[13] Küpfmüller K, Kohn G. Theoretische Elektrotechnik und Elektronik. 15th edition. Berlin Heidelberg New York, Springer Verlag, 2000, p. 233.
[14] Holm R. Electric Contacts – Theory and Applications. Berlin Heidelberg New York, Springer Verlag, Reprint of the fourth edition of 1967, 2000.
[15] Tietze U, Schenk C. Halbleiter-Schaltungstechnik, 12th edition. Berlin Heidelberg New York, Springer-Verlag, 2002.

[16]  Görlich P. Einhundert Jahre Wissenschaftliche Spektralanalyse. Berlin, Akademie Verlag, 1960.
[17]  Junkes SJ. Hundert Jahre chemische Emissions-Spektralanalyse. Laboratorio Astrofisico Della Specola Vaticana, Ricerche Spettroscopiche, Vol. 3, No. 1, Vatican state, 1962.
[18]  Gerlach W, Schweitzer E. Die chemische Emissions-Spektralanalyse. Leipzig, Verlag Leopold Voss, 1930.
[19]  Feussner O. Zur Durchführung der technischen Spektralanalyse. Archiv für das Eisenhüttenwesen, 1932/33, 6, p. 551.
[20]  Feussner O. Zeiß Nachrichten 1933, No. 4, p. 6 et seq.
[21]  Kaiser H, Walraff A. Gesteuerte Funkenentladungen als Lichtquelle für die Spektralanalyse. Zeitschrift für technische Physik 1938, No. 11, p. 399 et seq.
[22]  Kipsch D. Lichtemissions-Spektralanalyse. Leipzig, VEB Deutscher Verlag für Grundstoffindustrie, 1974.
[23]  De Galan L. Analytische Spectrometrie. Amsterdam, Agon Elsevier, 1972.
[24]  Slickers KA. Die Automatische Atom-Emissions-Spektralanalyse. Gießen, Brühlsche Universitätsdruckerei, 1992.
[25]  Laqua K, Hagenah W-D. Spektrochemische Analyse mit zeitaufgelösten Spektren von Funkenentladungen. Spectrochim. Acta 1962, 18, p. 183 et seq.
[26]  Skoog DA, Leary JJ. Instrumentelle Analytik. Berlin, Springer Verlag, 1996.
[27]  Kneubühl F K, Sigrist M W. Laser. Stuttgart, B. G. Teubner, 1989.
[28]  Kaiser H, Walraff A. Über elektrische Funken und ihre Anwendung zur Anregung von Spektren. Phys. 1939, 34, p. 297 et seq.
[29]  Weizel W, Rompe R. Theorie elektrischer Lichtbögen und Funken. Leipzig, Verlag Johann Ambrosius Barth,1949.
[30]  Moenke H, Moenke-Blankenburg L. Einführung in die Laser-Mikro-Emissionsspektralanalyse. Leipzig, Geest & Portig, 1966.
[31]  Mika J, Török T. Analytical Emission Spectroscopy. Budapest, Akademiai Kiado, 1973.
[32]  Cremers DA, Radziemski LJ. Handbook of Laser-Induced Breakdown Spectroscopy. Chichester, John Wiley & Sons, Ltd., 2013.
[33]  Wikipedia article fiber laser. https://de.wikipedia.org/wiki/Faserlaser, retrieved March 20th, 2019.
[34]  Dong L, Samson B. Fiber Lasers: Basics, Technology, and Applications. Boca Raton, London, New York, CRC Press, Taylor & Francis Group, 2017.
[35]  Nufern information video on Youtube: How a Fiber Laser Works. https://www.youtube.com/watch?v=ofEqFlqkiS0, retrieved 21.03. March 21st, 2019.
[36]  Meschede D. Optik, Licht und Laser, 3. reviewed edition. Wiesbaden, Vieweg+Teubner, GWV Fachverlage GmbH, 2008.
[37]  Homepage Femto Fiber Tec GmbH. https://www.femtofibertec.de/en/home#home, retrieved March 21st, 2019.
[38]  Eglseder G, Mertens R, Guerrero M, Dantonel D, Malengreau N, Ryman C, Solhed H, Beskow K, Fischer H, Falcone R, Flock J, Thurmann U, Laserna J, Garcia C. Development of techniques for the production of glassy standard material for the analysis of slags by spectroscopic methods. RFCS publication, European Commission, Brussels, 2006.
[39]  Wong DM, Bolshakov A, Russo RE. Encyclopedia of Spectroscopy and Spectrometry, 2nd Edition. Academic Press, 2010, p. 1281–1287.
[40]  Noll R. Laser Induced Breakdown Spectroscopy, Fundamentals and Applications. Heidelberg Dordrecht London New York, Springer Verlag, 2012.
[41]  Hemmerlin M, Paulard L, Hellberg P, Thurmann U, Reinhard R, Sturm V, Vrenegor J, Guio MJ, Baena JM, Rivas O, Quirós E, Laserna JJ, Palanco S, Vadillo JM, Romero D, García PL, & Conesa S. Fast analysis pf production control samples without preparation. RFCS publication, European Commission, Brussels, 2005.

[42] Werheit P, Fricke-Begemann C, Gesing M, Noll R. Fast single piece identification with a 3D scanning LIBS for aluminium cast and wrought alloys recycling. Journal of Atomic Emission Spectrometry 2011, 11.

[43] Bohling C, John A, Feierabend A. Inline Multielement-Analysesysteme für Prozessoptimierung, Qualitätssicherung und Präzisionsrecycling – LIBS im industriellen Einsatz. Fachausschussbericht (report of expert committee) number 3.025, 141. Vollsitzung des Chemikerausschusses (plenary meeting of the Chemists' committee), Stahlinstitut VDEH, Düsseldorf, 2016.

[44] Method and Device for Spectrometric Elemental Analysis. German patent application, DE102009018253 (A1), OBLF, Gesellschaft für Elektronik und Feinwerktechnik, 2010.

[45] Spark chamber for optical emission analysis. European patent, EP 2612133 (B1), Thermo Fisher Scientific Ecublens Sarl, 2010 (priority year).

[46] Nadkarni RA. Modern Instrumental Methods of Elemental Analysis of Petroleum Products and Lubricants. American Society for Testing & Materials, 1991.

[47] Niederstraßer J. Funkenspektrometrische Stickstoffbestimmung in niedriglegierten Stählen unter Berücksichtigung der Einzelfunkenspektrometrie. Duisburg, Dissertation, 2002.

[48] Schriever U. Untersuchungen zur Wirkungsweise der Elemente Bor, Titan, Zirkon, Aluminium und Stickstoff in wasservergüteten, schweißbaren Baustählen. Luxemburg, Research Report EUR 13503 DE, Commission of the European Communities, 1991.

[49] Kupfer und Kupferlegierungen – Nahtlose Rundrohre aus Kupfer für medizinische Gase oder Vakuum; German version EN 13348:2008. Berlin, Beuth-Verlag, 2008.

[50] Directive 2011/65/EU of the European Parliament and of the Council of 8 June 2011.

[51] Sircal Instruments. Company publication MP-2000 Rare Gas Purifier – The cost effective and reliable solution for rare gas purification. Downloaded from the internet May 20th 2017, www.sircal.co.uk.

[52] American Holographic. Company publication Concave Diffraction Gratings. Littleton, MA, USA Company publication American Holographic, 1986.

[53] Clark, GL. The Encyclopedia of Spectroscopy. New York, Reinhold Publishing Corporation, 1961.

[54] Spektrometeroptik mit nicht-sphärischen Spiegeln. German patent application DE102007027010, Spectro, 2008.

[55] Carl Zeiss GmbH. Company publication Rowlandkreis-Gitter. Oberkochen, 1992.

[56] Carl Zeiss GmbH. Company publication Prüfung von Beugungsgittern. Oberkochen, 1993.

[57] Carl Zeiss GmbH. Company publication Präzisions-Beugungsgitter. Oberkochen, 1993.

[58] Carl Zeiss GmbH. Company publication Abbildende Gitter. Oberkochen, 1993.

[59] Agilent Technologies. Company publication Concave Diffraction Grating Design Guide. Agilent Technologies, 2002 (Downloaded from the internet, April 25, 2002 www.agilent.com).

[60] Dobschal HJ, Kröplin P, Reichel W, Rudolph, K, Steiner M. Beugungsgitter und Hohlspiegel zugleich. München, F&M 100, Carl Hanser Verlag, 1992.

[61] Grobenski Z, Radziuk B, Schlemmer G. Perfect Marriage: Echelle Optics and Solid State Detectors. Modern Aspects of Analytical Chemistry, Proceedings 5th Argus, 1997.

[62] Okruss M, Becker-Ross H, Florek S, Franzke J, Koch J. NIRES – Ein NIR-Echelle-Spektrometer mit Flächen-CCD-Sensor für die simultane, hochauflösende Spektrenregistrierung zwischen 640 und 990 nm. Berlin, Institut für Spektrochemie und Angewandte Spektroskopie, 2001.

[63] Paul H. Lexikon der Optik. Heidelberg, Berlin, Spektrum Akademischer Verlag GmbH, 2003.

[64] Schröder G. Technische Optik. Würzburg, Vogel Verlag, 1974.

[65] Niedrig H (editor). Bergmann – Schäfer Lehrbuch der Experimentalphysik. Berlin, New York, Verlag W. de Gruyter, 2004.

[66]  Demets R, Bertrand M, Bolkhovitinov A, Bryson K, Colas C, Cottin H, Dettmann J, Ehrenfreund P, Elsaesser A, Jaramillo E, Lebert M, van Papendrecht G, Pereira C, Rohr T, Saiagh K, Schuster M. Window contamination on Expose-R. International Journal of Astrobiology 2015, 14 (1), 33–45.

[67]  Zu Stolberg-Wernigerode O. Neue deutsche Biographie, vol. 6. Berlin, Duncker & Humblot, 1964.

[68]  Elster J, & Geitel H. Über die Verwendung des Natriumamalgams zu lichtelectrischen Versuchen. Annalen der Physik und Chemie, 1890, NF 41, p. 161–165.

[69]  Elster J, Geitel H. Notiz über eine neue Form der Apparate zur Demonstration der lichtelectrischen Entladung durch Tageslicht. Annalen der Physik und Chemie 1891, NF 42, p. 564–567.

[70]  Elster, J., & Geitel, H. Lichtelectrische Versuche. Annalen der Physik und Chemie 1892, NF 46, p. 281–291.

[71]  Elster J, Geitel H. Lichtelectrische Versuche. Annalen der Physik und Chemie 1894, NF 52, p. 433–454.

[72]  Koller LR. Photoelectric Emission from Thin Films of Caesium. Phys. Rev 36, 1639, American Physical Society, December 1930.

[73]  Photomultiplier Tubes. Hamamatsu Photonics K. K., Company publication TPMZ0002E01. Iwata City, Japan, February 2016.

[74]  Hamamatsu Photonics, K.K. Company publication Photomultiplier Tubes. Shimokanzo, Japan, 1994.

[75]  Jennewein T. Charakterisierung von Photomultiplier Tubes hinsichtlich deren Verwendung in Flüssig-Xenon Zeitprojektionskammern. Bachelor thesis, Mainz, 2012.

[76]  Joosten HG. Verfahren zur automatisierten Übertragung von Emissionsspektrometer-Kalibrationen. Duisburg, Dissertation, 2003.

[77]  Cox WG. Use of the Diode Array Detector with the DC Argon Plasma – Echelle Spectrometer. Applications of Plasma Emission Spectrochemistry, Editor Barnes R M, Philadelphia, Heyden & Sons, 1979.

[78]  Krüger H. Entwicklung eines Detektorsystems zum schnellen ortsaufgelösten Nachweis von Einzelmolekülen. Dissertation, Universität Bonn, 1999.

[79]  Köstner R, Möschwitzer A. Elektronische Schaltungen. München, Carl Hanser Verlag, 1993.

[80]  Sweedler JV, Ratzlaff KL, Denton MB. Charge Transfer Devices in Spectroscopy. New York, VCH Publishers Inc., 1994.

[81]  Perkampus HH. Encyclopedia of Spectroscopy. Weinheim, VCH Verlagsgesellschaft GmbH, 1995.

[82]  Hynecek J. Virtual Phase Technology: A New Approach to Fabrication of Large-Area CCD's. IEEE Transactions on Electron Devices May 1981, Vol ED-28, No. 5.

[83]  Ninkov Z. Advanced Image Devices. New York, Center for Electronic Imaging, 2002, (www.cis.rit. edu).

[84]  Sony. Company publication CCD Camera Systems. 1999, (www.sony.net/products/SC-HP /Index.html).

[85]  Ninkov Z, Backer B, Corba M. Characterization of a CID-38 Charge Injection Device. New York, Center for Electronic Imaging, 1996, (www.cis.rit.edu/research/CID/SJ96/paper.html).

[86]  Göhring D. Digitalkameratechnologien, eine vergleichende Betrachtung CCD kontra CMOS. Humboldt Universität Berlin, 2002, (www.informatik.hu-berlin.de/~meffert/Seminararbeiten/ Weitere/cmos/ccd-cmos.pdf).

[87]  McCormick DT. Line Array Sensor Comparison Version 1.0. Publikation der Firma Advanced-MEMS, San Francisco, 2016, (www.advancedmems.com/pdf/AMEMS_LineSensorArraySum-mary_v1.pdf).

[88]  CMOS linear image sensor S11639-01. Hamamatsu Photonics K. K., Solid State Division, data sheet, Hamamatsu City, Japan, December 2016.

[89]  Pfundt HU. Beiträge zur Spektralanalyse der Leicht- und Schwermetalle mit einer direkten lichtelektrischen Anzeige. Dissertation, München, 1955.

[90]  Simultanes Doppelgitter-Spektrometer mit Halbleiterzeilensensoren oder Photoelektronenvervielfachern. German patent application DE19853754B4, Spectro, 1998 (priority year).

[91]  Engeln-Müllges G, Reutter F. Formelsammlung zur numerischen Mathematik mit C-Programmen. Mannheim/Wien/Zürich, B.I. Wissenschaftsverlag, 1987.

[92]  Saidel AN, Prokofjew, WK, & Raiski SM. Spektraltabellen. Berlin, VEB Verlag Technik, 1955.

[93]  Thomsen V. Modern Spectrochemical Analysis of Metals. Materials Park, OH, ASM International, 1996.

[94]  Schweitzer E. Eine absolute Methode zur Ausführung der quantitativen Emissionsspektralanalyse. Zweite Mitteilung. Z. Anorg. Allg. Chem. 1927, 164, 127–144.

[95]  Lührs C, Kudermann G. Funkenspektrometrie. Clausthal-Zellerfeld, Chemists Committee of the GDMB Gesellschaft für Bergbau, Metallurgie, Rohstoff- und Umwelttechnik, 1996.

[96]  Verfahren zur vollautomatischen Übertragung von Kalibrationen optischer Emissionsspektrometer. German patent DE10152679B4, Spectro, 2001 (priority year).

[97]  UV-Spektrometer mit positionierbaren Spalten und Verfahren zur vollautomatischen Übertragung von Kalibrationen zwischen mit solchen Optiken bestückten Spektrometern. European patent EP1825234B8, Spectro, 2004 (priority year).

# 4 Sampling and sample preparation

This chapter is intended to provide the reader with information about how to take and prepare samples to be analyzed. Sample taking, sample preparation and sample analysis are separate work steps. The operator responsible for the analysis is usually not directly engaged in sampling and frequently not involved with sample preparation, but he should know them well. Only with this knowledge is it possible to identify measurement uncertainties that are caused by faulty sampling and to arrange for the delivery of flawless samples.

The focus of this chapter is on iron and steel samples. Here sampling and sample preparation are complicated and there are, at the same time, high demands on sample quality. The principle procedure can often be transferred to other metal bases.

## 4.1 Basic requirements for spectrometer samples

Several aspects must be considered when taking samples for spectrometric analysis. Usable samples from molten metal can only be obtained when the following conditions are met:
- Attempts must be made to ensure the *homogeneity* of the chemical composition. It must be ensured, for example, for steel samples, that the sample originates from a defined zone of the melt below the slag.
- When sampling from semi-finished products, it should be noted that the surfaces often have a different chemical composition than the core material. Causes for this can be coatings, such as galvanizing, but other effects such as decarburization, carburization, or nitriding can also influence the composition on the surface.
- If the sample is taken from a melt, the *cooling process* should be carefully controlled to be able to obtain samples with identical structures. Rapid cooling is usually advantageous.
- The sample should be *free from inclusions, cavities, fissures, segregations and burrs or ridges*.
- The *size of the sample* must be sufficient so that multiple spectroscopic measurements can be conducted. For rotationally symmetrical samples, it has become customary to use only a ring parallel to the outer circumference for the analysis. Segregation in the center can often be tolerated here.
- The sample must be prepared with a suitable method. *Grinding, milling and turning* are possible. After preparation, the sample should be flat so that it tightly closes the spectrometer's spark stand opening.

   Particularly manual sample preparation with disc grinding machines requires practice. There is a tendency to grind samples "with crowning." More

https://doi.org/10.1515/9783110529692-004

sample material is then removed on the edges resulting in a slight pillow shape. The risk of this occurring is especially large for softer materials such as pure iron. Only in rare cases, for example, for quick tests on the surfaces of Cr/Ni steels, can sample preparation via machining be dispensed with.

- The sample surface must be *free of coatings, moisture, dirt or lubricants* prior to analysis. Materials that are susceptible to corrosion, for example, magnesium-based materials, tend to oxide formation on the surfaces. Here, the time period between sample preparation and analysis should be kept as short as possible. For the samples of some metal bases, such as those made of aluminum, it is advantageous to store standardization and control samples in a desiccator. This counteracts changes in the surface and prevents the buildup of moisture; thus, increasing the sample preparation intervals.
- Careful *labeling* prevents mixing up of the samples.
- Frequently, it is necessary to store the sample after analysis in order to meet *record keeping and archiving obligations*. It must, then, be kept safe and protected from contamination. The archived sample itself should be traceable as should the analytical results for it.

It is highly recommended to document the procedure for sampling and sample preparation in work instructions in order to always have samples with the same properties, independent from the sample taker.

Work safety must also be taken into account. Both sampling and sample preparation may be associated with safety risks. For this reason, all work steps should be reviewed by safety experts/officers and sample taker and sample preparer should be accordingly trained and provided with the appropriate personal protective equipment.

What needs to be considered for sample taking for semi-finished products and scrap is described in detail in Section 7.8. This subject is particularly important for incoming goods controls and outgoing inspections as well as in the secondary raw materials industry. Because these tasks are usually conducted with mobile spectrometers, the sampling of semi-finished products is included in Chapter 7 which deals with the design and operation of such spectrometer systems.

## 4.2 Sampling from liquid melts

Sampling and sample preparation of iron-based metals are described in detail in the *Handbuch für das Eisenhüttenlaboratorium* (Handbook for the Ironworks Laboratory) [1] and standardized in DIN EN ISO 14284 [2].

### 4.2.1 Sampling from pig iron

Samples of molten pig iron for steel making can be taken from various points of production:
– from the blast furnace during the tapping process
– directly from the pouring stream
– from transfer ladles

The following sampling procedures are used:
– sampling with a ladle
– sampling with the dipping mold
– sampling with the immersion probe
– sampling by suction

#### 4.2.1.1 Sampling with the ladle
In this type of sampling, the liquid pig iron is removed using a ladle and filled into a cold metal mold. A drawing of such a mold can be found in DIN EN ISO 14284 [2].

#### 4.2.1.2 Sampling with the dipping mold
Sampling with the dipping mold is done by dipping the mold into the melt and letting it fill. The influences of oxygen in the air and slag are prevented with this procedure. It must be ensured that the sample is quickly white solidified, that is, without the formation of graphite. Figure 4.1 shows a white solidified pig iron sample; Figure 4.2 one that is grey solidified and by which a portion of the carbon is present as graphite (see also Figures 5.1 and 5.2).

#### 4.2.1.3 Sampling with the immersion probe
Disposable dipping molds are used for this technique. The mold is placed in a cardboard tube and dipped into the molten metal with a lance.

A disk-shaped sample that is well suited to spectrometry is produced when using immersion probes. The prerequisite is a white structure in the sample. The process of sampling can vary: Immersion duration, angle and depth should, however, be kept the same once the optimal parameters have been found by experimentation.

The sample can be taken in the blast furnace runner or from the iron pouring stream. To do this, the probe is placed into the melt and the disposable mold fills after several seconds. Then the probe is pulled out of the melt and broken apart. The sample is removed and cooled. These immersion probes differ from the design of the probe for molten steel sampling by several design features, such as the inlet on the side, which enables taking the sample from the blast furnace runner.

Figure 4.1: White solidified pig iron sample.

Figure 4.2: Grey solidified pig iron sample.

## 4.2.2 Sampling from liquid steel

Sampling with the immersion probe has become the established method for sampling from liquid steel since the mid-1970s (see the *Handbuch für das Eisenhütten-Laboratorium* (Handbook for the Ironworks Laboratory) [1]). This development is explained by the advantages of the process:

- The process can be automated.
- The sample is easily taken with a robotic arm.
- Well-formed samples that do not require time-consuming sample preparation are generated.

Therefore, by using immersion samples, it is possible to shorten analytical run times, which is important for an acceleration of the production process.

Sampling with the immersion probe requires good knowledge of the procedure. Segregation in the sample must be avoided during the cooling process. Metal strips made of aluminum or zirconium are placed inside the probe to bind oxygen.

An important advantage to the immersion probe technique is that it makes it possible to directly (and thus quickly) determine the elements carbon and nitrogen in the sample with spark emission spectrometers.

The "purge and suction technology" is a special technique when working with the immersion probe:

- The immersion probe is flushed with argon during sampling until the slag zone has been broken through and the sample form is free from oxygen and nitrogen.
- Then suction is applied and the sample mold is filled.

The expenditure for equipment is high when using this technology.

Figure 4.3 shows the design of a suction probe. Figure 4.4 presents different sample shapes.

## 4.2.3 Sampling from molten cast iron

It is advantageous to to take two cast iron samples. This increases the probability of getting at least one homogeneous sample. Taking immersion samples is possible, but sampling with a ladle is more common. A graphite ladle or a ladle made of steel and coated with a refractory material is used.

The sampling process is as follows:

- The slag on the melt is stripped off, the preheated ladle is dipped into the melt and filled with iron. When taking the sample during the casting process, the ladle is held directly in the pouring stream.
- The contents of the ladle are poured into a mold that has good heat dissipation. The advantage to molds made completely of copper compared to steel molds is that they enable faster cooling. It must be ensured that the mold is not too hot before it is filled. The formation of a white solidified iron structure can be achieved with a cold mold made of a material with good heat conductivity.

MINKON
German Technology

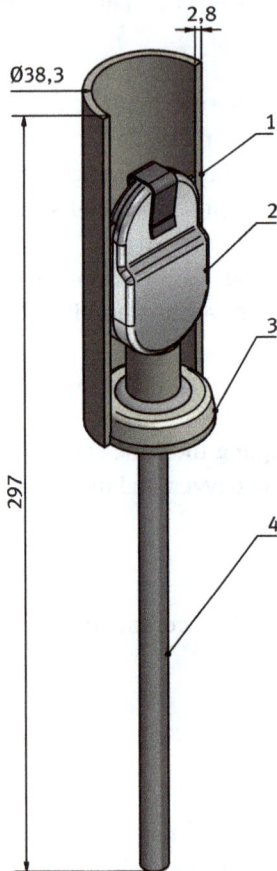

MINKON Saugsonde / Suction Sampler
Typ / Type:          SLC-79-200-NK-X
Art.Nr. / Partno:   111-6430

| 1 | Aufnahmehülse / Paper tube |
| 2 | Probenform Typ 79DM / Steel mould type 79DM |
| 3 | Keramikring / Ceramicring |
| 4 | Edelstahlrohr / Stainless steel pipe |

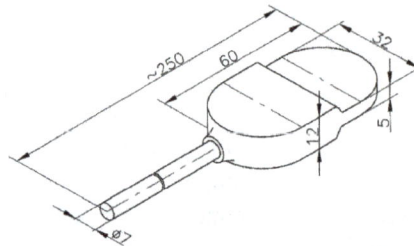

Minkon GmbH, Heinrich-Hertz-Straße 30-32, D-40699 Erkrath
Tel.: +49-211-2099080, Fax: +49-211-20990890

**Figure 4.3:** Suction sampler for liquid steel (printed with friendly permission of the company MINKON GmbH, Heinrich-Hertz-Str. 30–32, D-40699 Erkrath, Germany).

Kipsch [3] noted that white solidification on the one hand and the freedom from cavities, slags and striations on the other hand are contradictory demands. When in doubt, it is advantageous to generate samples that are sure to be white solidified. If the spark strikes, for example, a cavity, this can be seen immediately on the burn spot. Grey solidified samples can, in contrast, be accompanied by measurement errors for carbon that may go unnoticed.

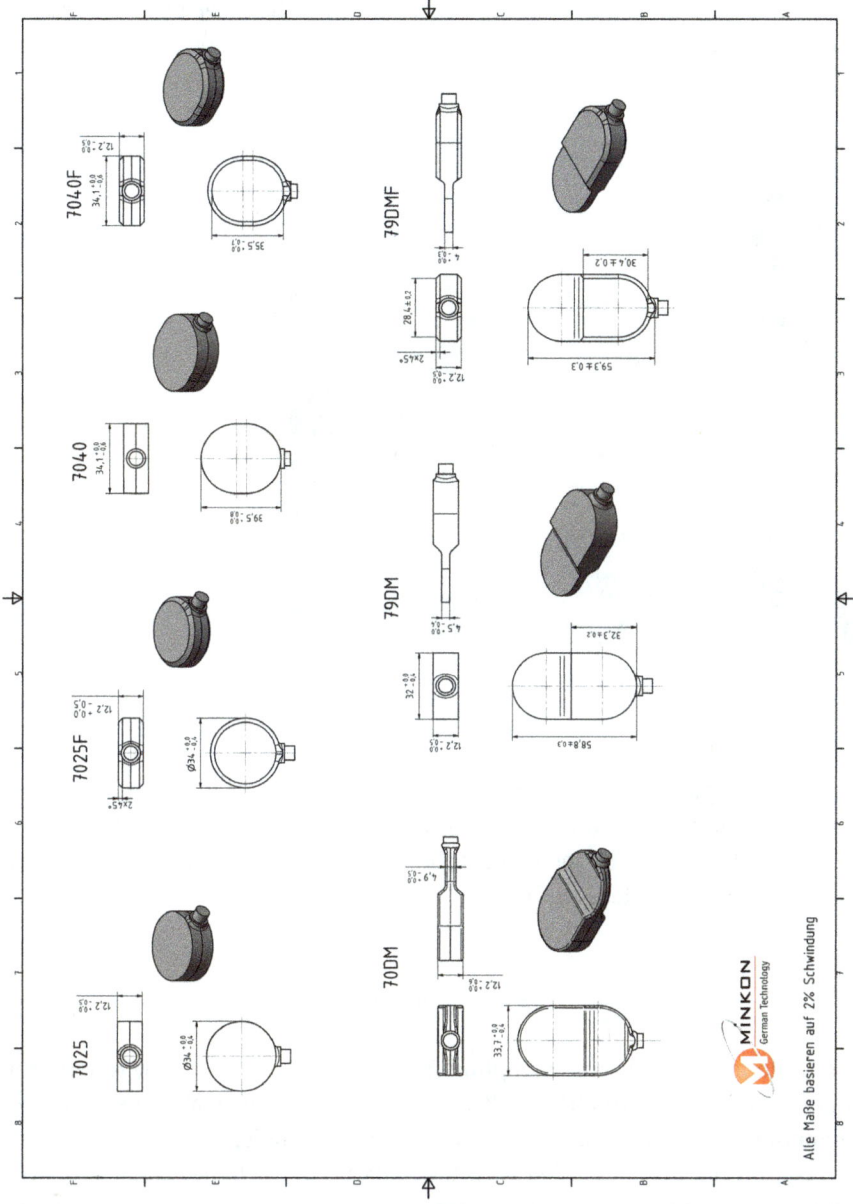

Figure 4.4: Different sample shapes (printed with friendly permission of the company MINKON GmbH, Heinrich-Hertz-Str. 30–32, D-40699 Erkrath, Germany).

A round sample with a thickness of 4–8 mm and a diameter of 35–40 mm is formed (see Figure 4.5). The sample is removed from the mold as soon as it has cooled.

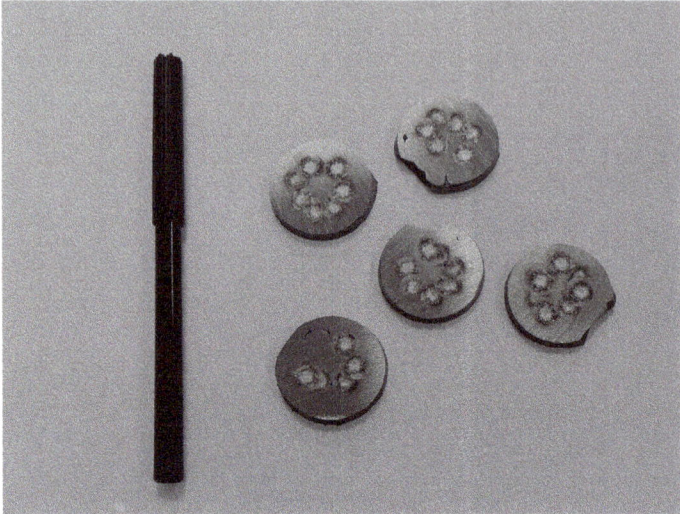

Figure 4.5: Typical cast iron samples.

Figure 4.5 shows typical cast iron samples after sample preparation and measurement. A drawing of a mold suited to the production of cast iron samples is presented in EN ISO 14284 [2]. The molds must be well cared for and prepared in order to obtain samples with flawless surfaces.

## 4.2.4 Sampling from aluminum melts and other metal bases

Lührs and Kudermann [4] discuss sampling from aluminum melts. They report that samples are usually taken from the melt with a ladle. The ladles must be prepared with a temperature-resistant coating, before being used. Molds, which are similar to those for cast and pig iron, are filled with the material in the ladle. Immersion and suction molds can also be used. They also mention that the optimal zone for sample taking should be empirically determined.

In copper-base, ladles as well as immersion probes are also used. For the low-melting metals, zinc, lead and tin-base, samples that are formed by filling molds with ladles (spoon samples) dominate. Figure 4.6 shows a zinc sample as an example.

Figure 4.6: Zinc sample.

## 4.3 Sample preparation

The following methods are commonly used for sample preparation:

– *Grinding with belt or disc grinding machines*

A wet pre-grinding can be used. At the end, however, it should be sanded dry. After grinding, the sample should no longer be cooled with water.

The abrasive material can lead to contamination. $Al_2O_3$, SiC and $ZrO_2$ are common. A negative influence on the measurement uncertainty must be expected in trace ranges for the elements Al, Si, C, Zr and O, if single spark evaluation is being carried out.

Manual grinding with disc grinding machines is widely used for steels, nickel and cobalt base alloys. In automated systems, sample preparation for such materials is performed with belt grinding machines, as longer service lives for the abrasive material can then be realized. However, milling machines are usually used in newer automatic systems.

The grain of the abrasive disc should not be too fine. Grinding discs for metallographic polishing should not be used. A surface that is too smooth is rather disadvantageous. Grain sizes of 60 or 80 have been proven to be suitable.

– *Grinding with swing grinding machines*

Cast iron samples are usually ground using a swing grinder. The sample is held in place with a magnetic clamping plate or a vise and a cup wheel with a diameter of about 150 mm is swung back and forth over the sample surface. The cup wheel is first lowered so that it just touches the sample surface. In the course of the grinding process, it can be lowered with a handwheel in increments of fractions of a millimeter. The grinding process is ended as soon as the entire surface is flat and has a bright metal finish.

Preparation of cast iron samples with disc grinding machines is not recommended. There is a risk of overheating. The sample then turns blue. This overheating of the sample can result in an increase in the measurement uncertainty. In addition, it is difficult to handle the thin "coin shaped" samples customary for cast iron on disc grinders. If a suitable tool is used that enables holding of the cast iron sample securely and flat on the grinding disc (e.g., a fitting magnetic sample holder) and if overheating can be avoided, it is, in principle, possible to use disc grinding machines.

– *Turning and milling*

For soft non-ferrous materials, turning or milling is superior to grinding, because the soft material "smears" the abrasive discs and can lead to the contamination of subsequent samples. Milling has the advantage compared to turning, that no burr remains in the center of the sample. Such a burr can prevent the sample from lying flat on the spark stand. This fact makes milling the preferred method for sample preparation. However, it cannot be used for very hard materials, such as titanium aluminides. In recent years, the milling technique has been constantly improved and an increasing number of alloys can be prepared using milling.

Lührs and Kudermann [4] point out that cooling with propanol can be advantageous when turning and milling. When using this coolant, it is necessary to ensure that there is a sufficiently sized exhaust system.

Table 4.1 is an overview of which sample preparation method is possible or recommended for which alloy group.

**Table 4.1:** Sample preparation methods possible for different alloy groups.

| Preparation method | Steel | Cast iron, Pig iron | Ni, Co, Ti-Base Alloys | Al, Mg, Zn, Pb, Sn | Cu-Base |
|---|---|---|---|---|---|
| Grinding (belt or disc grinder) | Suitable | Less suitable | Suitable | Not suitable | Conditionally suitable for some alloys |
| Grinding (swing grinder) | Unusual | Good | Unusual | Not suitable | Unusual |
| Milling and turning | Good for samples that are not too hard | Conditionally suitable | Good for samples that are not too hard | Good | Good |

# Bibliography

Standards that regulate sampling and sample preparation exist for some material grades. These standards are listed in the following bibliography [2, 5–8].

[1]     Handbuch für das Eisenhüttenlaboratorium, Band 5. Düsseldorf, Verlag Stahleisen, 2012.

[2]     DIN EN ISO 14284:2003-02: Stahl und Eisen – Entnahme und Vorbereitung von Proben für die Bestimmung der chemischen Zusammensetzung (ISO 14284:1996); German edition EN ISO 14284:2002. Berlin, Beuth Verlag, 2002.

[3]     Kipsch D. Lichtemissions-Spektralanalyse. Leipzig, VEB Deutscher Verlag für Grundstoffindustrie, 1974.

[4]     Lührs C, Kudermann G. Funkenspektrometrie. Clausthal-Zellerfeld, Chemikerausschuss des GDMB Gesellschaft für Bergbau, Metallurgie, Rohstoff- und Umwelttechnik, 1996.

[5]     ISO 1811-2:1988-10: Kupfer und Kupferlegierungen; Auswahl und Vorbereitung von Proben für die chemische Analyse; Teil 2: Probenahme von Kneterzeugnissen und Gussstücken. Berlin, Beuth Verlag, 1988.

[6]     DIN EN 14361:2005-02. Aluminium und Aluminiumlegierungen – Chemische Analyse – Probenahme von Metallschmelzen. Berlin, Beuth Verlag, 2005.

[7]     DIN EN 12060:1998-01: Zink und Zinklegierungen – Probenahme – Spezifikationen. Berlin, Beuth Verlag, 1997.

[8]     DIN EN 12402:1999-10: Blei und Bleilegierungen – Probenahme für die Analyse. Berlin, Beuth Verlag, 1999.

# 5 Analytical performance for the most important metals

The operation of modern emission spectrometers with spark excitation was described in Chapter 3. With these instruments, it is possible to determine the elemental contents of a large number of different metals. This chapter will give an overview of which elements can be meaningfully measured in which concentration ranges for the individual metal bases. The ranges indicate the span between a reasonable lower limit (the choice of this limit will be explained on the next page) and the highest possible concentration that can be determined with spark emission spectrometers according to the knowledge of the authors. The analytical performance with arc and laser excitation will not be covered in this chapter.

Only such elements as are commonly determined with arc/spark spectrometers will be discussed. However, with very few exceptions, this selection reflects the list of elements relevant to the metals concerned. Some important exceptions will be mentioned in the course of this chapter.

## Comments about the upper limits of the measuring range

The upper limits of the concentration range are based, on one hand, on the application requirements; on the other hand, the reference materials cannot always provide sufficient coverage. If the concentration range is not sufficient for practical purposes, then it may, in this case, be possible to employ secondary standards. Customer samples, for which the elemental content has been determined with other methods, are meant here as secondary standards. Such analysis methods include atomic absorption spectrometry, colormetry and optical emission spectrometry with inductively coupled plasma (ICP-OES). The homogeneity of secondary standards must be checked, whereby this control can be performed with a spark emission spectrometer using measurements of intensity ratios on different spots on the sample. The homogeneity of the sample is considered sufficient when no correlation between measurement position and measured value can be detected.

The upper concentration ranges listed in this chapter do not completely exclude secondary standards. However, they are only considered to a limited extent. The calibration range should only be extended beyond the range covered by reference materials if several secondary standards from different sources are available that can also provide a consistent calibration, that is, a calibration function with reasonable scattering.

If the value for the upper concentration limit is followed by a second value in parentheses, this means that calibration to the value given in parentheses is, in principle, possible, but it is difficult to cover the range between the upper limit and the value in parentheses with reference samples or a sufficient number of

https://doi.org/10.1515/9783110529692-005

secondary standards. The calibration range is completely enclosed in parentheses if the shortage of suitable reference materials extends over the entire range.

The upper limits for the alloying elements are limited to 50% in the tables in this chapter. In practice, however, higher upper limits are acceptable so that groups of metals with similar application areas do not have to be distributed over multiple methods. Then, strictly speaking, measurements are made in the metal base of the alloying element. For example: In the tin screening method, the calibration curve for lead is often extended beyond 60%. If an alloy with more than 60% lead is measured, it is not exactly a metal of the tin base that is being analyzed, but a lead alloy. But in this way, it is possible to measure solders with widely varying combinations of the elements lead and tin without having to change the method.

## Choice of the lower limits of the measuring range

It is difficult to set the lower concentration limits. It is sensible to couple these to the achievable detection sensitivity.

The detection limits, which can be achieved with larger spark laboratory instruments, and determined according to DIN 32645 [1] form an important criterion for setting the lower limits of the measuring range. A larger laboratory spectrometer is understood to be a system with a maximum reciprocal linear dispersion of 0.6 nm per mm. Many of the lower limits listed are only achievable when all technical possibilities have been fully utilized. Time resolved measurement, for example, is one such possibility for improvement of the detection limits. Detection limits can also be optimized by specially tailoring the measurement parameters to the analytical lines. Although such measures must be applied in order to reach the listed limits, the line-specific adjustments should be kept within limits. This means that each one of a larger number of measured elements (e.g., 40) is assigned to a much smaller number of excitation parameters (e.g., three). This procedure is useful. If every element was assigned its own measuring time with optimized excitation parameters, the total measuring time would be too long. The compromise solution described here is used in commercial spark emission spectrometers. By good analytical performance, measurement times below 30 s for a single measurement can be realized. Short measuring times are an important boundary condition in many cases, for example, melting bath control. However, sometimes it is still necessary to optimize individual elements because the application requires the lowest possible detection limits. Longer measuring times or other drawbacks must then be accepted. In such cases, the then achievable detection limit is set in parentheses in front of the measuring range.

It is not permissible to interpret a detection limit G determined according to DIN 32645 for an element E in such a way that when a sample P made of a random alloy is measured and a concentration G is calculated, that the element E is actually contained in the sample. This conclusion is only permissible if P is a pure sample of the base metal containing only traces of E. This assumption is also justified when the calibration functions exhibit no dispersion in the lower ranges. In practice, the achievable real detection limits for an entire group of metals depend on the dispersions of the calibration curves. However, it does not make sense to take this scattering as the basis for the listed lower limits, because it can be strongly reduced by narrowing the material group covered by a method. When the type recalibration described in Section 3.9.6.3 is used, the detection limits for the given alloy grade determined using the pure sample may also be achieved. However, this is only the case when the analytical line is not significantly overlapped by a interfering line from the alloying elements and the height of the spectral background is approximately the same as for the pure sample.

Several reasons for an increased dispersion cannot be compensated for with a more sophisticated calibration:
- Elements can be partially metallically dissolved and partially found in inclusions. This condition can vary among the standards used for the calibration.
- The measurement uncertainties for the standards can be high in the trace range. They are, particularly for older reference materials, not noted in the certificates. However, "even" values arouse suspicion. For example: If for the five standards in a set, the concentrations of a trace element are given as 1 ppm, 5 ppm, 10 ppm, 20 ppm and 50 ppm, it is not very probable that these are true values, as it is not possible to exactly achieve a target concentration. Most probably rounding was used.

The following should be noted about the lower limits for the measuring range specified in this chapter:
- The lower limits for the measuring ranges specified in the tables in this chapter can only be guiding values intended for rough orientation.
- They do not correspond to the real achievable detection limits for every alloy.
- Consultation with the instrument manufacturer is recommended for specific analytical applications. If the achievement of minimum detection levels is important for the application, they should be specified in the order and confirmed by the sales person. Compliance with the limits must be checked by the manufacturer before delivery to ensure that they are not exceeded despite the unavoidable series instrument variation. Control of the detection limit is not possible with just any samples. If a sample contains, for example, low levels of a trace element that is not homogeneously distributed, it is not suitable for verification purposes. The non-homogenous distribution leads to higher standard deviation, thus simulating higher detection limits.
- The detection limits, as well as other important analytical criteria, should be checked in regular intervals throughout the entire instrument lifetime.

Finally, it should be mentioned that more than one analytical line is usually required to cover the calibration ranges given in the subsections.

Sometimes, there are not enough reference materials for continuous coverage of the range between low and high concentrations. In these cases, two concentration ranges are listed for the elements concerned, between which there is a gap in which no reliable concentration determination is possible.

## 5.1 Analysis of iron-based materials

Table 5.1 shows the elements and their associated concentration ranges for which routine determination in steels, cast iron and other iron-based materials is possible.

Table 5.1: Concentration ranges for measurements in iron base.

| Element | Concentration range (%) | Element | Concentration range (%) | Element | Concentration range (%) |
|---------|-------------------------|---------|-------------------------|---------|-------------------------|
| Al | 0.0001–5.6 | La | 0.00005–0.02 (0.3) | Sb | 0.0004–0.3 |
| As | 0.0003–0.28 | Mg | 0.00005–0.4 | Se | 0.001–0.31 |
| B | 0.0001–1.3 | Mn | 0.0002–24 | Si | 0.0003–24 |
| Bi | 0.0002–0.2 | Mo | 0.0001–12 | Sn | 0.0001–0.4 |
| C | 0.0001–5.4 | N | 0.0006–1.1 | Ta | 0.001–0.8 |
| Ca | 0.0001–0.013 | Nb | 0.0003–4 | Te | 0.0008–0.06 (0.76) |
| Cd | 0.0002–0.015 | Ni | 0.0003–45 | Ti | 0.0002–3.6 |
| Ce | 0.0002–0.12 (0.66) | O | 0.002–0.06 | V | 0.0001–10 |
| Co | 0.0002–22 | P | 0.0002–2.9 | W | 0.002–25 |
| Cr | 0.0002–40 | Pb | 0.0002–0.4 | Zn | 0.0001–0.1 |
| Cu | 0.00005–10 | S | 0.0002–0.52 | Zr | 0.0003–0.3 |

Creation of the following sub-methods is useful for iron-based metals:
– Nonalloyed and low alloy steels
  (C < 2%, no alloying element > 5%, S < 0.12%, Pb < 0.1%)
– Nonalloyed and low alloy free-cutting steels
  (additionally 0.12%-0.5% sulfur and/or several tenths of a percent lead)
– Chromium and chrome/nickel steels
  (C < 2%, Cr 5–30%, Ni 0–30%, S < 0.12%)
– Chrome/nickel free-cutting steels
  (additionally > 0.12% sulfur)
– Nonalloyed and low alloy cast iron
  (C > 2%, no alloying element > 5%)
– Manganese steel/austenitic manganese steel
  (C < 2%, manganese > 10%, chromium 0–20%)

- High-speed steels
  (approx. 1% C, approx. 4% Cr, several percent of W, Mo, V and/or Co)
- Maraging steel
  (C < 0.05%, Ni 12–25%, Co up to 12%, Mo up to 5%, Ti up to 1.7%)
- Chrome chill casting
  (wear-resistant cast iron with up to 5% C, 12–32% chromium and 0–3% Mo)
- Austenitic cast iron
  (Ni–resist, up to 3% C, Si up to 6%, Ni up to 36%, Cr up to 4%, Mn up to 5%,
  plus one grade with up to 7.5% Cu)

Detailed information on the chemical composition and the properties of the alloys listed in the groups above can be found in the *Stahlschlüssel* (Key to Steel) [2]. The widespread works on material science can also help to obtain an overview on the common alloys, for example, the books from Domke [3], Merkel/Thomas [4], and Bargel/Schulze [5].

If steels, in which the assignment to one of the above groups is not clear from the outset, are to be analyzed, then a screening method can be used. For analysis using these methods, however, compromises in accuracy must be made. The analysis obtained is sufficient for many purposes. The presence of such a screening method is also helpful when there is no suitable sub-method available on the spectrometer system.

Far above the lower limits (i.e., above a hundred times the lower limits) listed in Table 5.1, it is possible to achieve a coefficient of variation for the measured values that usually lies between 0.1% and 0.7% for elements that are completely dissolved in the iron matrix. There is an especially low coefficient of variation if an internal standard can be found for an analyte line that reacts similarly to fluctuations in the plasma temperature. This is, for example, the case for the chromium line 298.9 nm.

For metallically dissolved elements, it is not the repeatability but the dispersion of the calibration functions and the stability of the system that is the limiting factor for the achievable accuracy. When elements are present completely or partially as non-metallic precipitates, the standard deviation of the determined concentrations are higher as the inclusions are often preferably hit by sparks. Here, a coefficient of variation of 1% to 3% are expected for concentrations far above the lower limits listed in Table 5.1.

## 5.1.1 Details for the analysis of steels

In addition to iron, carbon is the most important element in steels. The line at 193 nm is usually used for its determination. However, if very small carbon concentrations are to be determined, it is advantageous to use the more sensitive carbon line at 133 nm. To be able to use this line, however, the optical system must be equipped with components that are transparent for such short-wave radiation, which entails a certain amount of additional outlay. In this case, windows and lenses must be made of lithium fluoride, magnesium fluoride or calcium fluoride.

The analysis of free-cutting steel semi-finished products is problematic. Samples taken from the melt can be analyzed without great difficulty. A large part of the sulfur in rolled semi-finished products is present in the form of elongated manganese sulfide inclusions. The spark prefers to strike these inclusions, which leads to determination of sulfur and manganese concentrations that are too high in relation to the average contents in the sample. In addition to the sulfur and manganese concentrations, the excess depends on the form of the inclusions and, thus, from the mechanical history of the sample. Using a long pre-spark, whose duration can be up to a minute in extreme cases, helps. Nevertheless, the same accuracy achieved with other methods cannot be obtained.

Spark spectrometry enables the analysis of almost all elements of interest with abundant accuracy. Hydrogen is an exception. Although the spectral line at 121 nm is quite sensitive, the following factors hinder its use:
- The quantity degraded from the sample material by the spark is very small.
- Oxygen can be easily flushed out of the argon system and spark stand. In contrast, moisture adhering to the inner surfaces of the spark stand and argon system is only removed slowly. As a consequence, there is still water vapor present in the argon during the spark; resulting in an increase the hydrogen signal.
- Hydrogen atoms are mobile. The H-content on the surface, exactly where the material is ablated, can be reduced simply due to the sample warming that occurs during grinding.

According to the current state of technology, other procedures for the determination of hydrogen in steels are preferable to spark spectrometry, for example, carrier gas hot extraction. The method is described in the *Handbuch für das Eisenhüttenlaboratorium* (Handbook for Ironworks Laboratories) [6]. Contract laboratories offer carrier gas hot extraction for many different materials. Detection limits well below one ppm [7] are reported.

For the determination of nitrogen, it is necessary to take measures with regard to the purity of the spectrometer argon and the air-tightness of the argon system (see Section 3.4). Determination is possible if the necessary requirements have been fulfilled. Details can be found in the work of J. Niederstraßer [8].

### 5.1.2 Details for the analysis of cast iron

As in steel, carbon is by far the most important alloying element in cast iron. Unfortunately, carbon cannot generally be determined with spark spectrometry in finished cast iron products because it is precipitated as graphite. This is usually present as spheroidal or lamellar graphite in finished products.

Spheroidal graphite is preferably attacked during the pre-spark. If a graphite inclusion is hit, it leaves the sample by sublimation. During the actual measurement time, the carbon content is greatly reduced.

Even after the pre-spark time, lamellar graphite generates an excessive carbon signal. Here, an approximate analysis is possible if very long pre-spark times are used. Of course, the same, long pre-spark times must also be used during the creation of a method for the determination of carbon in semi-finished products.

Measurements on finished products succeed only in exceptional cases:
- Carbon is bound as carbide in chromium hard casting and, therefore, does not precipitate as graphite.
- If the surface is quenched during casting, the cementite phase ($Fe_3C$) is retained. In shell casting, this improves the wear resistance on the surface. Carbon can be determined on such surfaces.

Carbon measurements on samples that are taken from the melt during the production process are, in contrast, possible. However, it must be ensured that the sample cools suddenly. The best results are achieved by pouring the molten cast iron into a solid copper mold (see also Section 4.2.3). With this quenching, the carbon is then present exclusively as cementite. If graphite precipitates, the sample is useless for spark spectral analysis. This can, for example, happen when the mold is already too warm due to previous use. If the sample is broken apart, the fracture surface is slightly grey, which is easy to detect under a microscope. Metallographic section can also provide information about the sample quality. There are often graphite inclusions in the regions of the sample where feeding takes place. This may result in other carbon values there than on other parts of the sample surface where there is no graphite. The non-matching single measurements lead to confusion. There are algorithms in the spectrometer software that enable the identification of measurement on unproper sample positions. This is done by observing whether there is an excessive carbon signal at the beginning of the pre-spark. If the graphite precipitations and, thus, the intensity excesses moderate, a correction can be made. However, if the graphite content is too large, the sample is classified as unusable. A new sample must be taken. Such a procedure is described in the German patent DE102010008839B4 [9]. Figures 5.1 and 5.2 show two metallographic section images taken at different locations of the same sample surface. Each image represents an area of approximately 0.2 mm$^2$. The zone shown in Figure 5.1 is almost free of graphite precipitations, while there are many graphite precipitations in the form of dark spots in Figure 5.2.

Kipsch [10] pointed out that it is difficult to take a cast iron sample, that is, on one hand, white solidified and, on the other hand, free of cavities and slag particles. In recent years the production of white solidified samples has been further complicated by the use of preconditioned base iron which favors graphite formation. This is often desired by the foundries but it is detrimental to the production of usable samples.

Ductile cast iron types with spheroidal graphite, such as GJS 450-18, GJS 500-14 and GJS 600-10, have gained in importance in the last few years. These grades,

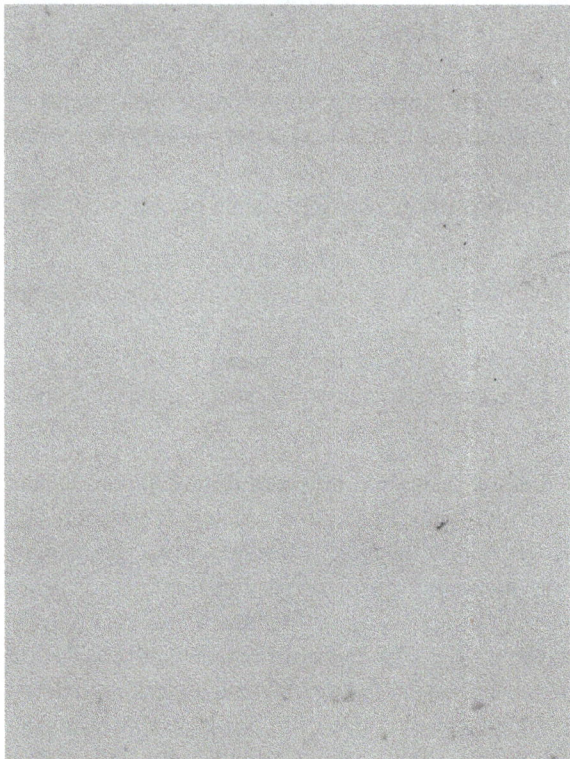

Figure 5.1: Section of the surface of a cast iron sample that is almost free of graphite inclusions.

standardized in DIN EN 1563 [11], exhibit both high values for tensile strength and elongation at break. In addition to good mechanical strength, they have the advantage that they can be plastically deformed before they break. This facilitates diagnosis in the event of damage. Fatigue fractures can be distinguished from those caused by mechanical overload. Monitoring for the correct silicon values is important for these materials. These should be within tight tolerances around 4.3%. The tensile strength is optimum for this Si value. If the value is exceeded, the elongation at break value drops abruptly; for silicon contents of about 4.8%, it drops to almost zero. Thus, the Si values are higher than for classical grades with spheroidal graphite and remaining within the limits is more critical. A description of the depicted problem can be found in Werner, Lappat and Ajrich [12].

## 5.2 Analysis of aluminum-based materials

The analysis of aluminum and its alloys with spark spectrometry is simple and provides good detection limits as the spectra, compared with those from iron or nickel base, have relatively few lines. The usable concentration ranges are summarized in

**Figure 5.2:** Section of the surface of a cast iron sample with graphite inclusions.

Table 5.2. However, it must be mentioned that the spectral range between 172 and 222 nm can only be used to a limited extent because a consistently high background signal is present throughout. Therefore, the use of some sensitive spectral lines in this range cannot be recommended. Such lines are, for example As 189.0 nm, B 182.6 nm and Sn 190.0 nm. It is better to select alternative lines outside the range mentioned for these elements. The determination of phosphorous, however, must be done with the line P 178.3 nm for lack of suitable alternatives. Here, a background correction helps in achieving the detection limits required to determine P.

Distribution into the following sub-methods is useful for aluminum base metals:
- Pure aluminum
  The main alloying elements are calibrated to a maximum of 0.05%, the calibration ranges for other elements usually end below 0.01%.
- Low alloy aluminum
  Si and Fe < 2%, Cu < 0.4%, Mn, Mg and Pb < 1.5%

- Aluminium/silicon
  Si < 30%, Fe < 5%, Cu < 6%, Ni < 3%, Mn < 1.5%, Zn < 4.5%, Mg < 2%,
  other elements each below 1%
- Aluminum/copper
  Cu < 10%, Si < 4%, Fe < 1.5%, Ni < 2.5%, Mn < 2%, Zn < 4%, Mg < 2%,
  other elements each below 1%
- Aluminum/magnesium
  Mg < 11%, Si < 4%, Mn < 1.5%, Zn < 1.5%, other elements each below 1%
- Aluminum/zinc
  Zn < 12%, Si < 10%, Cu < 10%, Mg < 4%, other elements each below 1%

Table 5.2: Concentration ranges for measurements in aluminum base.

| Element | Concentration range (%) | Element | Concentration range (%) |
|---|---|---|---|
| Ag | 0.00002–1.2 | Mo | (0.00002) 0.0005–0.2 |
| As | 0.0004–0.2 | Na | 0.00004–0.03 |
| B | (0.00002) 0.00005–2.8 | Nd | (0.0002–0.25) |
| Ba | 0.00005–0.02 | Ni | 0.00004–5 |
| Be | 0.00001–0.05 | P | (0.0002) 0.0003–0.035 |
| Bi | 0.0001–0.75 | Pb | (0.00003) 0.00008–2 |
| Ca | 0.0001–2.1 | Pr | (0.0003–0.08) |
| Cd | 0.00003–0.4 | Sb | 0.0002–0.6 |
| Ce | 0.0001–0.055 (0.6) | Sc | 0.00002–0.22 (0.8) |
| Co | 0.00005–10 | Se | (0.001–0.01) |
| Cr | 0.00003–0.8 | Si | 0.00005–43 |
| Cu | 0.00001–50 | Sm | (0.0002–0.02) |
| Fe | 0.00005–14 | Sn | 0.0001–22 (30) |
| Ga | 0.00003–0.1 | Sr | 0.00002–0.5 |
| Hg | 0.0002–0.01 | Ti | 0.00002–7 |
| In | 0.0002–0.12 | Tl | 0.00005–0.1 |
| La | 0.0001–0.04 (0.3) | V | 0.00005–0.5 |
| Li | 0.00001–12 | W | (0.001–0.01) |
| Mg | 0.00001–15 | Zn | 0.00006–12 |
| Mn | 0.00003–32 | Zr | 0.00005–0.8 |

Information about the chemical composition of the alloys in these groups can be found in the first volume of the *Aluminium-Taschenbuch* (Handbook of Aluminum) [13] and the standards dealing with the properties and composition of aluminum and its alloys [14, 15].

There are alloys outside the limits described above. For example, aerospace alloys containing lithium in percent ranges do not fall into any of the above ranges. It is, however, possible to create a powerful screening method for aluminum base.

The achievable accuracy is sufficient for many applications, for example, to verify the expected grade.

## 5.3 Analysis of copper-based metals

Distribution into the following sub-methods is useful for copper base metals:
- Pure coper
  All elements less than 0.05%
- Low alloy copper
  Si < 4%, Ni < 3%, Fe < 2%, Cr, Mn and Pb < 1.5%, Ag, Zn, Mg and Sn < 1%, P, As,
  Sb, Co and Zr < 0.4%, Te, Cd and Se < 0.2%, other elements each < 0.1%
- Beryllium bronzes and cobalt-beryllium alloyed copper
  Be < 3%, Co < 2.5%, Ni and Ag < 1.6%, Fe, Cr, Si and Al < 0.25%, Zn, Sn and
  Mn < 0.15%, other elements each < 0.1%
- Brass, special brass and tombac
  Zn 8–45%, Al < 8%, Mn, Bi and Si < 6%, Pb and Fe < 5%, Sn < 3%, Se < 2%,
  Sb < 1%, Co < 0.5%, other elements each < 0.3%
- Nickel silver
  Zn < 31%, Ni < 20%, Fe < 1.5%, Mn < 1%, other elements all below 0.3%
- Cupronickel
  Ni 8–35%, Mn, Cr, Fe and Nb < 3%, Zn, Ti and Si < 1%, Zr < 0.5%,
  other elements each below 0.2%
- CuZnSnPb(Ni) alloys
  Zn < 12%, Sn < 10%, Pb < 7%, Bi, Ni < 6%, Sb and Se < 2%, Fe < 1%, Co < 0.5%,
  Al, As, S, Mn and P < 0.25%, all other elements each < 0.1%
- Tin and lead bronze
  Pb < 23%, Sn < 15%, Zn < 8%, Ni < 3%, Sb < 1.5%, Fe and Mn < 0.5%,
  As, Al, Si and Bi up to 0.3%
- Aluminum bronze
  Al < 12%, Fe and Ni < 7%, Mn and Zn < 2.5%, Sn < 1.5%, Pb and Si < 1%,
  As < 0.5%, Cr, Mg and P < 0.25%, other elements each < 0.1%

Other sub-methods are possible. For example, the calibrations for low alloy copper and beryllium bronze can be combined. If alloys containing beryllium are measured, it is necessary to ensure that no metal condensate escapes into the environment, as beryllium is hazardous to health. In company operating instructions it can be specified that after loading the respective method and before measuring samples containing beryllium protective measures, for example control of the exhaust system and installing a suction system to the spark stand, must be taken. It is also important to be sure that the exhaust and suction systems are working when samples containing other hazardous elements such as Pb, Cd, Hg and As are measured.

In many cases, the lower concentration limits specified in Table 5.3 are not sufficient for the analysis of pure copper. A maximum of 400 ppm oxygen, 50 ppm lead and 5 ppm bismuth may be contained in electrolytic copper with the grade Cu-ETP. The sum of the contents of all other elements, except for silver, must remain

**Table 5.3:** Concentration ranges for measurements in copper base.

| Element | Concentration range (%) | Element | Concentration range (%) |
|---------|------------------------|---------|------------------------|
| Ag | 0.00005–2 | Ni | 0.00005–40 |
| Al | 0.00002–15 | O | 0.002–0.05 |
| As | 0.00005–1.5 (2) | P | 0.00005–10 |
| Au | 0.00005–0.1 | Pb | (0.0001) 0.0002–23 |
| B | 0.00002–0.04 | Pd | (0.00002–0.05) |
| Be | 0.00001–3 | Pt | (0.0001–0.06) |
| Bi | 0.00005–7 | Rh | (0.00005–0.03) |
| C | 0.001–0.07 | Ru | (0.00005–0.02) |
| Cd | 0.00002–0.5 (1) | S | 0.00005–0.3 |
| Co | 0.00005–3 | Sb | 0.0001–4.2 |
| Cr | 0.00002–3 | Se | 0.00005–1.5 |
| Fe | 0.00005–7.5 | Si | 0.00005–8 |
| Ir | (0.0001–0.03) | Sn | 0.0001–20 |
| Li | (0.00001–0.015) | Te | (0.0002) 0.0003–0.55 |
| Mg | 0.000005–0.3 | Ti | 0.00005–0.8 |
| Mn | 0.00002–25 | Zn | (0.00005) 0.0001–50 |
| Nb | 0.001–1.5 | Zr | 0.00005–0.4 |

under a limit of 300 ppm (source: Deutsches Kupferinstitut [16], German Copper Institute). In order to measure small concentrations in pure copper, it is possible to optimize the spectrometer system specifically for this application. Table 5.4 lists the detection limits that can be achieved for some elements with systems optimized in this way. However, the limits for some pure copper grades are so low that even with optimized detection sensitivity the performance limits of spark spectroscopy are reached. For the grade Cu-OTP, maximum concentrations of 2 ppm for the elements bismuth, tin, tellurium and selenium and a maximum value of 1 ppm for Cd and Zn

**Table 5.4:** Improved detection limits for systems optimized for measurements on pure copper.

| Element | Detection limit (%) | Element | Detection limit (%) | Element | Detection limit (%) |
|---------|---------------------|---------|---------------------|---------|---------------------|
| Ag | 0.00003 | Co | 0.00002 | Pb | (0.00008) 0.0002 |
| Al | 0.00001 | Cr | 0.00001 | S | 0.000035 |
| As | 0.00004 | Fe | 0.000035 | Se | 0.00003 |
| Au | 0.00003 | Li | (0.000004) | Si | 0.000035 |
| B | 0.00001 | Mg | 0.000002 | Sn | 0.00006 |
| Be | 0.000006 | Mn | 0.00001 | Te | (0.0001) 0.0002 |
| Bi | 0.00003 | Ni | 0.00002 | Ti | 0.00002 |
| Cd | 0.000015 | P | 0.000035 | Zn | (0.000025) 0.00005 |

must be monitored [17, 18]. It should be noted that the values listed in Table 5.4 are detection limits. Limits of quantitation are, according to DIN 32645 [1], a factor of 10/3 higher and are, thus, close to the concentrations to be monitored. The element oxygen, which is important in copper base, can be analyzed with the spectral line at 130 nm. The requirements for a gas-tight argon system and on a minimal level of oxygen in the spectrometer argon are even higher than for the determination of nitrogen in steels. This fact was already pointed out in 1992 by the esteemed Manfred Richter [19], who unfortunately died much too early. It should be remembered that a vpm of a gas in argon produces the same signal intensity as 20–50 ppm in the sample (see Section 3.4.1). A higher argon flow is generally used when analyzing pure copper to quickly flush out oxygen and moisture, which can enter the spark stand when changing the sample.

Comprehensive information about alloys of the copper base, but also about other non-ferrous metals can be found in volume 2 of the ASM Handbook [20].

Conducting analyses of copper and copper alloys with spark emission spectrometers is standardized in DIN EN 15079 [21].

## 5.4 Analysis of nickel and nickel alloys

Table 5.5 provides a rough overview of limits within which the measurement of nickel and nickel alloys is possible.

Distribution into the following sub-methods makes sense:
- Nonalloyed and low alloy nickel
  Co < 2%, C < 1.5%, Fe, Cr and Al < 1%, Si, Mn, Cu and Ti < 0.5%, Mg < 0.2%, other elements each < 0.1%
- Ni/Cu (Monel)
  Cu 20–40%, Al, Fe and Si < 5%, Mn < 3%, Ti < 2%, all other elements less than 1%
- Ni/Cr
  Cr 15–30%, Mo < 10%, Fe and Nb < 5%, Ti < 3%, Co, Si and Al < 2%, all other elements less than 1%
- Ni/Cr/Fe
  Cr 15–30%, Fe 5–40%, Mo < 10%, Nb < 7%, Co and W < 5%, Si < 3%, Al, Mn, Ti and Cu < 2%, other elements each < 1%
- Ni/Mo
  Mo 10–35%, Cr and Fe < 30%, Fe < 25%, W and Co < 5%, Mn and Si < 1.5%, all other elements less than 1%
- Ni/Cr/Co
  Cr 10–20%, Co 10–20%, W and Nb < 7%, Al and Ti < 5%, Ta < 4%, Fe < 1%, other elements each < 0.5%

- Monocrystalline Ni-base superalloys
  Cr 1–12%, Co < 16%, W < 15%, Ta < 14%, Re and Ru < 7%, Al 4–7%, Ti < 5%,
  Mo < 4%, V < 2%, Nb < 1%, Hf < 0.5%

**Table 5.5:** Concentration ranges for measurements in nickel base.

| Element | Concentration range (%) | Element | Concentration range (%) | Element | Concentration range (%) |
|---------|-------------------------|---------|-------------------------|---------|-------------------------|
| Al | 0.0001–10 | Mg | 0.0001–1 | Si | 0.0005–7 |
| As | 0.0003–0.015 | Mn | 0.0001–12 | Sn | 0.0001–19 |
| B | 0.0001–0.03 | Mo | 0.001–40 | Ta | 0.003–15 |
| C | 0.0001–1 | N | 0.0005–0.35 | Ti | 0.0005–10 |
| Co | 0.0005–35 | Nb | 0.0005–10 | V | 0.0002–0.85 (1.2) |
| Cr | 0.0005–50 | P | 0.0005–0.05 (2.4) | W | 0.0005–15 |
| Cu | 0.0001–40 | Pb | 0.0001–0.08 | Zn | 0.0001–0.02 (0.5) |
| Fe | 0.0002–50 | Re | 0.01–8 | Zr | 0.0002–0.3 |
| Hf | 0.005–4 | S | 0.0005–0.35 | | |

For the analysis of monocrystalline superalloys, the elements rhenium and ruthenium, which are rarely present in other metal bases, must often be determined with high concentrations. The accurate determination of carbon, boron and yttrium at trace levels is also often required. Superalloys are named for the fact that they can be used at temperatures above the operating temperatures of most steels. Monocrystalline superalloys are used for turbine blades in combustion chambers for gas turbines and enable operating temperatures of up to 90% of the melting point. The disks on which the turbine blades are mounted are usually made of polycrystalline superalloys that are measured in the methods Ni/Cr, Ni/Cr/Fe, Ni/Mo and Ni/Cr/Co. These alloys are operated at up to 80% of the melting point. Details about nickel base superalloys can be found in Reed [22]. Parts of aircraft turbines are safety-relevant and, thus, often require 100% testing. This is why there is such a high requirement for analyses for production of the material as well as for further processing up to the finished turbine. Laboratories involved with this must take precautions, for example, by periodically measuring control samples and maintaining quality control charts, to ensure that the specified measurement uncertainties are not exceeded.

# 5.5 Analysis of zinc and zinc alloys

Table 5.6 shows the concentration ranges within which zinc alloys can be measured. There are gaps in the concentration ranges due to a lack of standards.

Table 5.6: Concentration ranges for measurements in zinc base.

| Element | Concentration range (%) | Element | Concentration range (%) |
|---|---|---|---|
| Ag | 0.00005–0.05 | Mg | 0.00002–1.2 |
| Al | 0.00002–50 | Mn | (0.00004) 0.0001–0.25, 0.9–1.1 |
| As | 0.0002–0.005 | Ni | 0.00001–0.1, 1.9–2.1 |
| Cd | 0.00002–0.7, 30–32 | Pb | (0.00002) 0.0001–3 |
| Ce | (0.00001) 0.00005–0.08 | Sb | 0.0005–0.25, 0.8–4 |
| Cr | 0.0001–0.19 | Si | 0.00002–0.1, 1.2–3.3 |
| Cu | 0.00002–7 | Sn | (0.00002) 0.00005–3 |
| Fe | 0.00005–1.5 | Ti | 0.0001–0.35 |
| In | 0.00001–0.04 | Tl | 0.00001–0.07 |
| La | (0.00001) 0.00005–0.06 | Zr | 0.0001–0.02 |

Distribution into the following sub-methods makes sense:
–  Pure zinc
   All elements < 0.2%
–  Titanium zinc
   Ti 0.1-0.4%, Cu < 2%, other elements each < 0.2%
–  Commercial and secondary zinc
   Pb and Cd < 3%, Sn < 1%, other elements all < 0.2%
–  Cast alloys with Al contents up to 6%
   Al 1.5–6%, Cu < 4%, other elements all < 0.2%
–  Cast alloys with Al contents > 6%
   Al 6–15%, Cu < 4%, Cr and Ti < 0.4%, other elements < 0.2%
–  Alloys with high aluminum contents
   Al 15–50%, Cr and Ti < 0.4%, other elements each < 0.2%

The calibration ranges for the pure zinc method greatly exceed the permitted upper limits for real grades. The total concentration of impurities for refined zinc of the grade Z3 is maximum 0.05%. Only 0.005% is allowed for refined zinc of the grade Z1. The primary zinc grades Z1 to Z5 are standardized in DIN EN 1179 [23]; the secondary zinc grades ZSA, ZS1 and ZS2 in DIN EN 13283 [24]. The compositions of the most common cast zinc alloys can be found in DIN EN 1774 [25]. The process for sample taking is regulated in DIN EN 12060 [26]. Maximum concentrations for the elements Sn, Pb, Bi, Ni, Al and the sum of other elements in zinc baths for hot-dip galvanizing are specified in the DASt directive 022 [27]. With DIN EN ISO 3815–1 [28] there is also a standard for conducting spark spectral analysis. It also contains information about detection limits, however, those listed there are rather conservative.

In zinc base, it does not make sense to create an overview program. The content of the main alloying element, Al, has a strong influence on the excitation process. This can be seen from measurements of the burning voltages. Slickers [29] reported that for a fixed spark parameter, a spark burning voltage of 25 V was measured. For pure aluminum it was 35 V. The burning voltages for aluminum alloyed zinc alloys lie between these, namely 28 V for an Al content of 2% and 29 V for an Al content of 6%. Suitable methods must be created for alloys that are outside the limits listed in the sub-methods above. This calibration is usually performed by the instrument manufacturer on request, whereby it is often necessary for the end user to provide calibration standards.

Small Ti and Al contents (Ti between 100 and 2,000 ppm and Al between 20 and 200 ppm) are sometimes problematic. For reasons that are not understood, the scattering of the calibration curves and, thus, the measurement uncertainties are higher than those for other elements and also higher than for aluminum and titanium in other metal bases.

## 5.6 Analysis of lead and lead-based alloys

When analyzing the heavy metal lead, it is necessary to make sure that a properly working suction system is attached to the spark stand. The exhaust gas cleaning system, as described in Section 3.4.5, must also be checked regularly. Clogged filter cartridges are to be replaced and disposed of in accordance with safety aspects. Respiratory protection should be used when doing this work.

Table 5.7 shows the measuring ranges currently possible for lead base. The upper concentration ranges are mainly limited by the availability of standards. Trace elements in battery lead can negatively influence the properties of batteries, especially self-discharge. Therefore, an increase in the number of elements would be desirable but cannot take place due to a lack of suitable reference samples.

Unlike for zinc base, an overview method can be created with no problem for lead. Distribution into the following sub-methods makes sense:
- Pure lead
  Sn < 2%, Ag, Sb, Bi, Ca < 0.2%, other elements each < 0.1%
- Low alloy lead
  Ag < 4%, Sn < 3%, Cd, As, Bi < 1%, In and Cu < 0.5%, other elements all under 0.1%
- Hard lead
  Sb 0.5–21%, Sn < 12%, As < 1,7%, Sn < 1%, In < 0.5%, Bi < 0.3%, all other elements under 0.1%
- Pb-Sn
  Sn 1–50%, Sb < 17%, As < 1.7%, Cu < 0.6%, Bi < 0.25%, all other elements < 0.1%

Table 5.7: Concentration ranges for measurements in lead base.

| Element | Concentration range (%) | Element | Concentration range (%) |
|---------|------------------------|---------|------------------------|
| Ag | 0.00001–6 | Na | (0.000002) 0.00001–0.07 |
| Al | 0.00001–0.08 | Ni | 0.00001–0.05 |
| As | 0.00005–1.7 | P | (0.0003–0.025) |
| Au | 0.00005–0.18 (0.26) | Pd | 0.00005–0.008 |
| Ba | (0.000001) 0.00001–0.06 | Pt | 0.00003–0.01 |
| Bi | 0.00001–1.3. 10–50 | Rh | (0.00001–0.001) |
| Ca | 0.000005–1.1 | Ru | (0.00001–0.005) |
| Cd | 0.00001–18.5 | S | 0.0001–0.05 |
| Co | 0.00001–0.05 | Sb | 0.00003–21 |
| Cu | 0.000005–1.6 (7.5) | Se | 0.0002–0.05 |
| Fe | 0.00001–0.05 | Sn | 0.00005–50 |
| Hg | 0.0005–0.03 | Sr | (0.00001–0.005) |
| In | 0.000005–2.2, 15–30 | Te | 0.00004–0.1 |
| Mg | (0.0000005) 0.0001–0.2 | Zn | 0.00001–0.1 |
| Mn | 0.00001–0.0015 | | |

Concentration limits for common grades of lead and lead alloys are specified in the standards DIN EN 12659 [30] and DIN 17640–1 [31]. The composition of lead for cable sheaths and sleeves can be found in DIN EN 12548 [32], that of lead for construction in DIN EN 12588 [33]. The procedure for sample taking for analytical purposes is described in DIN EN 12402 [34]. A pre-standard for conducting analyses with spark emission spectrometers also exists (DIN V ENV 12908 [35]). The measuring ranges given there begin at a much higher levels than those listed in Table 5.7.

## 5.7 Analysis of tin and tin-based alloys

The safety precautions regarding the suction system and exhaust gas filtration mentioned in Section 5.6 must also be met for the analysis of tin alloys.

A sensible sub-method classification could look like the following:
- Fine and commercial tin and pewter
  Pb < 1.5%, Cu, As, Ge < 1%, Sb, Ag, Au, Hg, In < 0.2%, all other elements less than 0.1%
- Bearing metal
  Sb and Cu < 10%, As, Bi and Pb < 0.5%, Cd < 0.2%, all other elements < 0.1%
- Soft solder, cadmium-free special solder
  Pb < 50%, Sb < 6%, Cu and Ag < 5%, Bi and Au < 0.3%, other elements all less than 0.1%

Other tin-base materials can be analyzed in the overview method. Table 5.8 shows the concentration ranges possible for tin-base.

Table 5.8: Concentration ranges for measurements in tin base.

| Element | Concentration range (%) | Element | Concentration range (%) |
|---------|------------------------|---------|------------------------|
| Ag | 0.00001–4 (35) | Hg | 0.0002–0.15 |
| Al | 0.00001–0.07 (0.11) | In | 0.00001–0.11 (21) |
| As | 0.0004–0.6 (2.2) | Ni | 0.00001–1.3 (13) |
| Au | 0.00002–0.5 | P | 0.0006–0.03 |
| Bi | 0.00005–1.1 (3) | Pb | 0.00002–4 (50) |
| Cd | 0.00002–0.15 (2) | Pd | (0.00005–0.6) |
| Cr | 0.00001–0.005 (1.3) | S | 0.0001–0.005 (0.02) |
| Cu | 0.00001–12 (41) | Sb | 0.0001–8.5 (15) |
| Fe | 0.00001–0.06 (3.1) | Se | 0.0003–0.05 (0.07) |
| Ge | 0.0001–0.5 (1.1) | Zn | 0.00001–8.5 (50) |

The chemical concentrations of tin, tin alloys and soft solder are described in the standards DIN EN 610 [36], DIN EN 611 [37, 38] and DIN 1707-100 [39].

## 5.8 Analysis of titanium-based alloys

Table 5.9 shows the concentration ranges in which elements can be measured in titanium base.

Possible sub-methods:
- Pure titanium and Ti-0.2Pd
  W < 0.5%, Pd, Cr, O, Fe, Mo < 0.3%, Al, Sn, V, Cu, Ru, Mn, Ni < 0.2%, other elements < 0.1%
- Alloys with Zr and Mo
  Al < 7%, Zr < 6%, Mo < 15%, Mn < 7%, Fe and Cr < 4%, Nb and W < 1%, O < 0.5%, other elements each < 0.1%
- Alloys with niobium and vanadium
  Al < 8%, Sn < 4%, Nb < 7%, Mn < 5%, Mo, Fe and Cr < 1.5%, W < 1%, O < 0.5%, other elements each less than 0.1%

The compositions of titanium and titanium alloys are standardized in DIN EN 17850 [40] and DIN EN 17851 [41]. Interesting information about the use of different titanium alloys can be found in a monograph edited by E. W. Kleefisch [42].

Table 5.9: Concentration ranges for measurements in titanium base.

| Element | Concentration range(s) (%) | Element | Concentration range(s) (%) |
|---------|---------------------------|---------|---------------------------|
| Al | 0.0003–40 | Ni | 0.001–0.15 (20) |
| C | 0.001–0.05 (0.13), 0.9–1.1 | O | 0.005–0.4 |
| Co | 0.004–0.06 (0.3) | Pd | 0.001–0.3 |
| Cr | 0.001–5 | Ru | 0.01–0.05 (0.2) |
| Cu | 0.0001–3 | Si | 0.0005–0.75 |
| Fe | 0.0005–4 | Sn | 0.0005–3.1 (14) |
| H | 0.0001–0.013 | V | 0.001–14 |
| Mn | 0.0003–10 | W | 0.02–1.8 |
| Mo | 0.0003–16 | Y | 0.0005–0.004, 1.8–2.2 |
| N | 0.002–0.04 | Zr | 0.001–4.5 (50) |

# 5.9 Analysis of magnesium-based alloys

The following sub-methods make sense:
-   Pure magnesium
    Al < 0.1%, all other elements each less than 0.05%
-   Alloy groups magnesium-aluminum-zinc (AZ),
    magnesium-aluminum-manganese (AM), magnesium-aluminum-silicon (AS)
    and magnesium-manganese (M)
    Al < 12%, Zn < 7%, Mn < 3%, Si < 2%, Zr < 1%, Ca < 0.5%, Cu < 0.1%, other
    elements each < 0.1%
-   Alloy groups magnesium-zinc-rare earths (ZE), magnesium-zinc-thorium (ZH),
    magnesium-silver-rare earths-zirconium (QE), magnesium-zinc-copper (ZC) and
    magnesium-zinc-zirconium (ZK)
    Zn < 7%, Ag, Th, Nd and Ce < 4%, Cu < 3%, Zr and La < 1%, Pr < 0.5%, Cu, Mn,
    Al, Pb < 0.2%, other elements each less than 0.1%

The above list of sub-methods does not take several alloy groups into account. For
example, the group magnesium-yttrium-rare earths (WE) is not included. The main
obstacle here is the lack of standard samples with a high yttrium content.
Table 5.10 shows the typical concentration ranges for the analytes in magnesium
base.

The use of an electrode brush with molybdenum wire bristles is recommended
for magnesium base. The reason for this is the fact that small concentrations of
only a few ppm of the elements Fe and Ni must be determined. The abrasion from
the bristles can be sufficient for contamination. It would also be desirable to be
able to determine the rare earth elements scandium, samarium, europium, terbium,

Table 5.10: Concentration ranges for measurements in magnesium base.

| Element | Concentration range (%) | Element | Concentration range (%) |
|---------|-------------------------|---------|-------------------------|
| Ag | 0.00005–4 | Ni | 0.0001–0.03 |
| Al | 0.00001–14 | P | 0.0003–0.007 |
| Ca | 0.00003–0.5 | Pb | 0.0002–0.1 |
| Cd | 0.00003–0.08 | Pr | (0.0005) 0.0015–0.5 |
| Ce | 0.0001–3.3 | Si | 0.0003–2 |
| Cu | 0.00005–0.4 (3) | Sn | 0.0003–0.11 |
| Fe | 0.0001–0.05 | Sr | 0.0001–2.5 |
| Gd | 0.002–0.11 (0.3) | Th | 0.01–4 |
| La | 0.0005–1.5 | Ti | 0.0001–0.011 |
| Mn | 0.00003–3 | Y | 0.001–0.04 |
| Na | 0.00005–0.01 | Zn | 0.0001–8 |
| Nd | 0.0005–3.5 | Zr | 0.0001–0.9 |

dysprosium, holmium, erbium and ytterbium since mischmetal (a mixture of rare earth elements and small percentages of other metals) is used in the melting of alloys containing rare earths. Cerium mischmetal consists, for example, depending on the origin, of 45–60% cerium, 20–30% lanthanum, 15–21% neodymium and 4–6% praseodymium as well as 1–4% samarium, yttrium and terbium. In addition, approximately 1% each iron and magnesium can be found in cerium mischmetal. For a long time, it was difficult to separate the rare earth elements from each other. This is why they were used as mischmetal. However, today neodymium is often separated out because this element is used for the production of permanent magnets. Thus, the concentrations found for Ce, La and Pr, are often higher than they were in the past. The measurement of previously not analyzed rare earth elements should be pursued so that the total concentration of the rare earths can be more accurately determined. At the moment, however, the required reference samples are not available. In the past, the elements were calibrated to the following concentrations: Dy to 0.35%, Er to 0.15%, Eu to 0.01%, Ho to 0.07%, Sm to 0.1%, Tb to 0.05% and Yb to 0.1%. However, it is currently hardly possible to obtain the necessary standards. Details about the spark spectrometric determination of the rare earths can be found in the *Handbook of Rare Earth Elements* [43].

The element thorium is slightly radioactive. Therefore, magnesium alloys containing thorium are rarely used today. The determination of thorium usually serves to ensure that the alloys produced or processed do not contain this element. If, however, alloys containing thorium are to be determined, it is important to take the safety precautions specified in Section 5.6 into account.

DIN 1729 standardizes the composition of common magnesium-aluminum-zinc and magnesium-manganese wrought alloys [44]. Comprehensive information about the composition of the common magnesium alloys and their properties, current trends in development and about European and international standards can be found in the *Magnesium Handbook* [45]. *Smithells Light Metals Handbook* [46] is also helpful in this respect. In addition, it contains useful information about the aluminum and titanium bases.

## 5.10 Analysis of cobalt-based alloys

Table 5.11 lists the ranges covered with spark emission spectrometry for cobalt-based alloys. Cobalt-based alloys are hard and wear-resistant due to embedded carbides. An informative diagram depicting the important Co-based alloys with the hardness and corrosion resistance along the axes of a coordinate system is available in a company publication of the Deutsche Edelstahlwerke [47]. In the past, Co-based alloys were developed for aircraft turbines. Today, nickel-based super alloys are preferred for this. However, according to Reed [22], nickel-based turbine blades are coated with cobalt alloys, for example, with the alloy Co – 25% Cr – 14% Al – 0.5% Y, to increase the corrosion resistance.

**Table 5.11:** Concentration ranges for measurements in cobalt base.

| Element | Concentration range (%) | Element | Concentration range (%) |
|---------|------------------------|---------|------------------------|
| Al | 0.001–1.2 | Ni | .001–36 |
| B | 0.00005–0.14 | P | 0.0001–0.05 |
| C | 0.0002–3.5 | Pb | 0.00005–0.03 |
| Cr | 0.05–41 | S | 0.0001–0.05 |
| Cu | 0.0003–0.15 (0.4) | Si | 0.0001–2.2 (4) |
| Fe | 0.0001–50 | Sn | 0.0001–0.11 |
| La | 0.0003–0.06 | Ta | 0.003–4 (12) |
| Mn | 0.0001–3.5 | Ti | 0.0002–1 (6) |
| Mo | 0.0001–10 (35) | V | 0.0002–0.05 (11) |
| N | 0.003–0.2 | W | 0.001–16 (23) |
| Nb | 0.0002–3 (15) | | |

The following sub-methods are possible for cobalt-based alloys:
- Low alloy cobalt
  Fe < 2.5%, Si < 1%, Cr < 0.1%, all other elements < 0.1%
- Co-Cr alloys with low nickel contents
  C and Si < 1%, Cr < 25%, W < 12%, Mo < 10%, Ni, Fe and Nb < 3%, Mn < 2%, Al < 1%, Sn and Ta < 0.2%

- Co-Cr alloys with high nickel contents
  C < 3%, Cr < 35%, Ni 2–30%, W < 16%, Nb < 5%, Mo and Si < 4%, Fe < 3%, Mn < 2%, V < 0.5%, Al < 0.3%, N < 0.2%, B < 0.1%
- Co-Cr alloys with high molybdenum contents
  Si < 4%, Cr 5–20%, Mo 25–35%, Fe < 4% Ni < 1%
  Concentrations that exceed the limits of the sub-methods above can be measured with an overview method.

## 5.11 Analysis of precious metals

In the field of precious metal analysis, spark spectrometry is mainly used to determine impurities in pure metals. Table 5.12 lists proven concentration ranges for the elements silver, gold, palladium and platinum. Spark spectrometry is less suited to the determination of the main components of precious metal alloys because, with high concentrations of the alloying components, better absolute accuracies can be achieved with other methods, for example, X-ray fluorescence spectrometry. An overview of the precious metal alloys can be found in the *Metals Handbook* [48] or in the previously mentioned more comprehensive version of the *ASM Handbook* [20]. Information about methods suitable for the analysis of precious metals are given in the monograph *Determination of the Precious Metals* [49].

## 5.12 Other metals

### 5.12.1 Refractory metals

Pure zirconium and grades alloyed with approximately 1.5% tin are examined for the zirconium base. In addition to tin, iron, chromium, nickel, copper and niobium are also determined. There are few standards on the market. Sorting of the main grades via intensity ratios (see Section 7.7.3) is usually possible. The same applies for the common materials of the tungsten and molybdenum base. Here, material ablation is only very low due to the high melting points; this limits the detection sensitivity.

### 5.12.2 Low-melting heavy metals

Bismuth, indium, thallium and cadmium are low-melting heavy metals and, thus, related to tin and lead. There are very few commercially available standards, most of them for the calibration of small concentrations of Ni, As, Sb, Sn, Cu, Pb, Tl and

**Table 5.12:** Concentration ranges for the determination of traces in precious metals.

| | Concentration range (ppm) | | | |
|---|---|---|---|---|
| | **Ag** | **Au** | **Pd** | **Pt** |
| Ag | – | 0.1–200 | 0.5–100 | 0.2–300 |
| Al | – | 0.1–500 | 0.5–100 | 0.5–30 |
| As | – | 1–100 | – | – |
| Au | 0.1–250 | – | 0.5–1000 | 0.5–30 |
| Bi | 0.4–200 | 1–100 | 0.5–100 | 1–100 |
| Ca | – | 0.05–100 | – | – |
| Cd | 0.1–200 | 0.1–50 | – | – |
| Co | – | 0.2–50 | – | – |
| Cr | – | 0.1–100 | – | – |
| Cu | 0.2–200 | 0.2–200 | 0.1–200 | 0.2–200 |
| Fe | 0.3–20 | 0.5–200 | 2–1000 | 0.5–1000 |
| Ir | – | – | – | 3–1000 |
| Mg | – | 0.1–100 | 0.1–20 | 0.1–20 |
| Mn | – | 0.1–800 | 0.5–100 | 0.2–200 |
| Ni | 0.5–200 | 0.1–100 | 0.5–50 | 0.5–100 |
| Pb | 0.8–200 | 0.3–100 | 0.2–50 | 0.5–20 |
| Pd | 0.2–200 | 0.3–200 | – | 0.5–1000 |
| Pt | 1–200 | 1–200 | 2–1500 | – |
| Rh | – | 0.2–20 | 2–1000 | 0.2–400 |
| Ru | – | – | 1–1000 | 0.5–500 |
| Sb | 1–200 | 1–100 | 2–300 | 3–200 |
| Se | 1–200 | 2–30 | – | – |
| Si | – | 0.5–50 | 2–300 | 1–100 |
| Sn | 1–200 | 1–100 | 1–300 | 2–200 |
| Te | 2–100 | 0.5–50 | – | – |
| Ti | – | 0.1–20 | – | – |
| Zn | 0.1–200 | 0.2–100 | 0.5–300 | 0.3–150 |

Ni in cadmium base. The limited analytical experience with these heavy metals leads to the assumption that the detection sensitivity is similar to that of tin or lead base.

Chapter 7 of the *Handbook of Rare Earth Elements* [43], written by Jörg Niederstraßer, deals with the determination of the rare earths in various metal bases using spark spectrometry.

The monograph *Alloys* [50], published by F. Habashi, provides a good overview of the metal alloys and also deals with refractory metals and low-melting alloys.

# Bibliography

[1]   DIN 32645:2008-11. Chemische Analytik – Nachweis-, Erfassungs- und Bestimmungsgrenze unter Wiederholbedingungen – Begriffe, Verfahren, Auswertung. Berlin, Beuth Verlag, 2008.

[2]   Wegst M, Wegst C. Stahlschlüssel – Key to Metals. Marbach, Verlag Stahlschlüssel Wegst, 2016, ISBN 978-3922599326.

[3]   Domke W. Werkstoffkunde und Werkstoffprüfung, 10th edition. Düsseldorf, Verlag W. Girardet, 1986.

[4]   Merkel M, Thomas KH. Taschenbuch der Werkstoffe. München, Wien, Fachbuchverlag Leipzig im Carl Hanser Verlag, 2003.

[5]   Bargel HJ, Schulze G. Werkstoffkunde. Düsseldorf, VDI-Verlag, 1988.

[6]   Handbuch für das Eisenhüttenlaboratorium, Vol. Band 2, Analyse der Metalle, Teil 2, Neue Verfahren. Düsseldorf, Verlag Stahleisen, 1998.

[7]   Trägergasheißextraktion, from the internet Internet: http://www.revierlabor.de/fileadmin/ Resources/Public/pdf/Flyer-_Traegergasheissextraktion_09.14.pdf. Publication of the company Revierlabor, Essen, reviewed July, 22nd, 2017.

[8]   Niederstraßer J. Funkenspektrometrische Stickstoffbestimmung in niedriglegierten Stählen unter Berücksichtigung der Einzelfunkenspektrometrie. Dissertation, Duisburg, 2002.

[9]   Van Driel R, Van Stuivenberg B. Verfahren zur Bestimmung von Kohlenstoff in Gusseisen. German patent DE 102010008839 B4, granted April 21st, 2016.

[10]  Kipsch D. Lichtemissions-Spektralanalyse. Leipzig, VEB Deutscher Verlag für Gundstoffindustrie, 1974.

[11]  DIN EN 1563:2012-03. Gießereiwesen – Gusseisen mit Kugelgraphit. Berlin, Beuth Verlag, 2013.

[12]  Werner H, Lappat I, Ajrich B. Mischkristallverfestigte GJS-Werkstoffe für Groß- und Schwergussteile. Mering, Werkstoffzeitschrift, HW-Verlag, edition 2, 2014.

[13]  Kammer C. Aluminium-Taschenbuch, Volume 1, Grundlagen und Werkstoffe, 16. Edition. Düsseldorf, Aluminium-Verlag Marketing & Kommunikation GmbH, 2002.

[14]  DIN EN 573-3:2013-12. Aluminium und Aluminiumlegierungen – Chemische Zusammensetzung und Form von Halbzeug – Teil 3: Chemische Zusammensetzung und Erzeugnisformen; German edition EN 573-3:2013. Berlin, Beuth Verlag, 2013.

[15]  DIN EN 1780-1:2003-01. Aluminium und Aluminiumlegierungen – Bezeichnung von legiertem Aluminium in Masseln, Vorlegierungen und Gussstücken – Teil 1: Numerisches Bezeichnungssystem; Deutsche Fassung EN 1780-1: 2002. Berlin, Beuth Verlag, 2003.

[16]  Datenblatt Cu-ETP. www.kupferinstitut.de/fileadmin/user_upload/kupferinstitut.de/de/ Documents/Shop/Verlag/Downloads/Werkstoffe/Datenblaetter/Kupfer/Cu-ETP.pdf. Düsseldorf, Deutsches Kupferinstitut, reviewed July 20th, 2017.

[17]  DIN EN 13601:2013-09. Kupfer und Kupferlegierungen – Stangen und Drähte aus Kupfer für die allgemeine Anwendung in der Elektrotechnik. Berlin, Beuth Verlag, 2013.

[18]  Datenblatt Cu-OFE. https://www.kupferinstitut.de/fileadmin/user_upload/kupferinstitut.de/ de/Documents/Shop/Verlag/Downloads/Werkstoffe/Datenblaetter/Kupfer/Cu-OFE.pdf. Düsseldorf, Deutsches Kupferinstitut, reviewed July 20th, 2017.

[19]  Richter M. Erfahrungen bei der Bestimmung von Stickstoff und Sauerstoff in Metallen mittels optischer Emissionsspektrometrie, Beitrag im Tagungsband: Nichtmetalle in Metallen '92. Oberursel, DGM Informationsgesellschaft mbH, 1992.

[20]  ASM Handbook. Properties and Selection: Nonferrous Alloys and Special-Purpose Materials (ASM Handbook), Vol. 2, 10th Edition. Metals Park, Ohio, USA, 1992, ISBN-13: 978-0871703781.

[21]  DIN EN 15079:2015-07. Kupfer und Kupferlegierungen – Analyse durch optische Emissionsspektrometrie mit Funkenanregung (F-OES). Berlin, Beuth Verlag, 2015.

[22]  Reed RC. The Superalloys: Fundamentals and Applications. Cambridge (GB), Cambridge University Press, 2006, ISBN 978-0-521-07011-9.

[23]  DIN EN 1179:2003-09. Zink und Zinklegierungen – Primärzink. Berlin, Beuth Verlag, 2003.

[24]  DIN EN 13283:2003-01. Zink und Zinklegierungen – Sekundärzink. Berlin, Beuth Verlag, 2003.

[25]  DIN EN 1774:1997-11. Zink und Zinklegierungen – Gusslegierungen – In Blockform und in flüssiger Form. Berlin, Beuth Verlag, 1997.

[26]  DIN EN 12060:1998-01. Zink und Zinklegierungen – Probenahme – Spezifikationen. Berlin, Beuth Verlag, 1997.

[27]  Deutscher Ausschuss für Stahlbau. Feuerverzinken von tragenden Stahlbauteilen, DASt-Richtlinie 022. Düsseldorf, Institut Feuerverzinken GmbH, 2009.

[28]  DIN EN ISO 3815-1:2005-08. Zink und Zinklegierungen – Teil 1: Optische Emissionsspektrometrie an festen Proben (ISO 3815-1:2005). Berlin, Beuth Verlag, 2005.

[29]  Slickers K, Iten J, Frings S. Spektrometrische Zinkanalyse in Argon. GIT, Labor-Fachzeitschrift, GIT-Verlag, October 1977.

[30]  DIN EN 12659:1999-11. Blei und Bleilegierungen – Blei. Berlin, Beuth Verlag, 1999.

[31]  DIN 17640-1: 2004-02. Bleilegierungen für allgemeine Verwendung. Berlin, Beuth Verlag, 2004.

[32]  DIN EN 12548:1999-11. Blei und Bleilegierungen – Bleilegierungen in Blöcken für Kabelmäntel und Muffen. Berlin, Beuth Verlag, 1999.

[33]  DIN EN 12588:2007-03. Blei und Bleilegierungen – Gewalzte Bleche aus Blei für das Bauwesen. Berlin, Beuth Verlag, 2007.

[34]  DIN EN 12402:1999-10. Blei und Bleilegierungen – Probenahme für die Analyse. Berlin, Beuth Verlag, 1999.

[35]  DIN V ENV 12908:1998-01. Blei und Bleilegierungen – Analyse durch Optische Emissionsspektrometrie (OES) mit Funkenanregung. Berlin, Beuth Verlag, 1998.

[36]  DIN EN 610:1995-09. Zinn und Zinnlegierungen – Zinn in Masseln. Berlin, Beuth Verlag, 1995.

[37]  DIN EN 611-1:1995-09. Zinn und Zinnlegierungen – Zinnlegierungen und Zinngerät – Teil 1: Zinnlegierungen. Berlin, Beuth Verlag, 1995.

[38]  DIN EN 611-2:1996-08. Zinn und Zinnlegierungen – Zinnlegierungen und Zinngerät – Teil 2: Zinngerät. Berlin, Beuth Verlag, 1996.

[39]  DIN 1707-100: 2016-12. Weichlote – Chemische Zusammensetzung und Lieferformen. Berlin, Beuth Verlag, 2016.

[40]  DIN 17850:1990-11. Titan; Chemische Zusammensetzung. Berlin, Beuth Verlag, 1990.

[41]  DIN 17851:1990-11. Titanlegierungen; Chemische Zusammensetzung. Berlin, Beuth Verlag, 1990.

[42]  Kleefisch EW (editor). Industrial Applications of Titanium and Zirconium. Philadelphia USA, American Society for Testing and Materials, 1981.

[43]  Golloch A (editor). Handbook of Rare Earth Elements – Analytics. Berlin Boston, DeGruyter, 2017.

[44]  DIN 1729-1: 1982-08. Magnesiumlegierungen; Knetlegierungen. Berlin, Beuth Verlag, 1982.

[45]  Kammer C. Magnesium-Taschenbuch. Düsseldorf, Aluminium-Verlag, 2000.

[46]  Brandes EA, Brook GB (editor). Smithells Light Metals Handbook. Oxford, England, Butterworth-Heinemann, 1998.

[47]  Schematischer Stammbaum „Celsite". https://www.dew-stahl.com/fileadmin/files/dew-stahl.com/documents/Publikationen/Werkstoffdiagramme/dew_stammbaum_celsite_de_131031.pdf. Deutsche Edelstahlwerke, reviewed July, 24th, 2017.

[48]  Boyer HE, Gall TL (editor). Metals Handbook – Desktop Edition. Metals Park, Ohio, USA, American Society for Metals, 1984.

[49]  Van Loon JC, Barefoot RR. Determination of the Precious Metals. Chichester, England, John Wiley & Sons, 1991.

[50]  Habashi F (editor). Alloys. Weinheim, Wiley-VCH, 1998.

# 6 Analysis of inclusions in metal

## 6.1 Introduction

The visual appearance of metallic materials gives the impression that they are homogenous, single-phased substances. The experience of hundreds of years of metal production and processing, however, shows that metals – regardless of the main element – are complex multi-phased materials that also contain non-metallic compounds. These are rarely perceptible without technical aids. A well-known example of this is so-called Damascene steel, whose different phases are visible to the naked eye.

The technological properties of metallic materials are determined on one hand by their chemical composition and on the other hand by the production parameters. Particularly the temperature profile during solidification and the temperature regime during processing are deciding factors in the formation of the desired crystal structures and, thus, the material properties. For the worldwide most important metallic material, steel, in addition to the two iron crystal forms, ferrite and austenite, based on the iron-carbon diagram, there are, for example, the phases pearlite and cementite. An almost incalculable number of different steel grades can be produced by modification of the chemical composition and formation of different structures.

In addition, almost all metallic materials contain so-called inclusions or segregations. Exogenous inclusions are the compounds that are introduced into the material from outside during the melting process. Endogenous inclusions or segregations are formed during reactions within the melt or during processing of the material. The term "degree of purity" is used as the cumulative parameter for both types of inclusions in metallurgy.

Further explanations here are limited to examples dealing with iron based materials because there is a very wide variety of possible types of inclusion and because there is, in terms of analysis, long-term experience available for use. However, the interpretations can be analogously applied to other materials, such as aluminum, zinc and tin.

## 6.2 Inclusions and their origins

During production processes, the molten metal comes into contact with various process media. These include the refractory linings of the different aggregates, but also input materials such as casting aids or covering compounds. Components of these materials can be absorbed into the melt. These are predominantly oxidic compounds like $Al_2O_3$, $ZrO_2$, etc. With optimum process control, these rise due to the lower specific gravity compared to the melt and merge with the slag phase floating

https://doi.org/10.1515/9783110529692-006

at the top. In rare cases, these exogenous inclusions can remain in the melt. Because of their particle size (>10 μm), they almost invariably have a negative influence on the material properties. These inclusions result in surface defects during the production of thin sheets for exposed panel applications in the automotive industry. Such inclusions can also be the starting point for cracks on severely deformed components. Aging and signs of fatigue in cyclically stressed components are also possible.

Both in the melt phase and during processing, different non-metallic reactants are available to the various alloying elements to form non-metallic inclusion compounds and segregations in chemical reactions. Nitrogen, oxygen, carbon and sulfur are typical reactants. Table 6.1 lists a selection of important non-metallic inclusions that may be formed on the basis of these reactions. These particles are much smaller compared to the exogenous inclusions (see Figures 6.1 and 6.2). The particle sizes reach from several micrometers down to the nanometer range. The probability that a given reaction will occur depends strongly on the composition of the material and the affinity to the respective reaction partners. Titanium and aluminum have, for example, a high affinity to oxygen and nitrogen so that, when they are present, the oxide or nitride are preferably formed. Calcium has a high affinity to oxygen and sulfur. Inclusions based on chromium, vanadium, niobium, etc., are predominantly found only in high alloyed materials. That is, when the composition of the melt is known, it is possible to predict which inclusion compounds will be formed based on known plausibilities. In contrast to exogenous inclusions, endogenous inclusions can have both positive and negative influences on the material properties. As an illustration, the influences of inclusions on the materials are presented in the following examples.

During steel production, carbon present in the pig iron is removed in the converter process by blowing in oxygen. Thus, the pig iron contains up to 1,000 $\mu g\ g^{-1}$ dissolved oxygen by the end of the process. The melt must be deoxidized before any further metallurgical treatment is carried out [3, 4]. Aluminum, for example, is added to the melt for this purpose. Aluminum oxide is formed, which rises due to the lower specific gravity compared to the melt and then merges with the slag phase found at the top. However, this process doesn't proceed completely so that particles remain in the melt. The ratio between the metallic aluminum which functions as the alloying element and the inclusion compounds aluminum oxide and aluminum nitride must be monitored and adjusted during process control. The electromagnetic permeability of transformer sheets, for example, depends on the targeted adjustment to a predetermined value for the ratio of the metallic phase and the inclusions. If the oxide proportion is too high, qualitative problems could occur during the subsequent processing steps.

So-called IF (interstitial free) steel is an example for the calculated precipitation of inclusions to improve the material quality. These are micro-alloyed steels with a very low carbon content. By alloying with titanium or niobium, the free nitrogen or carbon atoms are precipitated as nitrides or carbides during the last hot forming thus hardening the material.

Table 6.1: Inclusion compounds depending on the reactants.

| | Si | Fe | Al | Mn | Mg | Ca | B | Ti | V | Cr | Zr | Nb |
|---|---|---|---|---|---|---|---|---|---|---|---|---|
| **O** | $SiO_2$ | $FeO$<br>$Fe_2O_3$<br>$Fe_3O_4$ | $Al_2O_3$ | $MnO$<br>$Mn_2O_3$<br>$MnO_2$ | $MgO$ | $CaO$ | $B_2O_3$ | $TiO$<br>$TiO_2$ | $VO$<br>$V_2O_3$<br>$V_2O_5$ | $CrO$<br>$Cr_2O_3$<br>$CrO_3$ | $ZrO_2$<br>$ZrOS$ | $NbO_2$ |
| **N** | $Si_3N_4$ | $Fe_2N$<br>$Fe_4N$ | $AlN$ | $Mn_3N_2$ | | | $BN$ | $TiN$<br>$TiN_xC_{1-x}$ | $VN$<br>$V(C,N)$ | $CrN$<br>$Cr_2N$ | $ZrN$ | $NbN$<br>$Nb(C,N)$ |
| **C** | $SiC$ | $Fe_3C$<br>$Fe_7C_3$<br>$Fe_{23}C_6$ | | $Mn_3C$ | | | | $TiC$<br>$TiC_xN_{1-x}$<br>$Ti_4C_2S_2$ | $VC$<br>$V(C,N)$ | $Cr_3C$<br>$Cr_7C_3$<br>$Cr_{23}C_6$ | $ZrC$<br>$Zr_4C_2S_2$ | $NbC$<br>$Nb(CN)$ |
| **S** | | $FeS$<br>$FeS_2$ | $Al_2S_3$ | $MnS$ | | $CaS$ | $Fe_3(C,B)$ | $TiS$<br>$Ti_2S$<br>$Ti_3S_4$ | | | $ZrS_x$ | |

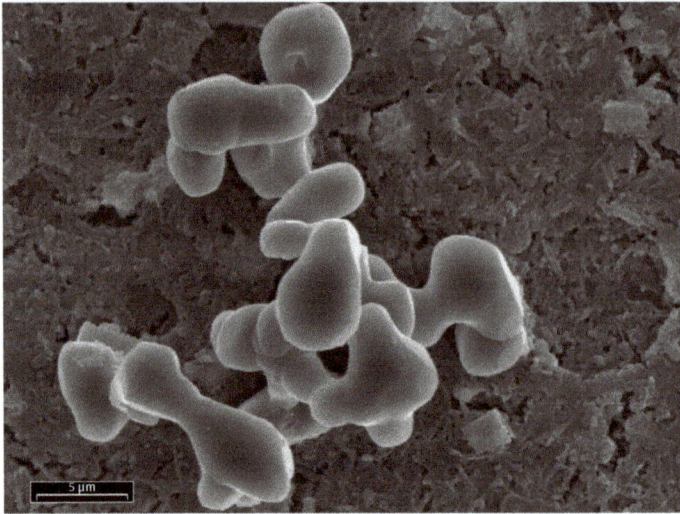

Figure 6.1: Isolated $Al_2O_3$ inclusion (with permission of the Rubikon Verlag Rolf Kickuth, Bammentaler Straße 6–8, 69251 Gaiberg, Germany) [1].

Figure 6.2: Isolated AlN inclusion (with permission of the Rubikon Verlag Rolf Kickuth, Bammentaler Straße 6–8, 69251 Gaiberg, Germany) [1, 2].

These two examples show that with endogenous inclusions or precipitations, the material properties can be influenced in both a positive and a negative sense. In any case, real-time monitoring of the degree of purity and thus the inclusions contained in the material is required for optimal control of the metallurgical processes.

## 6.3 Reference methods for the determination of inclusions in steels

As a relative measurement method, the respective appropriate standard procedures, enabling internal laboratory quality control, are required for quality-assured utilization of spark emission spectrometry. Several chemical and instrumental analytical methods used as reference methods are presented in the following. The descriptions are limited to the fundamentals. Further information can be found in the respective literature.

### 6.3.1 Classical analytical procedures

The first work step for all chemical testing methods is exposure of the isolate, that is, complete removal of the metallic matrix. Here it must be distinguished between chemical and electrochemical dissolution processes. Chemical dissolving processes only deliver information about the individual inclusion compounds in a short amount of time. Electrochemical dissolution processes require more time but provide information about a wider range of inclusion types.

#### Chlorination
For treatment with chlorine gas in an oven, the Fe matrix is transformed at higher temperatures into $FeCl_3$ in the gas phase. The residues are analyzed using inductively coupled plasma optical emission spectrometry (ICP-OES) and X-ray diffraction (XRD). All oxides are quantitatively detectable [5].

#### Dissolution process with iodine-methanol or bromine-methanol solutions
The sample materials are refluxed in the respective solution. The matrix goes into solution. The residues are analyzed using ICP-OES and XRD. All oxides as well as BN are quantitatively detectable. Carbides can only be qualitatively identified [6].

#### Determination of AlN according to the Beeghly method
The Fe matrix is dissolved with an iodine-methyl acetate or bromine-methyl acetate solution and the AlN is converted to ammonia with a sodium hydroxide fusion and subsequently determined with UV-VIS spectrometry [7].

#### Galvanostatic electrolysis
The specimen is dissolved with galvanostatic electrolysis and the residue is filtered off. The residue is analyzed using ICP-OES, carrier gas hot extraction and XRD. The

method provides information about the oxides, several nitrides as well as qualitative information about carbides and sulfides contained. In addition, the mass share of all inclusions in the sample is obtained as a cumulative parameter [8, 9].

### Potentiostatic electrolysis

The procedure is analogous to the galvanostatic electrolysis. The residue provides information about all types of inclusion in the sample [8, 9]. If water-soluble inclusions (CaO or MgO) are to be detected, an anhydrous electrolyte must be used. This procedure is still being used in various laboratories today.

### Determination of acid soluble aluminum using AAS or ICP-OES

This is a time-optimized testing method for process control with which it is possible to distinguish between acid-soluble ($Al_{sl}$) and acid-insoluble or precipitated ($Al_a$) inclusions containing aluminum. However, the analytical information in respect to the characterization of the type of inclusion is not clear. The acid-soluble fraction includes fine-grained AlN particles and the metallic aluminum present. As an inverse conclusion, the acid-insoluble fraction includes aluminum oxides as well as coarse-grained AlN particles. The sample material can be dissolved either with acid mixtures [10, 34] or electrochemically [11, 12].

## 6.3.2 Microscopic testing methods

### Light microscopy

Light microscopy is the most commonly used method for the determination of inclusions and the purity of steels. The standards ISO 4967, EN 10247, ASTM E45 and JIS G 0555 are very similar to identical in terms of implementation and evaluation [13–16]. The basis for these standards is the comparison of image panels with light microscopic images of prepared samples. Based on this, information about the types of inclusions (shape), inclusion size and frequency can be obtained. The degree of purity is given as the surface fraction of the non-metallic inclusions per 1,000 $mm^2$. Light microscopy can be automated by using image analysis software and sample changers.

### Scanning electron microscopy

Scanning electron microscopy has long been one of the standard methods in metallography and offers, compared to light microscopy, in addition to higher resolution, the advantage that elemental analyses can be simultaneously conducted using wavelength dispersive or energy dispersive X-ray fluorescence spectrometers [9]. It must, however, be mentioned that in particular carbon, nitrogen and oxygen cannot be quantitatively determined.

**Reference materials**

The use of the methods presented here is of particular importance, as there are no reference materials available for spark emission spectrometry that are certified regarding type of inclusion and inclusion ratio. Production fails largely due to the fact that the primary material with the required uniform distribution of inclusions is not available or the homogeneity cannot be guaranteed.

For this reason, Mittelstädt, among others, has developed a method for producing this type of reference material based on hot isostatic pressing (HIP) and has applied for a patent. Hereby, pure iron powder is intentionally doped with inclusion compounds, homogenized and pressed into a cylindrical specimen. These binary standard samples can then be used for quality control [17].

## 6.4 Influence of the inclusions on the spark discharge

The technical possibilities for optimization of the precision and accuracy using spark discharge for the analysis of metallic material have been discussed in detail in Chapter 3. In addition to the obvious requirements on the sample material, for example, homogeneity and a flawlessly flat surface, the spark behavior is influenced by the microstructures and inclusions. Especially changes in the spark behavior depending on the analyte species can, under certain conditions, be analytically detected and quantified.

In the thirties of the last century, H. Winter already investigated the evaporation processes on metallic electrodes [18]. Using metallographic investigations, he was able to prove the dependence of the ablation processes for light metal alloys on chemical reactions on the sample surface and on their reaction products. The term "stationary state" as a prerequisite for the analytically usable phase of the spark process was coined here.

Commercialization of spark spectrometry began in the fifties and it found its way into the laboratories of the primary industries. Accordingly, interest in clarifying ablation processes and optimizing the key analytical figures increased among both academic and industrial users. Both Laqua, among others [19], and Höller, among others [20–22] were able to show in the sixties of the previous century that the signal increase at the beginning of the spark process was due to inclusions in the sample material. Figure 6.3 qualitatively shows the spark curve for steel samples with and without aluminum oxide inclusions.

These effects could be proven for different types of inclusion. Koch, among others, examined the influence of manganese sulfide inclusions on spark processes [23]. The lateral distribution of the spark impacts as a function of the positions of the inclusions was also considered here. Figure 6.4 schematically shows the distribution of the impacts after a shorter and a longer spark duration. The boundaries between the inclusion and the metal matrix are preferred by the first impacts. Only

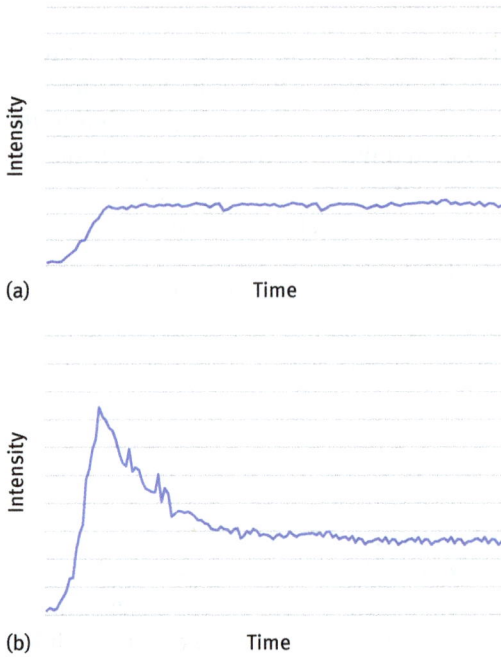

Figure 6.3: Spark curves for steel samples with an $Al_2O_3$ fraction <0.001% (a) and 0.049% (b).

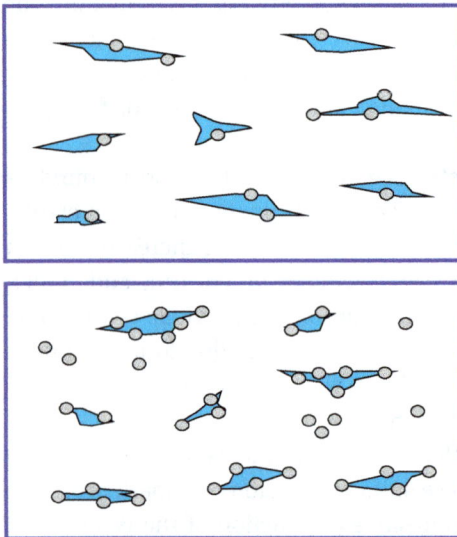

Figure 6.4: Schematic representation of the spark distribution in the initial spark phase (above: low number of sparks; below: increased number of sparks) [23].

after a certain amount of time are other areas of the sample hit by the sparks. Of course, particularly in the initial phase, this leads to strongly increased signal intensities for the elements involved in the inclusions and, thus, explains the spark profile shown in Figure 6.3b.

Increased field strengths in the boundary area between the inclusion and the matrix were named as an explanation for the effect. A physical explanation is possible with the band model of semiconductors (see Figure 6.5). The transition between inclusion and matrix is comparable to the situation in a diode. In the metallic matrix, the energy difference between the valence and the conduction band is very low or valence and conduction band are at the same level. Electrons can move freely in the conduction band. As insulators, inclusions have a larger band gap. Therefore, the electrons can only enter the conduction band with external energy input. A band shift occurs between metal and oxide resulting in a space charge. This leads to an increase in potential here making it possible for spark impacts to prefer this area [24].

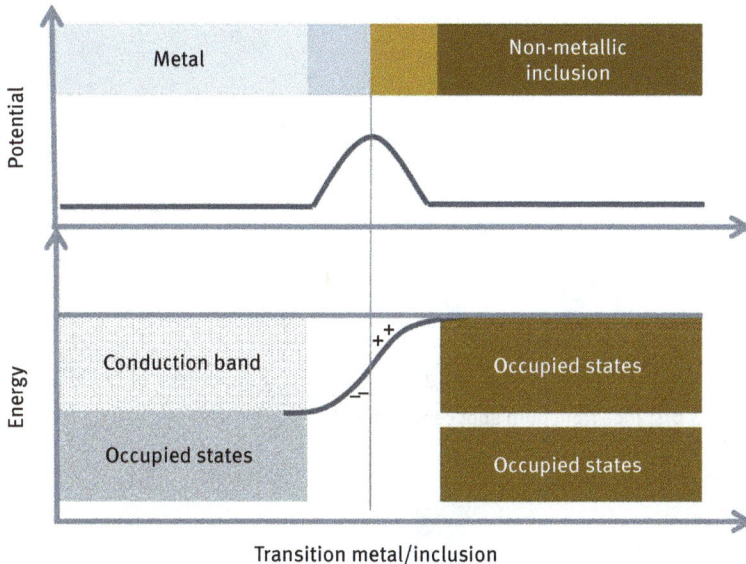

Figure 6.5: Band model for ablation processes for spark discharge (printed with the permission of D. Poerschke) [24].

K. Slickers called the initial phase of the spark, "Einfunken" (spark-in) [25, 26]. This is followed by a homogenization phase in which the sample surface is melted and homogenized by concentrated discharges. The existing inclusions are dispersed in the melt. At the end of the homogenization phase, the spark process enters a stationary phase and the signal intensities equalize. Material is ablated during this phase, so that deeper layers can be reached and new inclusions can be detected.

The crater from a single spark depends on the spark parameters selected and usually has a diameter between 15 and 20 μm. The term "spark parameter" is understood to be the course of the current strength produced by the excitation generator over the duration of the spark. The spark crater is a few μm deep. Figure 6.6 shows the cross section of a burn spot after 2,000 discharges with an effective depth of about 50 μm. The diameter of the burn spot is greatly dependent on the spark parameters set, the spark repetition frequency and the electrode shape.

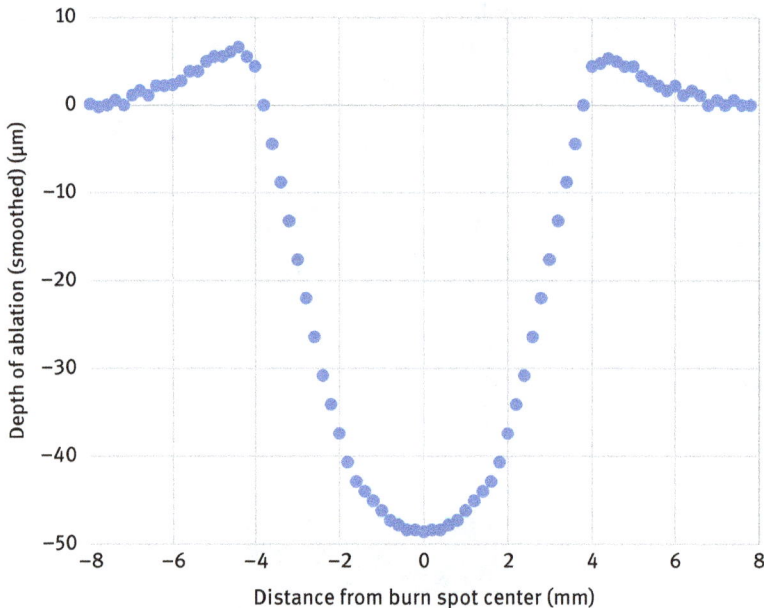

Figure 6.6: Cross section of a burn spot.

## 6.5 Measuring methods and evaluation

Until the mid-seventies of the last century, commercially available spectrometers offered few possibilities of switching multiple integration windows or recording the signal intensities of single spark discharges. Only improved spark generators, faster evaluation electronics and, particularly, powerful computer systems enabled time-resolved measurements to determine quantitative information about the inclusions in steels. Various measuring and evaluation methods are presented in the following together with an assessment of their results.

### 6.5.1 Peak integration method and related methods

Slickers and Gruber described the so-called "peak integration method" based on the phases of the spark process explained in Section 6.4 using the determination of oxidic aluminum in steel as an example [36]. The basis of this model is the assumption that the edges of the oxide particles are preferably hit at the beginning of the spark. The sample material is melted to a depth of 50 μm and inclusions are broken down and dispersed during the homogenization phase. Metallic and bound aluminum are captured alike in the stationary phase of the sparking process.

With regard to the characteristic parameters of the acid-soluble aluminum ($Al_{sl}$) obtained with the established reference method, the intensity ratio of the integrals of the pre-spark phase and the stationary phase was defined as the ratio of $Al_{sl}$ to the total aluminum ($Al_g$). The $Al_{sl}$ fraction can be determined using the normal calibration function $f(I)$ for $Al_g$ (eq. 6.1). The difference between $Al_g$ and $Al_{sl}$ gives the fraction of precipitated aluminum $Al_o$ (see Figure 6.7).

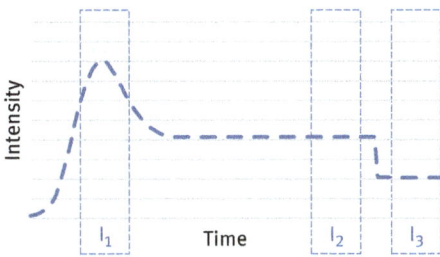

Figure 6.7: Diagram of the course of the spark.

$$Al_g = f(I_2); \quad Al_{sl} = \frac{I_2}{I_1} c(Al_g) \tag{6.1}$$

Based on this scheme, variants, which better suited the respective production processes, were created by different users [34, 38]. These approaches are still used today in many production companies. An additional integration phase, among other things, with a lower spark energy was introduced.

Steel producer 1 $$Al_g = f(I_3); \quad Al_{sl} = \frac{I_2}{I_1} * c(Al_g) \tag{6.2}$$

Steel producer 2 $$Al_g = f(I_3); \quad Al_{sl} = f\left(\frac{I_2}{I_1}\right) * c(Al_g) \tag{6.3}$$

Steel producer 3 $\quad$ $Al_g = f(I_2); \; Al_{sl} = f\left(\dfrac{I_2}{I_1}\right) * c(Al_g)$ $\hspace{2cm}$ (6.4)

As part of a European research project, Wittmann postulated that only metallic aluminum remained in the melted crater by the end of the sparking phase [27]. Therefore, for this model, the fraction $Al_g$ was determined by integration during the entire spark phase. This approach was ruled out with the single spark analysis described below, because elevated signals caused by inclusions are still recorded, even at the end of the sparking process.

The "peak integration method" is an empirical approach that can be used for many operational applications, but it has its limitations. Sample properties have a large influence on the quality of the results. These include not only the usual requirements such as homogeneity, freedom from surface defects, etc. but also differences in the sample origin (production stage) sampling probes and immersion depth during sampling.

## 6.5.2 Pulse distribution analysis

Pulse distribution analysis (PDA) was developed by Onodera among others at the Nippon Steel Corporation in the 1970s. The prerequisite for this method is that the intensities for the element-specific emissions can be recorded and evaluated for each spark [28, 29]. At that time, this was not possible with the commercially available spectrometers. The integral over the individual signals is used to determine the total concentration for each of the elements observed. The results of all spark discharges are divided into intensity classes represented as a histogram. For the evaluation, it is assumed that the emissions of the metallic components show a normal distribution, which shows an asymmetry in the direction of the higher signals due to the higher signals of the non-metallic inclusions (see Figure 6.8).

Various mathematical approaches for the deconvolution of the frequency distribution are described in the literature listed in the bibliography [30–32]. For samples without inclusions, a frequency distribution that can be described with a normal Gaussian distribution is obtained. For samples with inclusions, a normal distribution is calculated with respect to the signal class with the largest incidence. The difference between the integral of the normal distribution and the integral of the total frequency distribution gives the fraction of signals caused by inclusions. Both area fractions are proportional to the element shares present as metallic and inclusions. The area fractions can be converted into the respective portions using the normal calibration function determined with the total concentration of the analyte.

The characteristic parameters for evaluation of the degree of purity of the steel can be determined based on the results thus obtained. Wintjens and Falk used the

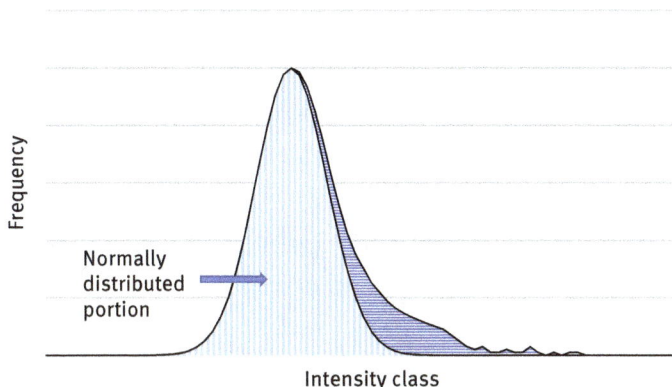

Figure 6.8: PDA frequency distribution.

ratio of the integrals of the normal distribution and the total frequency distribution for this [31]. According to eq. (6.5), the $K$ value for samples without inclusions approached zero; the value increases for materials with higher inclusion fractions.

$$K_{PDA} = 1 - \left( \frac{A_{Normal\_distribution}}{A_{Total}} \right)$$

(6.5)

Comparable approaches have also been described by other working groups [32, 33]. Here, it has also been shown that changes in the degree of purity correlate with operational events.

### 6.5.3 Single spark analysis

Today, modern spectrometer systems enable time resolved detection and storage of the radiation emitted by all elements of interest [35, 39, 37]. This makes it possible to determine dependencies between individual analytes within a very short time. For this purpose, the adjusted average of the intensities for $n$ spark discharges is determined. In the first step, the average of all signals and the associated standard deviation are calculated. In the next step, all values above the limit 3s are no longer considered when a new average is calculated. This procedure is repeated up to ten times. Measured values that are about 3s above the adjusted average are assigned to inclusions. The intensities for the elements Ca and Mg are plotted in chronological order using a production sample as an example in Figure 6.9. If signals occur above the 3s limit for only one of the elements, it can be assumed that the respective oxide is present. If these signals occur simultaneously for both elements, then there is a corresponding mixed oxide. Since the determination of the oxygen content is hardly possible, the decision as to the type of inclusion must be made based

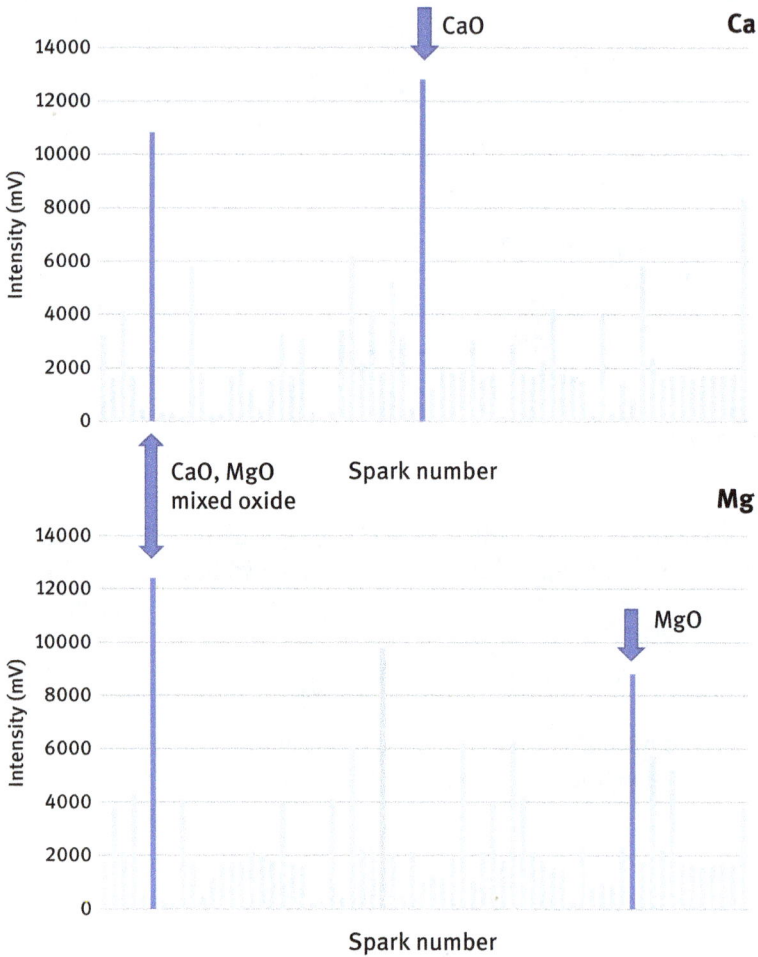

Figure 6.9: Correlation between elements with single spark analysis using an exogenous inclusion as an example [24].

on knowledge of the respective metallurgical process. For other binding partners, such as sulfur, the corresponding correlations to the detectable element concentrations of the single spark must be available.

The element concentrations can be calculated according to eq. (6.6).

$$c_{ie} = \frac{\sum I_{>3s} - n_{>3s} * \overline{I}_{\leq 3s}}{\sum I_{total}} * c_i \qquad (6.6)$$

$I_{>3s}$    Is the intensity of the single spark above the 3s limit

$n_{>3s}$    The number of signals above the 3s limit

$\bar{I}_{\leq 3s}$     The average of the signals $\leq 3s$

$I_{gesamt}$     The intensity of all single sparks

$c_i$     The total concentration of the analyte

$c_{ie}$     The concentration of the analyte $i$ in the inclusion type $e$

As already stated, determination of the oxygen content in the material steel is hardly possible using spark emission spectrometry. Assuming that the larger part of the inclusion compounds are oxides, it is, however, possible to calculate the theoretical oxygen content using eq. (6.7). In this case, the concentrations of the determined inclusion compounds of the elements i, j are converted into the respective oxygen content according to the stoichiometric composition, whereby i and j stand for element symbols.

$$c_{O\,cal} = c_{ei} * \frac{c_{Oi\,rel}}{100} + c_{ej} * \frac{c_{Oj\,rel}}{100} + \ldots \tag{6.7}$$

$c_{O\,cal}$     Is the calculated oxygen content

$c_{ei}, c_{ej}$     The fraction of the inclusion type i and j

$c_{Oi\,rel}, c_{Oj\,rel}$  Fraction of oxygen in inclusion type $i$ and $j$

This calculated oxygen content is merely a value for orientation and not comparable to the results established with the usual determination methods, such as carrier gas hot extraction, or with the measurement data from an solid electrolyte cell used in the context of process control. For in-house quality control, this value can be used to monitor the quality of the samples or sample taking.

## 6.5.4 Determination of the particle size

As seen from the information in Section 6.4, light emission excited by a spark discharge depends on the ratio between steel matrix and inclusion mass in the effective area. Assuming that there is only one inclusion in the recorded region, its mass can be calculated based on the concentration determined for the analyte in the inclusion compound. The approach given in eq. (6.8), can be solved for $m_e$ so that the mass of the inclusion can be calculated with eq. (6.9) [31].

$$c_{if} * m_f = c_{ie} * m_e + c_{ig} * (m_f - m_e) \tag{6.8}$$

$$m_e = \frac{m_f * (c_{if} - c_{ig})}{c_{ie} - c_{ig}} \tag{6.9}$$

$c_{if}$     Is the concentration of the analyte per spark discharge in m/m %

$m_f$     The mass of the sample excited per spark discharge

$c_{ig}$     The total concentration of the analyte in m/m%

$c_{ie}$     The concentration of the analyte in an inclusion

$m_e$     The recorded mass of the inclusion

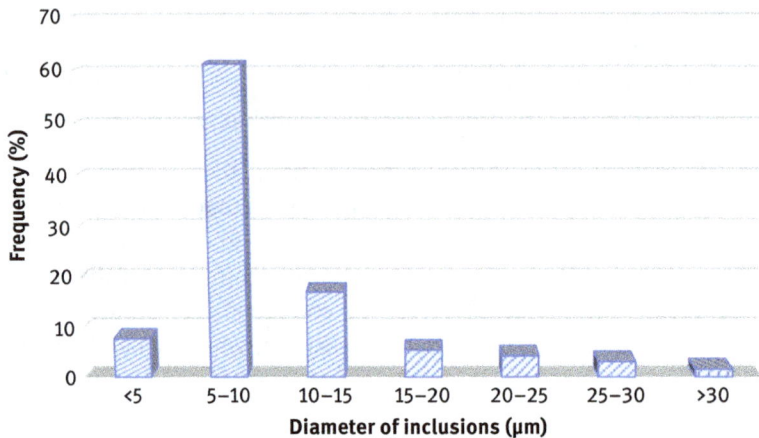

**Figure 6.10:** Size distribution.

Assuming that the inclusion is spherically shaped, its diameter is obtained using the specific gravity of the respective type of inclusion.

Calculation according to eq. (6.8) requires knowledge of the sample mass ablated by a single spark discharge. These values are typically between 60 and 100 ng for steels.

Figure 6.10 shows the size distribution for the $Al_2O_3$ inclusions detected with 2,000 spark discharges. The data for particles smaller than 5 μm should be assessed critically in this illustration as they can barely be detected with spark excitation. The mass that an inclusion must have so that its contribution to the emitted radiation is larger than the 3s limit can be calculated from the 3s criteria for distinguishing between inclusions and the metallic components. For an ablated sample mass of 60 ng, the minimum inclusion size for an $Al_2O_3$ particle is approximately 3 μm. If the lower excitation probability for $Al_2O_3$ is also taken into account, there is an even larger minimum inclusion size. Smaller inclusions can also be detected only for inclusions with alkaline earth elements due to the higher excitation probability. In addition, some of the types of inclusion presented in Table 6.1 frequently have diameters in the range of 1 μm and smaller. Several small inclusions in the effective area of a spark discharge cannot be distinguished from one larger inclusion. This is especially likely in segregation zones.

The size distribution of the inclusions determined in this way is not comparable with the size distribution determined using imaging reference methods such as light microscopy and scanning electron microscopy. In some cases, the deviations are around 50%. Even though the results currently have only a qualitative nature, determination of the particle size still represents additional information that may be included in the future in process control for assessment of the degree of purity in a melt.

### 6.5.5 Possibilities and limitations of inclusion analysis

Inclusion analysis using spark emission spectrometry is a powerful testing method that can effectively support the quality-oriented production of modern metallic materials. This method can be integrated into existing spectrometer systems without additional analytical effort; providing measurement data ideally suited as control indicators.

However, there are a number of limitations that must be considered when evaluating results. The particle size is one important aspect. As mentioned above, smaller inclusion particles cannot be detected. But also, physical effects, such as the splitting of a discharge into two or more subchannels, can falsify the results as, in this case, any eventually detected inclusion compounds do not receive the entire energy from the discharge. For system inherent reasons, the repeatability limits are worse for some analytes, for example, for the element nitrogen. Accordingly, the 3s limit for distinguishing between analytes dissolved in the matrix and the fraction contained in inclusions is significantly increased and the particles cannot be detected. Therefore, the relevant literature reports almost exclusively about the identification of oxides in metallic matrices. Inclusions such as $TiC_xS_x$ or $TiC_xN_x$, cannot be quantified at all with spark emission spectrometry due to their size (<1 μm), so that imaging methods and isolation procedures are indispensable here.

# Bibliography

[1]   Flock J. Geyer J, Sommer D. Elektrolytische Isolierung von Gefügebestandteilen in Stählen. CLB Chemie in Labor und Betrieb 1998, 49, 91–95.
[2]   Pappert E. Studien zur Bestimmung der Bindungsformen von Al, Ti, Zr und Zn in Feststoffen mittels Röntgenspektrometrie für Anwendungen in der Stahlindustrie. Dissertation, Universität Dortmund, 1998.
[3]   Philbrock WO. Oxygen reactions with liquid steel. Int. Met. Rev. 1977, 22, 187–201.
[4]   Gittins A. Effect of oxygen on hot workability of steel. Int. Met. Rev. 1977, 22, 213–221.
[5]   Dickens P, König P. Zur Ermittlung des Rückstandsgehaltes von Stählen durch unmittelbare Chlorierung. Archiv für das Eisenhüttenwesen 1968, 39, 453–456.
[6]   Bohnstedt U. Brom-Methanol als Hilfsmittel bei der Oxidisolierung aus Stählen und verwandten Legierungen. Fresenius Z. Anal. Chem. 1964, 199, 109–117.
[7]   Beeghly FF. Determination of combined Nitrogen. A rapid method. Analytical Chemistry 1949, 21, 1513–1519.
[8]   Koch W. Metallkundliche Analyse. Düsseldorf, Verlag Stahleisen, 1965.
[9]   Mittelstädt H, Müller G, Nazzikol C. Handbuch für das Eisenhüttenlaboratorium Band 1, Teil 2 – Methoden zur Analyse von Einschlüssen in Stählen, 2nd edition. Düsseldorf, Verlag StahlEisen, 2016.
[10]  Handbuch für das Eisenhüttenlaboratorium, Band 2, Teil 2, Test zur Ermittlung des säurelöslichen Aluminiums, Calciums und Magnesiums in niedriglegierten und unlegierten Stählen. Düsseldorf, Verlag StahlEisen, 1994.

[11]  Flock J, Ohls K. On-line-Stahlanalyse mit der ICP-Emissionsspektrometrie nach elektrolytischem Lösen. Fresenius Z. Anal. Chem. 1988, 326, 408–412.
[12]  Braas S, Lostak T, Pappert E, Flock J, Schram J. On-line electrolytic dissolution for soluble aluminium determination in high-silicon steel/electrical sheet by ED-FIA-ICP-OES. Talanta 2018, 189, 489–493.
[13]  ISO 4967:2016(E). Steel – Determination of content of nonmetallic inclusions – Micrographic method using standard diagrams. Geneva, ISO, 2016.
[14]  DIN EN 10247:2017-09. Metallographische Prüfung des Gehaltes nichtmetallischer Einschlüsse in Stählen mit Bildreihen. Berlin, Beuth Verlag, 2017.
[15]  ASTM E 45:2013. Steel – determination of content of non-metallic inclusions – Micrographic method with standard diagrams. IHS, Englewood, 2013.
[16]  JIS G 0555:2013. Microscopic test method for the non-metallic inclusions in steel. Japanese Standards Association, Tokyo, 2013.
[17]  Mittelstädt H, Müller G, Nell U. Lagerstabile Standardproben. German patent DE 10 2008 010 176 B3, granted 12.11.2009.
[18]  Winter H. Zur quantitativen Spektralanalyse: Untersuchung über den Verdampfungsvorgang an metallischen Elektroden unter der Einwirkung des Hochspannungsfunkens. Z. Metallkunde 1937, 29, 341–351.
[19]  Laqua K, Hagenah W. Spektrochemische Analyse mit zeitaufgelösten Spektren von Funkenentladungen. Spectrochimica Acta 1962, 18, 183–199.
[20]  Höller P. Zustandsänderung der Probe im Abfunkbereich bei der optischem Emissionsspektralanalyse von Stahl. Spectrochimica Acta 1967, 23B, 1–14.
[21]  Herberg G, Höller P, Köster-Pflugmacher A. Abbauvorgänge und Interelementeffekte bei der optischen Emissionsspektralanalyse von Stahl – I. Spectrochimica Acta 1967, 23B, 101–116.
[22]  Herberg G, Höller, P, Köster-Pflugmacher A. Abbauvorgänge und Interelementeffekte bei der optischen Emissionsspektralanalyse von Stahl – II. Spectrochimica Acta 1968, 23B, 363–371.
[23]  Koch W, Dittmann J. Picard K. Abhängigkeit der spektrometrischen Schwefelbestimmung von der Anisotropie der verformten Sulfide im Stahlgefüge. Fres. Z. Anal. Chem. 1967, 225, 196–203.
[24]  Poerschke D. Grundlagen und Anwendung der Funken-OES-PDA zur Stahlreinheitsbestimmung. Steinfurt, Anwendertreffen Röntgenfluoreszenz- und Funkenemissionsspektrometrie, 2018.
[25]  Slickers K, Gruber J. Spektrometrische Bestimmung von metallischen und nichtmetallischen Elementanteilen in Stahl. Stahl u. Eisen 1984, 104, 293–298.
[26]  Slickers K, Iten J. Abbauvorgänge bei der spektrometrischen Metallanalyse mit Funkenentladungen in Argon. Spectrochimica Acta 1987, 42B, 791–805.
[27]  Wittmann A, Willay G. Determination by OES of soluble and insoluble Aluminium in steel. ECSC 7210-GA-321.
[28]  Onodera M, Nishizaka K, Saeki M, Sakata T. A new method of metallographic analysis of a small quantity of Aluminum present in steel using a vacuum emission spectrometer. Nippon Steel Techn. Rep. 1977, 9, 73–77.
[29]  Imamura N, Fukui I. Pulse Distribution Analysis method for spark emission spectrochemical analysis. Cambridge, Colloq. Spec. Int. 1979, XXI (paper 225).
[30]  Mizukami K, Ohashi W, Tsuji M, Sugiyama M, Mizuno K. Identification of Acid-soluble Components and Acid-insoluble Inclusions in Spark OES Pulse-height Distribution Analysis. Nippon Steel Techn. Rep. 2011, 100, 63–71.
[31]  Falk H, Wintjens P. Charakterisierung von Nichtmetallen in Metallen mittels Einzelfunken-OES. Münster, Proceedings Nichtmetalle in Metallen, 1998, 189–192.

[32]  Weert H, Senk D. Online-Bestimmung des Stahl-Reinheitsgrads mittels Einzelfunkenspektrometrie (PDA-OES) im Rahmen der Pfannenmetallurgie. Aachen, Proceedings 27. Aachener Stahlkolloquium, 1998, 189–192.

[33]  Fronk S, Ho HT, Senk D. Aktuelle Methoden für die Bestimmung von Seigerungsphänomenen und dem Reinheitsgrad in Stahlproben. Aachen, Proceedings 30. Aachener Stahlkolloquium, 2015, 172–184.

[34]  Bewerunge J, Flock J, Marotz RK, Thiemann E. Möglichkeiten der Bestimmung von "säurelöslichem" und "säureunlöslichem" Aluminium im Stahl. Steel research 1988, 59, 239–245.

[35]  Kuß HM, Müller G, Flock J, Lüngen S, Mittelstädt H, Thurmann U. Vergleich von funkenemissionsspektrometrischen Methoden zur Bestimmung von Einschlüssen in der Stahlmatrix. Proceedings Nichtmetalle in Metallen, Münster, 2000, 29–41.

[36]  Slickers K. Die automatische Atom-Emissions-Spektralanalyse. Giessen, Buchvertrieb K. A. Slickers, 1992.

[37]  Pissenberger A, Illie S, Mayr M, Schuller M. Charakterisierung von Einschlüssen und Seigerungen in Brammen- und Blechproben. stahl u. eisen 2008, 128, 63–67.

[38]  Güldner D. Möglichkeiten der emissionsspektrometrischen Bestimmung von Oxiden im Stahl. Neue Hütte 1991, 36, 70–78.

[39]  Kuss HM, Lüngen S, Müller G, Thurmann U. Comparison of spark OES methods for analysis of inclusions in iron base matters. Anal. Bioanal. Chem. 2002, 374, 1242–1249.

# 7 Mobile spectrometers

Chapter 7 describes the state of development of modern mobile spectrometers. This chapter should be seen as an extension of Chapter 3. For Chapter 7 it should not go unmentioned that the authors describe this state, but do not claim the intellectual authorship for the equipment and procedures presented here. If the authors themselves have made significant contributions to the state of technology, references to their patent applications and publications are indicated in the relevant locations in the chapter.

Mobile spectrometers are designed for measurements directly on the workpiece. They are equipped with basic components similar to the laboratory instruments. However, there are considerable differences in the design of mobile spectrometers because the requirements on mobile spectrometers differ from those on laboratory spectrometers:

– Mobile spectrometers must be as *small and light* as possible so that they are easily brought to the sample. For example: If 100% testing is required in material storage, then the spectrometer must be moved from one storage shelf to another.
– Frequently, outdoor testing is required, so systems must operate over a *wide temperature range*. The resistance to ambient humidity also plays a larger role than for laboratory instruments.
– After transport, the instruments must be ready for use as quickly as possible (catchword: *readiness on the spot*). Delay times for tempering the optical system should be as short as possible. The optical systems for the measurement of spectral lines in the vacuum UV are flushed with argon, since the ambient air is not transparent for these wavelengths. The time lost for flushing the optic after storage or transport must also be minimized.
– Mobile spectrometers must enable not only mains power but also *battery operation*. The battery capacity must be designed so that it is possible to work for several hours without recharging or changing the battery.
– The *work pieces* to be tested are often *bulky* and cannot be placed directly on a spark stand. For this reason, mobile spectrometers (see Figure 7.1) usually have a pistol-shaped spark probe, the front side of which consists of a miniature spark stand. Thus, it is possible to hold the probe on a suitable spot on the work piece for measurement. The radiation generated in the *mini-spark stand* during the measurement is either passed with a direct light path into an optical system also built into the probe or it is transmitted through a fiber optic cable into a larger optic in the mobile spectrometer's main housing. Spark probe and main

https://doi.org/10.1515/9783110529692-007

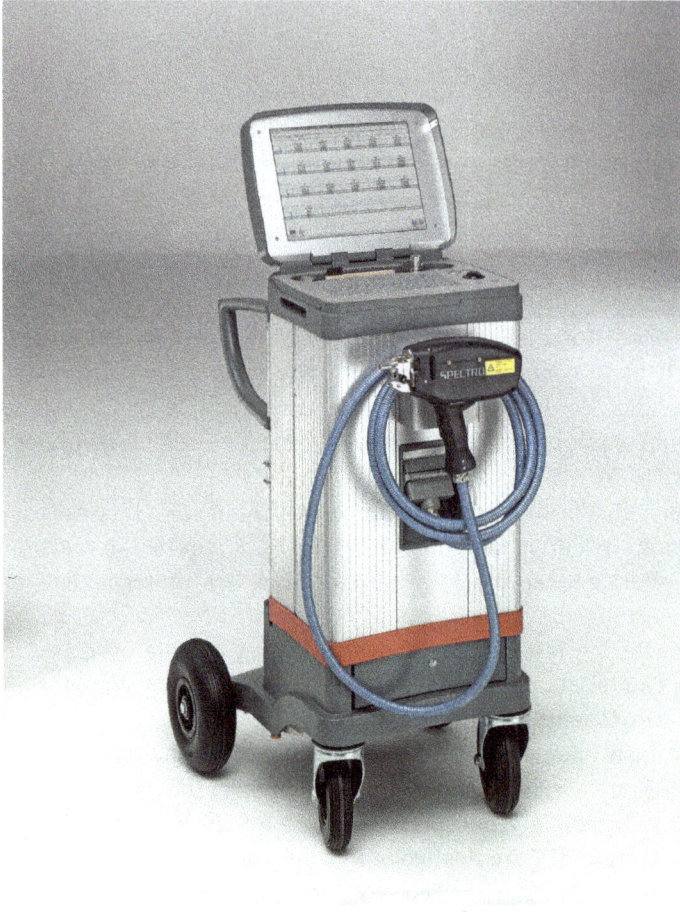

**Figure 7.1:** Modern mobile spectrometer, printed with the friendly permission of SPECTRO Analytical Instruments GmbH, Boschstr. 10, 47533 Kleve, Germany.

unit are, then, connected by a hose that is usually several meters long, which contains the signal cables and the supply lines between the excitation generator and the spark stand in addition to the fiber optics. The combination of optics built into the probe and optics in the main housing coupled via fiber optics is being built, because fiber optics are not transparent for the radiation of the spectral lines in the vacuum UV range.

– The software is usually not limited to simple analysis. *Special test procedures* for rapid and efficient sorting are specified. In addition, it is, in most cases, possible to analyze for compliance with target specifications and to identify unknown materials.

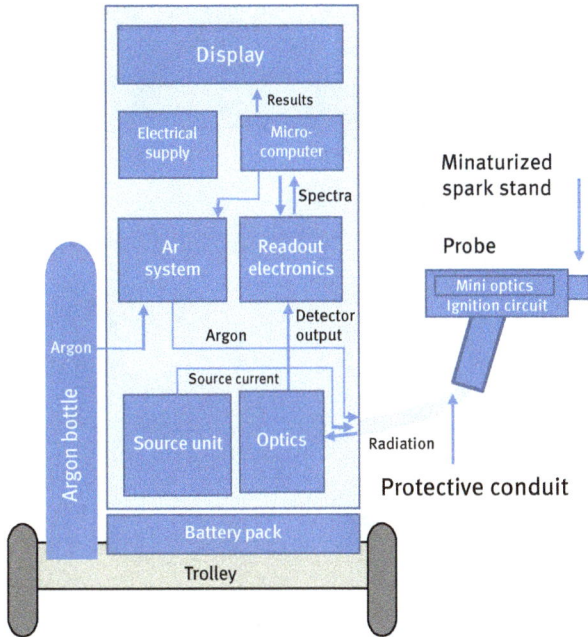

Figure 7.2 is a schematic diagram of the construction of mobile spectrometers consisting of a main unit connected via hose to the testing probe. There are also hand-held instruments for which either all components are housed in the measuring probe (see Figure 7.3a) or for which the components are distributed between the measuring probe and a small shoulder bag (Figure 7.3b).

The solution with the shoulder bag has the advantage that the measuring probe can be made lighter and more manageable, as the batteries required for the power supply have a not inconsiderable weight.

The design features of the mobile spectrometer components are discussed in detail in the following sections.

# 7.1 Excitation generators in mobile spectrometers

Mobile spectrometers usually offer the possibility of choosing to work with arc excitation in air or with spark excitation in an argon atmosphere. The circuits presented in Sections 3.1.1 and 3.1.2 are combined in one component, whereby several circuit components are used for both arc and spark operation.

The hardware costs for a combined arc/spark generator are not much higher than those for a generator with only one function.

(a)

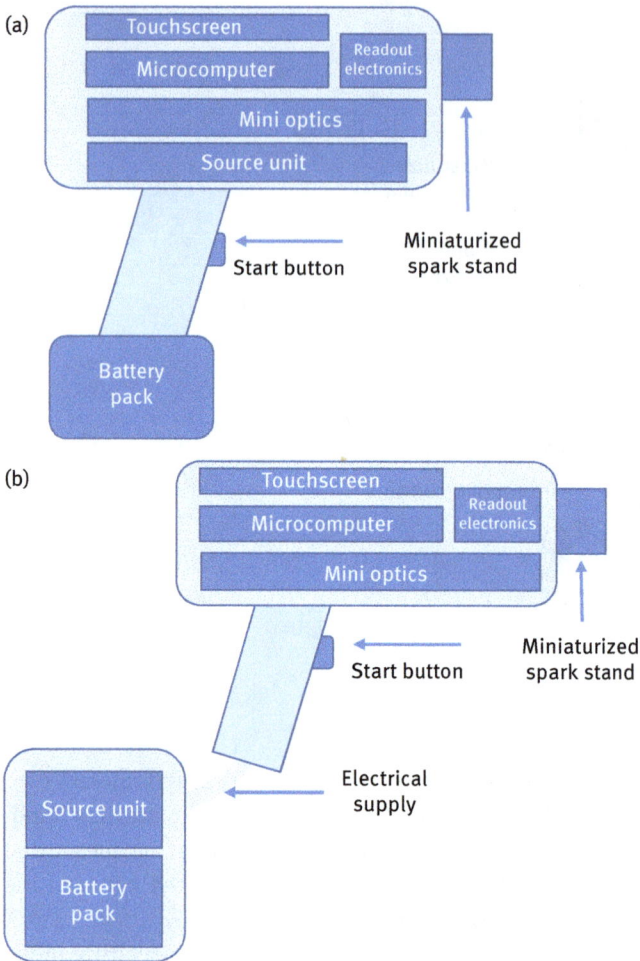

(b)

Figure 7.3: Construction of a handheld spectrometer.

## 7.1.1 Features of the ignition

The ignition unit is one such partial circuit that is used equally for arc and spark. It is usually built into the measuring probe and must, for ergonomic reasons, be constructed as small and light as possible. For 100% testing of larger lots, the user must often place the probe on the test pieces and hold it steadily for each measurement. Testing one piece takes only a few seconds, but checking a large lot may last for several hours.

Building the ignition into the main unit instead of into the probe is unfavorable due to reasons of electromagnetic compatibility. It is the job of the ignition electronics to make the gas between the tip of the electrode and the sample surface, the so-called gap, conductive with low-resistance. To do this, a high voltage impulse is generated that increases to the breakdown voltage. The breakdown voltage reaches values up to 12 kilovolts. A capacitor of several 100 pF is wired in parallel to the gap. After the breakthrough, it discharges. This capacitor is necessary to make the gap so low-resistant that the medium voltage for the arc or spark can then be introduced. The discharge takes place within a time of typically 100 ns. With this discharge time, a capacity of 200 pF and a breakdown voltage if 10 kV, the average current for the ignition phase is around 20 A. The peak current is significantly higher. Pulses with high current and voltage changes generate a broadband high-frequency spectrum, which can lead to radiofrequency emissions. Shielding measures are required to keep the RF-emissions within the permitted limits. In Europe, the basic requirements for permitted emissions are regulated in the directive 2014/30/EU on electromagnetic compatibility [1]. Compliance is required for the CE Declaration, which is in turn necessary for placing instruments on the market in Europe. The so-called "notified bodies" can, in exceptional cases, authorize violation of limiting values. However, this is only granted when all possibilities for the state of technology have been exhausted.

## 7.1.2 Arc generators for mobile spectrometers

Arc excitation sources for mobile spectrometers generally use the switching regulator principle described in Section 3.2.1.2, as energy efficiency is particularly important for battery operation. Besides, this design can be realized in the smallest space. Mostly arc currents between one and three amps are used. Electrode erosion and contamination are lower at low currents than at high ones. However, a minimum current, which is usually between one and two amps, is required for a stable arc. The value of the minimum current depends on the material to be analyzed. A current of 1.2 A is sufficient, for example, for tin bronzes, while aluminum bronzes require at least 1.8 A. High currents of three amps are needed only in special cases, for example, when carbon is to be determined using the CN molecular band. Here, the high arc current favors formation of the cyanide radical.

## 7.1.3 Spark generators for mobile spectrometers

The spark generators constructed for mobile spectrometers are not very different from those in spark laboratory instruments. Again, in modern instruments, the discharge energy curves are usually formed by electronically switched inductances.

However, there are several differences:
- Laboratory spectrometers use spark parameters with high current strengths for the excitation of spectral lines in the deep vacuum UV. Since these lines are not measured with mobile spectrometers, the spark generators do not need to be designed for such high currents. Inductance and resistance of the meters-long probe lines also complicate the realization of high currents.
- While the spark stand ground is galvanically connected to protective earth (PE) in laboratory instruments, the ground for the measuring probe spark stand is isolated from PE. The reason for this lies in the different handling: With the laboratory spectrometer, the sample is placed onto the spark stand, the sample clamp fixed and the measurement started. It is ensured that the user only has contact with metallic surfaces (spark stand, sample, sample holder) that are safely connected to PE.

   With mobile spectrometers, in contrast, it sometimes happens that the test piece is held with one hand and the measuring probe with the other. Although the operation manuals generally forbid this form of handling, this rule is often ignored in practice. Now, it may happen that the sample briefly loses contact with the spark stand during the sparking procedure but the distance between the tip of the electrode and the sample is so small that spark-over occurs. If the generator ground was connected to PE and the tester had bodily contact with earth, a body current would flow. If the excitation generator is completely galvanically isolated from PE, this case cannot happen.

   The problem described generally concerns more comfort than safety. It has been endeavored to design the generators so that they are not dangerous in cases like that described above. In Europe, the *DIN-EN 61010 – Sicherheitsbestimmungen für elektrische Mess-, Steuer-, Regel- und Laborgeräte* [2] (Safety requirements for electronic measuring, control and laboratory devices) regulates, which body currents must be considered as dangerous under which conditions. In order to guarantee the above-mentioned isolation requirements for the spark generator with respect to protective earth – even in the event of a fault – mobile spectrometer spark generators frequently measure the isolation resistance. If the isolation resistance is not big enough, sparking is inhibited.

## 7.1.4 Laser sources for handheld instruments

In recent years, handheld laser instruments have been brought onto the market by several manufacturers. These handheld instruments unite lasers, optical systems with detectors, readout electronics and the display in the probe. The power supply is provided by rechargeable batteries, which are usually fastened to the handle and are easy to change. The compact design forces miniaturization of all components, which is associated with compromises in terms of analytical performance. Section 7.4.2 discusses what this means for the optical system. The limited space also narrows the possibilities for the laser.

However, useful compromises can be found:
- Solid state lasers, for example, based on Nd:YAG crystals are frequently used.
- Frequency multiplication is neither common nor useful. Nd:YAG crystal lasers use pulses with the fundamental wavelength 1,064 nm.
- In most cases, systems use low-energy pulses. Plasma ignition and material ablation are already possible with pulse energies beginning at about 20 µJ. The typical pulse duration is on the order of some ns. However, there is at least one manufacturer who offers handheld units that are able to deliver pulses in the mJ range.
- In order to obtain sufficiently high signal intensities for systems with low energy pulses, repetition rates in the kilohertz range are common. The use of a large number of low individual energy pulses has an additional advantage: Pulses with high energy in the mJ range, produce peak temperatures on the sample surface that result in high thermal radiation that overlaps the usable signal. In addition, a continuum caused by electron recombination is emitted near the sample surface (see Section 2.5.2). The detection sensitivity is significantly worsened if the background radiation is present. If pulses with higher energy are used, either the sample surface must be masked, or detectors that enable blocking of the radiation at the beginning of the laser pulse for the duration of several 100 ns must be used. Both solutions are complex, which is why avoidance of thermal radiation is advantageous.
- Laser diodes present themselves as the pump source: In contrast to gas discharge lamps, they enable a pulse rate into the kilohertz range and can be implemented in a compact design. For diode pumped crystal lasers the abbreviation DPSS is common.
- Metals have a grain structure even though they appear to be homogenous. It is not enough to observe a spatially limited area of a few micrometers diameter when a statement as to the average contents is to be made. Therefore, the spot at which the laser pulse hits the metal surface must be varied. This can be done, for example, with deflection mirrors. Hereby, there can be the need for adjusting the focus of the laser beam. It should maintain the same distance to the sample surface when the spot of ablation is moved, so that the individual pulses ablate and excite the material with conditions that are as equal as possible.

In extreme cases, defocusing can cause the energy density to be too low for plasma ignition. This pulse cannot, then, contribute to the usable signal. Figure 7.4 shows the trace of laser craters that was generated during a measurement with shifting of the beam on the surface of a steel sample. This chain of craters has a length of only about three millimeters.

Figure 7.4: Chain of laser craters on the surface of a steel sample.

–   In most cases, the measurement is conducted in air. For measurements in an argon atmosphere, the results would be better in respect to detection sensitivity and reproducibility, but it would be necessary to carry along a bottle of compressed gas, whereby the manageability of the whole system would suffer. However, there are handheld systems with very small argon cylinders that are integrated in the probe.

Handheld laser devices have conquered the aluminum scrap sorting niche. There are three reasons for this:
–   The competing methods have weaknesses here: The determination of light elements with *handheld X-ray instruments* is difficult. Magnesium, silicon, lithium and titanium are, however, very important in aluminum base. Although silicon and titanium can be determined with handheld XRF instruments, the measuring times must be rather long. Magnesium is one of the most important alloying elements. Even with contents in the ppm range, lithium alters the grain structure of aluminum. However, there are alloys that contain several percent lithium. An entire heat can be ruined if scraps of this material get into the melt unintentionally. Despite long measuring times, the determination of magnesium is difficult and that of lithium hardly possible with handheld XRF instruments. This usually prevents the use of handheld X-ray devices for this application. *Arc instruments* are also unsuited for aluminum base. The reproducibility is in most cases not good enough. *Spark units* are the closest competitors to laser. The reproducibility, achievable accuracy and detection limits are very good and, in most cases, clearly superior to those with laser. However, the sample must be ground and the measuring times are comparatively long. The effort required for testing might be then, in regard to rapid presorting, too high.

–  The spectrum for aluminum base materials has few lines compared with that from, for example, iron base materials. In the wavelength atlas from Saidel and Prokofiew [3], only 116 aluminum lines are listed; the number of iron lines listed there is 3144. This reduces the problem of having to separate an analytical line from an adjacent line of the base metal, which is beneficial considering the limited resolution of the optical system in handheld instruments (see Section 7.4.2).

Aluminum wrought alloys are usually labelled with a four-digit number, whereby the first number provides information about the main alloying element (see Table 7.1). There is a similar system for aluminum casting alloys (Table 7.2). These are labelled with a three-digit number. For scrap sorting, assignment to the correct group according to Table 7.1 or 7.2 is often sufficient. Such sorting makes no special demands on detection sensitivity and accuracy. Distinctions within the main groups are also frequently possible, but rarely necessary in the recycling industry. Details about the classification of wrought and casting alloys can be found in the literature, for example, in the *Aluminium-Taschenbuch* [4] (Aluminum Handbook). The designations and compositions of wrought alloys are standardized in DIN EN 573-3 [5]; the designations for casting alloys in DIN EN 1780-1 [6].

Table 7.1: Classification of aluminum wrought alloys.

| Group | Main alloying element(s) | Group | Main alloying element(s) |
|---|---|---|---|
| 1XXX | Pure aluminum with Al > 99% | 2XXX | Copper |
| 3XXX | Manganese | 4XXX | Silicon |
| 5XXX | Magnesium | 6XXX | Magnesium and silicon |
| 7XXX | Zinc | 8XXX | Miscellaneous (e.g., lithium) |

Table 7.2: Classification of aluminum casting alloys.

| Group | Main alloying element(s) | Group | Main alloying element(s) |
|---|---|---|---|
| 1XX | Pure aluminum | 2XX | Copper |
| 3XX | Silicon with copper and/or magnesium | 4XX | Silicon |
| 5XX | Magnesium | 7XX | Magnesium, Zinc |
| 8XX | Tin | | |

Handheld laser instruments are also offered for other metal bases, like iron, nickel and copper. The spectra here are more line rich. For Fe-base, measurement in an argon atmosphere is desirable, because otherwise important elements such as phosphorous and sulfur cannot be determined as the main lines for detection are in the vacuum UV range and are absorbed by the ambient air.

## 7.2 Spark probes

Measuring probe and hose are the most distinctive distinguishing features between laboratory and mobile spectrometers. As already described above, the probe contains not only a miniaturized spark stand but also often further components in addition to the above-mentioned ignition:

- The *start button* is pressed to trigger the measurement. For safety reasons, the sparking process is interrupted when the button is released.
- There is usually a second button, that is frequently referred to as the *reset button*. This button is particularly important when working with the sorting mode. In this case, larger lots of workpieces are checked for identification. The testing procedure is blocked when a workpiece does not conform. This prevents the deviation from being mistakenly ignored. Pressing the reset button starts the testing procedure once again.

  In the "analyze," "material control" and "grade identification" testing modes, the button can be used for other purposes, for example, for averaging in the analysis mode or to search for a fitting grade if deviations are found during material control.
- A usually green *indicator light* is lit after agreement with a reference material has be determined during sorting. A red light is lit for deviations. This way, the user obtains the essential result without actually having to look at the display on the main unit, which may be several meters away.
- In some mobile spectrometers, especially handheld devices, a *display* is integrated into the probe instead of the indicator lights. This has the advantage that, in addition to a simple yes/no statement, it is also possible to display elements that deviate from the target values, analytical results, fitting grades, etc. directly on the probe. However, color displays are usually less readable in strong sunlight. In addition, the display increases the weight of the probe, which is quite relevant for applications where the probe must be held for hours.
- Part of the *argon supply system* is normally located in the probe. Argon is used to flush the spark stand during the measurement. A small argon flow is also maintained in the standby mode, especially to keep the spark stand free of water vapor, which cannot be removed with a short pre-flush even though the spark stand has a small volume. If the probe contains an optic for the wavelength range below 185 nm, this must also be constantly flushed with a low argon flow to maintain transparency in the vacuum ultraviolet range.

  Larger mobile spectrometers have optics for the wavelengths above 185 nm in the main unit. Coupling takes place via fiber optic cable. The transmission capabilities of fiber optics change with intense UV irradiation. This effect is referred to as "solarization." The UV radiation is extremely intense during the pre-burn time that precedes the measurement itself. A so-called shutter blocks the light path to the fiber optic during the pre-burn to protect the fiber optic.

The shutter usually consists of a miniature pneumatic cylinder operated with argon. Electromechanical components have proven themselves to be less effective here.

A separate gas line could be routed through the probe hose for each of these functions. However, that would increase its diameter and the weight on the probe-side of the hose, which is, of course, carried with the probe. The manageability of the probe would also be affected. A further drawback is the fact that, for reasons of easy repair, pluggability of the probe hose to the main unit is desired. Multiple argon connections would make this plug large and difficult to realize. For these reasons, there is only a reduction of the input pressure in the main unit together with any filtration of the argon. A single argon line is led to the probe. A block with needle valves or fixed shutters and miniature magnetic valves is located there; it allows switching of the above-mentioned gas flows.

- Magnetic valves, start and reset buttons, indicator lights and probe display could be controlled from the main unit. However, this would mean that the probe plug would have to have a large number of contacts and multicore cables would also have to be routed through the probe hose. It is more favorable to delegate the switching of signals to a separate microprocessor, the so-called *probe controller*. For safety reasons, activation of the excitation is conducted over a separate line to remain independent from probe controller failures.
- Commercially available fiber optics are not usable for wavelengths below 185 nm because they are not sufficiently transparent for short wave radiation. Some elements, such as phosphorous and sulfur, only have usable detection lines below 185 nm. The main determination line for the most important element in iron materials, carbon, lies with a wavelength of 193.0 nm above the 185 nm limit, but the requirements for accuracy are extremely high. The fiber optic gets in the way here. It clearly diminishes this wavelength and also changes its transmissivity under UV irradiation. For these reasons, small *optical systems* covering only the wavelength range between 170 nm to 200 nm are built into the probe.

The spark stand on the front of the spark probe forms the interface to the sample. It can be equipped for spark or arc operation by plugging different adapters. Arc and spark adapters for flat samples were already shown in Figure 3.44.

During spark operation, the sample area to be measured must form a gastight closure between the inner chamber of the spark stand and the external atmosphere. Small openings are favorable for a tight seal but make it more difficult to clean the tip of the electrode. Openings of about 8 mm are a practical compromise. If it is necessary to analyze small parts that do not have a sufficiently large flat surface, a removable disk made of a non-conductive material can be placed into the top of the spark stand. The diameter of the opening can be reduced to four millimeters. A non-conductive material must be used because otherwise the spark would strike the edge of the opening and the spark stand material would be ablated. Boron nitride and mica have proven to be effective as materials for the non-conductive disks. If the small parts to be analyzed are shaped so that no reliable galvanic contact to the spark stand can be ensured, this can be done using a clamp (see Figure 7.5a).

Tungsten electrodes with four or six millimeter diameters are commonly used for spark operation; occasionally, spectrally pure silver is used for the electrode material. The electrodes are usually sharpened to an angle between 90° and 140°. The gap, with two to three millimeters, is usually somewhat smaller than that for laboratory instruments. This makes it easier to achieve small burn spots but makes the correct shutter setting slightly more critical (see Section 3.3.1). The electrode remains relatively cold when sparking under argon and undergoes virtually no wear. Lifetimes of several 100,000 measurements are possible for tungsten electrodes.

In contrast, such lifetimes are not achieved in arc mode; the electrode tips undergo substantial erosion. Silver or copper electrodes are used. About 1.5 mm is common for the gap width. Electrode diameters of 6 mm are usually chosen. Thinner designs have not proven to be effective, as there is considerable heat development on the counter electrode for arc under air. If this heat is not sufficiently dissipated, burn-off or even deformation may occur. Arc electrodes are also sharpened. Occasionally, angles beginning from 60° can be found for stabilization of the arc position, but usually the same blunt electrodes as those for spark are used because sharp angles require more frequent readjustment of the electrode gap. For low alloy steel, it is necessary to change the electrode or at least readjust the gap after several hundred measurements to compensate for the burn-off. Tungsten cannot be used as an electrode material for an arc burning in air, as this material oxidizes on the tip of the electrode and the tip burns down even faster than silver or copper quickly increasing the gap. In addition, the resulting tungsten oxide is toxic.

As in spark operation, it is also common to work with different adapters when using arc depending on the testing application and nature of the samples. A tube-shaped attachment is used for flat samples. On one hand, it shields the eyes of the user from harmful UV radiation that occurs during arc excitation. On the other hand, it enables simple cleaning of the electrode. The adapter is also suited to the determination of carbon when flushing with air from which $CO_2$ has been removed. The same airtightness as with a spark adapter is not achieved but this is not necessary: While in spark mode, argon contamination leads to poor burn spots and adverse effects on the repeatability, the arc adapter must only ensure a rough separation of the purified air from the ambient atmosphere.

In addition to the standard adapter, the following adapter types are used for arc operation:

(a)                           Adapters for small parts

(b)                        Adapters for tubes and wires

(c)                               Wire adapters

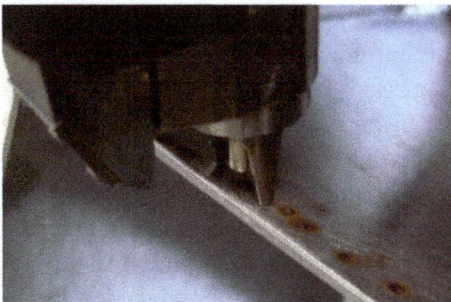

(d)        Adapter for fillet weldings

**Figure 7.5:** Spark and arc adapters for special sample shapes, printed with the friendly permission of SPECTRO Analytical Instruments GmbH, Boschstr. 10, 47533 Kleve, Germany.

- Wires can be inserted into wire adapters so that they are centered with the correct gap over the tip of the electrode (see Figure 7.5b and c).
- Tube adapters do the same for tubes and pipes. They have a circular recess perpendicular to the electrode axis. The diameter of this recess forms the upper limit of the diameter of the tubes that can be measured with this adapter. Tube adapters with recesses with different radii are frequently available (see Figure 7.5b).
- Fillet weld adapters are used to accurately determine the composition of welds (see Figure 7.5d).

## 7.3 Probe hoses

Except for handheld devices, mobile spectrometers always have a probe hose containing all electrical, pneumatic and optical connections between the main unit and the measuring probe.

Taking the design described as advantageous in the previous section into account, the probe hose contains the following components:
- It carries the *fiber optic cable* that transports the radiation from the probe spark stand to the main unit. It makes sense to provide the fiber optic cable with additional protection by surrounding it with a plastic tube inside the probe hose. This helps to prevent damage to the sensitive quartz fiber.
- In spark instruments, the *argon lines* supply the argon block located in the probe, which in turn uses the argon for the shutter as well as to flush the spark stand and mini-optic. For arc instruments that are designed to analyze carbon under air that has been freed from carbon dioxide, such a tube is used for air transport. Probes for arc and spark operation usually use the same lines for air and argon. PTFE tubes are used.
- The *electrical supply lines* for spark or arc currents must have a sufficient cross section and comply with the insulation requirements specified in DIN-EN 61010 [2]. The same is true for the supply for the ignition electronics.
- As explained in the previous section, the probe controller manages the gas flows as well as the display elements and informs the instrument computer as to when the start and reset buttons are pressed. Probe controller and main instrument are connected via a *serial bus*. If the probe has a mini-optic for the vacuum UV range, the measured values can be transmitted by the probe controller. Alternatively, the detector connections can be directly transferred to the main unit after an impedance conversion or the transmission can take place with a separate microcontroller. In these cases, additional *signal lines* are required; they must meet the same requirements as those for connection to the probe controller.

–   *Additional electrical lines* may be required to power detectors, analog electronics and microcontrollers as well as for deactivation of the excitation generator when the start button is released.

Especially the probe hose is susceptible to mechanical defects during operational use. Industrial trucks or forklifts frequently damage the hose when the main unit is several meters away from the operator. A design of flexible metal protective hose with a plastic sheath has proven itself to be effective (see Figure 7.6).

Figure 7.6: Probe hose with plastic sheath, printed with the friendly permission of SPECTRO Analytical Instruments GmbH, Boschstr. 10, 47533 Kleve, Germany.

Probe and probe hose generally form a unit. However, the end of the probe hose is often connected to the main unit with a plug.

Several reasons support the use of a pluggable connector:
–   As already mentioned, the probe hose is exposed to mechanical stress and hazards. It is advantageous to be able to easily replace the probe and/or hose in case of a defect.
–   As described in Section 7.2, the determination of the vacuum UV elements requires an optical system in the probe. Probes with such optics are heavier than those without. Therefore, it makes sense to use the heavy probe only when it is necessary, that is, when the elements P and S need to be determined. A light probe would then be plugged on when, for example, a lengthy sorting control in arc mode is to be conducted.

–  The probe plug must be able to establish a safe electrical contact, to couple argon to the gas lines in the probe hose and to transport the radiation at the end of the fiber optic cable to the optics within the instrument housing with as little loss as possible.
–  The electrical contacts are the least difficult. Here, there are standard solutions that can be used. There are also standard solutions for the gas connections.

The end of the fiber optic cable can directly connect to the end of another that attends to further transport of the radiation inside the main unit. Alternatively, it can be in the focus of a lens located on the end of a tube (see Figure 7.7). The lens holder tube can then directly illuminate the entrance slit to the main optic. Inlet via lens holder tube has the advantage over the solution of fiber optic on fiber optic that there is no loss of radiation at the transition between the fiber optic cables. Also, minor scratches and contamination on the lens disturb less than damage or contamination on the ends of the fiber optic cables, which usually have an optically active area of less than a square millimeter. A drawback to coupling via light tube is that there is no longer a choice as to the position in which the optic is installed.

Coupling of fiber optic to fiber optic can be combined with electrical contacts and gas connections in common plug systems. A proprietary plug connection must be constructed for a light tube solution or separate plugs must be used for the different types of connections.

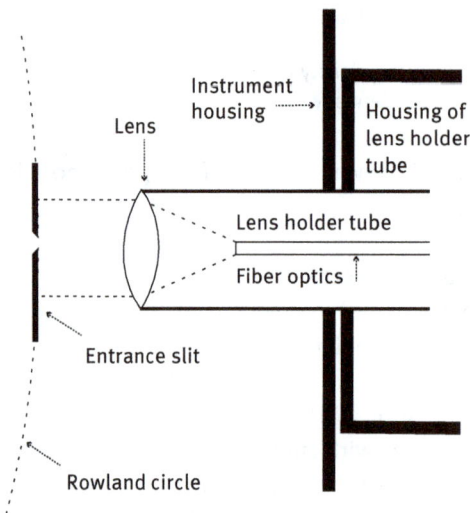

**Figure 7.7:** Lens holder tube for direct illumination of the entrance slit.

## 7.4 Optical systems for mobile spectrometers

It must be distinguished between three types of optical systems:
- Handheld instruments combine all components in the measuring probe, this also includes the optics. Only optics with small focal lengths and, therefore, limited resolution can be used due to space limitations.
- For larger mobile spectrometers, the radiation is generated in the spark probe and transported via fiber optic cable to the optical system located in the main unit. The optics no longer adversely affect the size and weight of the probe. Larger focal lengths and, thus, higher resolution become possible.
- If the elements phosphorous and sulfur need to be measured, an additional optic in the probe is required, as the radiation of the detection lines for these elements is in the vacuum UV range. Fiber optic cables are not or not sufficiently transparent for this radiation.

### 7.4.1 Optics coupled via fiber optic cable

The optics for measuring wavelengths above 187 nm are located in the main unit for larger mobile spectrometers. This lower limit results from the fact that being able to measure the carbon line C 193 nm together with the associated internal standard Fe 187 nm even with probes without UV optics is often desired. The optics are usually equipped with a Paschen-Runge mount. Photomultiplier tubes (PMTs) that are mounted behind the exit slits located on the focal curve are still used as detectors in older instruments. Modern mobile spectrometers use detector arrays with CMOS or CCD technologies. A large number of spectral lines can be measured inexpensively and in the smallest of spaces with detector arrays. A mobile spectrometer designed to analyze the alloys of all common metal bases generally uses a hundred or more spectral lines. So many PMTs are difficult to accommodate in a compact mobile spectrometer optic, even when small-sized photomultiplier tubes are used.

When designing the optical body, it must be considered that a mobile spectrometer, and thus with it, its optics, are frequently operated in different positions. The spectrum must not be shifted when the location is changed.

Mobile spectrometers are often operated outdoors. Usually they may only be operated at a temperature range between 5 and 40 °C. However, the distances entrance slit to grating and grating to detector must remain constant to a few tenths of a millimeter to prevent line broadening and, thus, a loss in resolution. For a total light path of one meter, a 35 °C temperature increase and aluminum as the optical mount, the optical path lengthens by 0.8 mm. If one considers only the influence of this elongation on the distance between grating and detector (about half of the one meter light path) and a grating illumination of 20 mm, the line broadening can be estimated using the theorem of intersecting lines known from high school geometry. The result is 16 µm. In addition, a widening of the same order of magnitude results from the extension on the light path between the entrance slit and the grating.

The material expansion of the optical mount in the above-mentioned temperature range can lead to significant line broadening and therefore represents a problem.

One solution to this problem is to heat the optic to the maximum operating temperature. The energy requirements for the heating, which must come from the limited resources of the battery during battery operation, is the disadvantage of this solution. In addition, after switching on the instrument, it is necessary to wait until the operating temperature has been reached. However, it is then possible to be sure that all the optical components, including the detectors, are always at the same, constant temperature during measurements. Other errors that are also caused by temperature fluctuations could be avoided in this way.

Another possibility for offsetting temperature fluctuations is to compensate for the light path elongation with a counter reaction of the same size. For example, using a bimetallic strip, the entrance slit could be moved along the path towards the grating exactly the same distance as the optical mount lengthens when there is a temperature increase.

## 7.4.2 Optical systems for handheld devices

Optics for handheld instruments must cover the entire relevant wavelength range with one optical system that is not much larger than the palm of a hand. Due to space limitations, such optics use only semiconductor detector arrays for radiation measurements. The wavelength range of laboratory instruments extends from H 121 nm to K 766 nm. Since oxygen, hydrogen and small nitrogen contents cannot be measured with mobile instruments anyway and potassium is almost never determined, it would be desirable to at least be able to measure between P 178 nm and Li 671 nm without gaps. If the wavelength range begins somewhat earlier, nitrogen contents above 800 ppm, as found in duplex steels, could also be determined using the nitrogen line at 174 nm. However, such a wide wavelength range can hardly be covered with a single concave grating optic only used in the first order of diffraction. This can be easily calculated using the grating equation (eq. (3.8)). If, for example, 39° is selected as an angle of incidence, then the wavelengths between 174 nm and 384 nm are available in the first order between grating normal and the entrance slit. Although angles above 39°, that is, beyond the entrance slit, can be used, image errors increase. Angles above 60° are not very practical. This results in a spectral range from 174 nm to 415 nm, whereby the range around 384 nm is missing because the entrance slit is located there and this space is, thus, not available for a detector. This limited wavelength range is satisfactory for many analytical applications, but it remains a compromise.

Disadvantageous is, for example, that the main detection lines for Na (589 nm) and Li (670 nm) are not available. It should be noted that only with compromises is it possible to accommodate a sufficient wavelength range in a single concave grating optic, as are normally used in arc/spark spectrometry. A further difficulty is to obtain a resolution that is good enough in the wavelength range available. Different approaches are conceivable:

1. It would be possible to miniaturize the optic described in Section 3.5.2 by scaling down the slit width, grating focal length and pixel width or exit slit width by a constant factor. However, there are limits to these efforts, as falling below the smallest optimum slit width described in Section 3.5.5.3 leads to no improvement of the resolution.

2. Optics with concave grating mounts can be shortened by almost half by folding the light path by installing a mirror (Figure 7.8b, the thick dotted line shows the segment of the Rowland circle used, the thin dotted line indicates the space required). The length of the optic can be almost halved by folding. If only a short wavelength range has to be covered, folding is a feasible solution. However, the length of the focal curve is not shortened by using this measure. If grating angles between 0° and 50° and a Rowland circle diameter of 400 mm is used, the focal curve forms a circular arc with a length of 349 mm. This is too long for a pistol-shaped measuring probe. The resulting optics would also be too heavy for a handheld device.

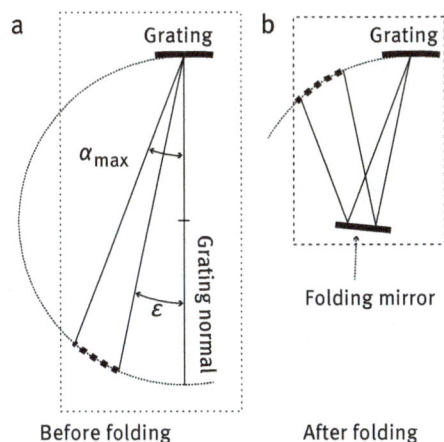

Figure 7.8: Folded Paschen-Runge optic.

3.  A further approach would be to use an Echelle optic, which has been described in Section 3.3. It is advantageous that the grating's diffraction orders provide short overlapping spectral sections that can be detected by a two-dimensional detector. If 100 diffraction orders are mapped on a detector with a width of 10 mm, then a spectrum with a length of 1,000 mm can be recorded in the smallest of spaces. In principle, an Echelle optic could solve the problem of wavelength coverage at sufficient resolution. However, Echelle optics have not established themselves in arc/spark spectrometry for the reasons explained in Section 3.5.4. At present, the authors are not aware of a handheld arc/spark instrument equipped with an Echelle optic.

4.  In a similar direction, the idea is to place several flat optics on top of one another. This makes it is possible to use a narrow range of angles in each optical section, which would also accommodate folding.

    Example: A wide spectral range can be recorded with three optical sections. 9° as the angle for the shortest recorded wavelength ($\alpha_{min}$), 29° as the position for the entrance slit ($\varepsilon$) and a grating with 3,600 grooves per mm results in a detectable spectral range from 178.1 nm to 269.3 nm. The combination $\alpha_{min} = 9°$, $\varepsilon = 29°$ and using a 2400 groove grating results in a usable spectrum between 267.2 nm and 404.0 nm. The spectrum from 392.1 nm to 572.2 nm is obtained with an 1,800 groove grating, $\alpha_{min} = 11°$ and $\varepsilon = 31°$. The numerical values are easily calculated with the grating eq. (3.8). When the spectrum is divided into three such modules the spectrum is, thus, distributed among three focal curves, one on top of the other, which helps to reduce the system dimensions. With such a design, it is also easier to cover the entire relevant wavelength range. A further advantage is that the resolution can be adjusted to the spectral range. In general, a high resolution is required especially for short wavelengths. The example takes this into account: The shortest wavelength module covers 91.2 nm, whereas that with the longest wavelengths covers 180.1 nm for the same angular difference. The shortest wavelength optic module has a direct light path; the other modules can be coupled to the spark stand with short fiber optic cables. The disadvantage to the system lies in the increased effort required and the higher complexity: Multiple gratings, several entrance slits and several coupling fiber optic cables, all of which must be installed and adjusted, are required.

5.  If the wavelength range is distributed among several optical modules, designs other than the classical concave grating arrangement can be selected for the individual modules. One such arrangement, the so-called crossed Czerny-Turner mount, is shown in Figure 7.9. The entrance slit is located at the focal point of a first concave mirror. A parallel bundle of radiation results from this concave mirror. The bundle hits a plane grating and is diffracted there according to wavelength. A second concave mirror focuses the spectrum sharply onto a

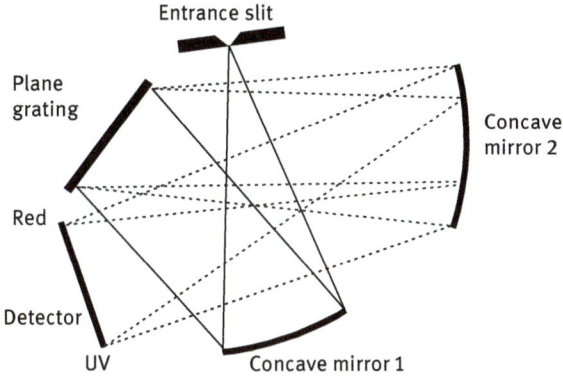

Figure 7.9: Czerny-Turner optic with crossed radiation path.

detector. Figure 7.9 clearly illustrates that small dimensions can be realized by crossing the radiation paths. The complexity of the component is increased once again compared to the solution with folded concave grating optics.

### 7.4.3 Probe optics for the VUV spectral range

In larger mobile spectrometers that have an optic in the main unit as described in Section 7.4.1, the probe optic only serves to measure the wavelengths between 170 nm and 200 nm. When a 3,600 groove grating and an entrance slit angle of 21° are used, the wavelengths 171 nm to 199 nm lie between 15° and 21°, that is, in an angular range of only 6°. Folding is easily possible. The arrangement could be designed similar to that in Figure 7.8. With the folding, the use of a 300 mm grating is possible without making the measuring probe unmanageably large. Such probe optics had already been used with photomultiplier tubes before semiconductors became established as detectors for mobile spectrometers [7]. Semiconductors facilitate a space-saving design. The curvature of the spectral curve is so low for gratings with 300 mm Rowland circle diameters that a linear detector with a length of 30 mm can be used without further corrective measures. It is not absolutely necessary to use a flat field grating. As already stated in Section 7.4.2, compact optics can be realized using the so-called "crossed Czerny-Turner mount", which is also suited to vacuum UV with a direct light path. Alternatively, the Wadsworth mount can be used. This arrangement is shown in Figure 7.10. The exit slit E is located in the focus of the mirror S and generates a parallel radiation beam that irradiates a concave grating G at an angle ε. The spectrum is detected in the focus of the grating on the focal curve F immediately left or right of the grating normal. Figure 7.11 shows a folded mini-optic for installation in a measuring probe.

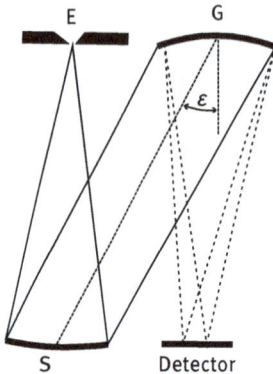

Figure 7.10: Wadsworth mount.

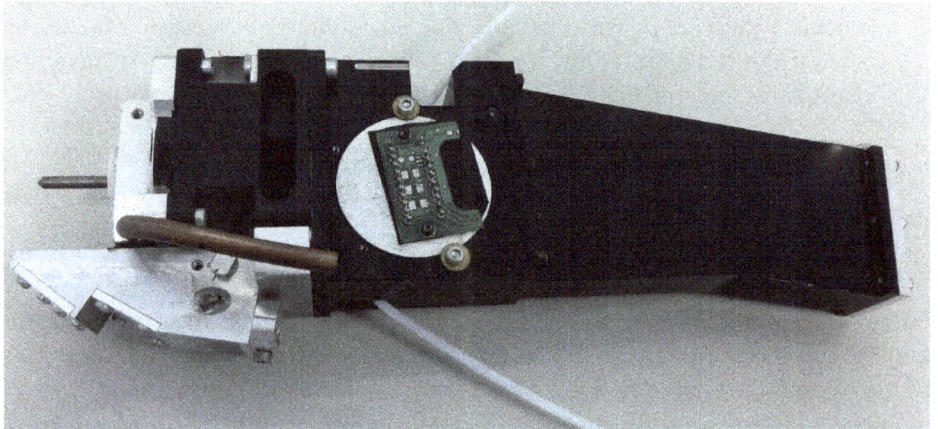

Figure 7.11: Hollow profile optic, printed with friendly permission of SPECTRO Analytical Instruments GmbH, Boschstr. 10, 47533 Kleve, Germany.

## 7.5 Housing, transport trolley and battery

The design of the instrument housing, the accompanying transport trolley and the selection and integration of the rechargeable batteries represent a considerable challenge in the development of mobile spectrometers.

The housing is the interface to the user. It must meet *ergonomic requirements*:
-   The display must be readable, even outdoors in direct sunlight. The user often stands several meters away from the instrument; it must also be possible to read the display from this distance.
-   In the testing room, pumps and instrument ventilators must not be so loud that they are distracting.

- The rechargeable battery must be easy to change.
- The instrument must be easy to place into the trunk of a car and easy to attach to the transport trolley.
- It should be possible to disassemble the trolley without tools and to stow it together with the instrument in the trunk of a car.
- For handheld devices, it is useful to offer a suitable case that accommodates the instrument and any accessories.
- A fixture on which the probe hose can be coiled and the probe can be held must be available to protect the probe hose and probe during transport.
- Of course, it is also important to find a pleasant shape and color scheme.

The *safety requirements* for laboratory equipment throughout Europe are specified in the standard DIN EN 61010 [2]. Mobile spectrometers are also covered by this standard. Among other things, it is important to consider the following properties here:
- Live parts must not be accessible from outside the instrument. What is considered to be "not accessible from outside," is precisely regulated in the standard. A so-called standard test finger, with which no galvanic contact to dangerous accessible surfaces may be made, is described. The term "dangerous to touch" is defined in DIN EN 61010 [2].
- This safety against contact must continue to exist even after damage to the housing due to fall or impact. No danger to the user may arise in the event of damage, for example, no dangerous glass splinters may fly around if the display shatters. Exact test procedures are also described for these hazards.
- Fire safety must be established. For this purpose, the instrument is surrounded with cotton, which is then ignited. The fire must not ignite the instrument, the transport trolley or the battery.
- All gas lines and other components of the gas system, such as valves, pressure reducers, etc., must withstand at least twice the maximum permissible operating pressures.
- The instrument must stand stably. It must not tip over when it is pulled with a fixed percentage of the instrument weight. This applies for every pull direction and every instrument configuration, that is, for the instrument alone, the instrument with battery and the instrument with battery and transport trolley.
- The requirements for internal insulation must be met. Certain clearance and creepage distances must be observed. Mobile spectrometers are designed for operation in industrial environments. Therefore, the creepage and clearance distances coupled to the contamination class are normally longer than for laboratory spectrometers.
- There are various labeling requirements. An abrasion resistant nameplate has to be installed. Places on the instrument that may be dangerous – for example, hot – must be marked with appropriate stickers. A meaningful pictogram must

be visible on the stickers. If there is no fitting symbol available, an exclamation mark must be displayed instead and the danger described in the operating instructions.

-   Risks must be prevented with constructive measures. Hereby, predictable operating errors must also be considered. If an instrument poses a residual risk, a hazard analysis that weighs the probability of danger and its consequences must be carried out. Of course, probable dangers with serious consequences are unacceptable.
-   The results of the tests outlined above must be documented. There are retention requirements for test documentation, for hazard analyses and for CE conformity declarations. Even ten years after the last instrument of a series has been placed on the market, these documents must be available on short notice upon request.

The design of the housing must also consider aspects of electromagnetic compatibility (EMC) with regard to emissions and immunity to interference. As has been mentioned in connection with the ignition in Section 7.1.1, the basic requirements in Europe are specified in Directive 2014/30/EU on electromagnetic compatibility [8]. In Germany, this European directive was transposed into national law by the "Elektromagnetische-Verträglichkeit-Gesetz" (Electromagnetic Compatibility Act) on December 14, 2016 [9].

Neither the European directive nor the EMCA contain the limiting values to be observed themselves. The standard DIN EN 55011 [10] specifies the permissible interference emissions, while DIN EN 61000-4 [11–15] group of standards determines which interferences must be tolerated by the instrument. An overview of the immunity requirements can be found in section one of the group of standards [16].

The required tests can be classified as follows:
1.  Electrostatic discharge immunity test (ESD-Test). Here, high voltage of a few kilovolts is discharged on the housing and on exposed connections (limiting values are defined in DIN EN 61000-4-2 [11]). The function of the instrument must not be impaired by these pulses. The same applies to the tests described in points 2–4.
2.  Test for immunity against high-frequency field emissions. The limiting values are specified in DIN EN 61000-4-3 [12].
3.  Electrical fast transient/burst immunity test (burst test). Here interference fields are introduced through instrument supply lines using a capacitive voltage probe or coupling network (limiting values are defined in DIN EN 61000-4-4 [13]).
4.  Test for immunity against voltage surges (surge test, limiting values: DIN EN 61000-4-5 [14]) and conducted interference induced by high-frequency fields (limits in DIN EN 61000-4-6 [15]).
5.  Test for electromagnetic emissions and interference voltage on the power supply connection (limiting value specifications in DIN EN 55011 [10]).

Other requirements must be met in addition to the above-mentioned standards for safety and electromagnetic compatibility, for example the so-called EU ROHS Directive [17]. This specifies the limiting values for heavy metals, such as, Cd, Hg, Pb, hexavalent chromium, halogenated flame retardants and some phthalates contained in components used in the spectrometer system. Regarding mobile spectrometers, compliance with these regulations is not more difficult than for laboratory systems.

A declaration of conformity may only be issued when an instrument complies with all regulations. The Declaration of Conformity combined with a CE mark is, in turn, required for marketing an analytical instrument within the EU. Similar regulations apply in other regions of the world, but fulfillment of the European regulations is a major step towards the meeting the standards in other regions. Of course, the local regulations must be checked and any additional tests carried out and supporting documents provided. In some countries like Russia, permission of the local authorities is required before a device may be placed on the market. Normally, this will be granted if the above listed requirements are met.

Rechargeable batteries of different technologies are offered for some mobile instruments. Older technologies, for example, lead gel batteries, can be found in addition to modern Li-ion or LiFePo batteries. Table 7.3 shows the weights, relative prices (as of 2016) and lifetimes for various battery technologies, all standardized to an energy of 360 Wh. After studying this data, it may come as a surprise that batteries based on lead are still available in addition to the modern lithium-based batteries. The reason lies in the regulatory provisions of another kind. Batteries that contain lithium above a certain limit are considered dangerous goods in accordance with international transport regulations. The batteries used in mobile spectrometers

Table 7.3: Key data for battery packs of various technologies based on a 360 Wh capacity.

| | Li-Ion | Lithium iron phosphate | Nickel metal hydride | Lead gel |
|---|---|---|---|---|
| Weight (kg) | 1.8 | 3.5 | 5.9 | 9 |
| Volume (l) | 0.9 | 2 | 1.9 | 3.1 |
| Charging cycles until the capacity falls for 70% | 500 | 4000 | 600 | 200 |
| Approximate suggested price in € for branded goods for industrial use (as of 2016, only cells, without housing and charging technology) | 180 | 410 | 360 | 85 |
| Strength | High energy density | Long lifetime | High energy density with fewer transport restrictions compared to batteries containing Li | Low price |

usually have a capacity of more than 100 watt-hours and are therefore classified as class 9A dangerous goods. Thus, according to IATA regulations they may not be transported in passenger aircraft, but only in special cargo planes. Specially approved safety packaging, which must undergo, for example, fall tests, is required. The packaging is subject to precisely regulated labeling guidelines. It is forbidden to transport fully charged lithium batteries. Before being placed on the market, the batteries must pass a so-called UN test. This involves a height simulation and tests for vibration, shock, short circuit, impact, overload and forced discharge.

In conclusion, it should be noted that the presentation of the regulations to be observed have been provided in Section 7.5 to the best of the author's knowledge. The presentation is certainly not complete and no guarantee can be given for the correctness. Also, the regulations are subject to constant changes.

## 7.6 Inert gas systems in mobile spectrometers

While laboratory instruments usually use only high-purity copper or stainless steel lines, mobile spectrometers also use plastic tubing, as the required flexibility of the probe hose excludes the use of metal.

It has already been mentioned that carbon analysis of steels with the aid of the cyanide radical is possible in arc mode. This cyanide radical CN is formed by the carbon in the sample and nitrogen in the surrounding atmosphere. However, the carbon dioxide contained in the atmosphere disturbs here. Every ppm contained there generates a signal for the cyanide band at 386 nm corresponding to a multiple of the signal of one ppm carbon in the sample. The $CO_2$ from the ambient air increases not only the background signal, the $CO_2$ concentration can also fluctuate depending of the location. The US patent US 7,227,636 B2 [18] shows a way to solve this problem:

The ambient air is passed through filling of NaOH or $Ca(OH)_2$ using a membrane pump. The reaction takes place as follows when $Ca(OH)_2$ is used:

$$Ca(OH)_2 + CO_2 \rightarrow CaCO_3 + H_2O \qquad (7.1)$$

Thus, the carbon dioxide is converted to calcium carbonate and water with the aid of calcium hydroxide. The arc plasma must burn in a closed chamber to separate it from the ambient air, which contains $CO_2$. The water formed in the reaction according to eq. (7.1) can lead to liquefaction of the strongly hygroscopic hydroxides. This effect is undesirable because alkaline liquids can leak out. Liquefaction can be prevented using a suitable substrate onto which the hydroxides are applied instead of pure hydroxide granules. The gas tubing in the probe hose makes up the largest volume fraction of the gas system in a mobile spectrometer. If this has an inner diameter of 2 mm and a length of 5 m, its volume is 15.7 ml, which is easy to calculate.

If the output of the membrane pump is 2 l/min, then the air in the gas system can be exchanged with a pump time of about 0.5 s.

## 7.7 Software for mobile spectrometers

The software for mobile spectrometer frequently differs from that of laboratory instruments. There are also additional functions that are not common for laboratory instruments.

### 7.7.1 The analysis mode

Calculation of the analysis is usually the same for mobile and laboratory instruments.

However, the presentation of the data is very different from that for laboratory instruments. For handheld devices, the display on the measuring probe offers little space. It is not possible to display all the elements and all the single measurements for a measurement series together with the current average values, standard deviations and coefficients of variation at a readable size. The display is restricted either to the most important elements together with the results of a small number (typically three) of the last measurements or to the result of the last spark for all the elements. Usually it is possible to switch to the less important elements or to the results of the previous measurements by pressing a button.

A similar problem exists for the larger mobile spectrometers. Although the display is larger here, the information shown must remain readable even from several meters away. Thus, the options are, in principle, the same as for the handheld instruments.

When using mobile spectrometers, the base metal of the sample is often unknown. This is a common problem, for example, in the recycling industry. In order to not have to find the fitting analytical method by trial and error, the spectrometer software has algorithms that automatically select the base metal and the fitting analytical method within that base. It is a great deal of effort to separately standardize all the methods in all the bases on an instrument before using it. A complete spectrum recalibration is more convenient (see Section 3.9.6.2).

An obvious algorithm for automatic method determination could appear as follows:
1. First, a method that only contains spectral lines for high contents of all the base metals to be measured is loaded. This method contains excitation parameters that are a compromise for all the sub-methods.

2.  A short measurement is conducted. Since it is convenient to work with complete spectrum recalibration, essentially the same intensities are measured for the same sample on all the instruments from a product series.
3.  The intensities of the base element spectral lines are now compared with a lower limit assigned to each line representing a metal base. The element of the line for which the ratio of the line intensity and the lower limit is maximum is selected as the matching metal base.
4.  The overview method for the metal base is loaded. If the measuring parameters deviate from those loaded for base selection, at least one further measuring phase is required. After this has been carried out, the contents are calculated using the calibration functions for the overview method.
5.  A set of concentration limits is stored for each method of a metal base. (Example for concentration limits: The method for low alloy steel is loaded if $C < 2\%$ and each of the remaining elements is smaller than 5%). The most suitable sub-method is selected and loaded using the results from the overview method.

It is usually sufficient to apply the calibration functions of the now loaded sub-method to the existing spectra (there is one spectrum per measuring phase) to determine the contents, since the sub-methods usually use the same excitation parameters as the overview method. If this is not the case, the deviating parameters are sparked subsequently. The automatic method search described here can be carried out before every measurement. Alternatively, it is possible to activate method search and then remain in the sub-method found for further measurements. In this way, the time required for the measuring phases used for identification can be avoided for the following measurements.

This simple algorithm is, however, only suitable for use with spark excitation. Use in combination with arc excitation is less advisable for two reasons:
–   The amounts ablated vary widely within a metal base, making it difficult to compare the base element spectral lines and, thus, to select the correct base.
–   It is not possible to create an overview method with enough accuracy for every metal base.

For these reasons, it is better to work with the so-called fingerprint algorithms in the arc mode. The principle is presented in the following section.

### 7.7.2 Calculating the contents using fingerprint algorithms

As the name implies, fingerprint algorithms use the spectrum obtained from the measurement of a sample as a spectral fingerprint. This fingerprint is compared with every entry in a file of fingerprints. Ideally, this file should contain at least one spectrum of a reference sample for every common grade.

The following additional information is stored for every reference sample in the spectra library (the so-called reference sample library):

1. The designation of the reference sample
2. A set of well-suited excitation parameters and measuring times
3. For every relevant element: the wavelength of an analytical line and that for a matching internal standard
4. A calibration function for each of these line pairs
5. Concentration ratios for these elements as well as the intensity ratios (see 3.9.4 and 3.9.5) measured with the matching excitation conditions (2)

Five process steps are processed sequentially to calculate the concentrations according to the fingerprint algorithm:

1. First, a measurement parameter that is suited as a compromise for all the alloys to be measured is set. For measurements in arc mode, the set of parameters consists of a current strength and a measuring time – in spark mode a set consisting of energy curve, frequency and measuring time is selected.
2. A measurement is conducted with the given set of parameters and a complete spectrum is recorded.
3. The complete spectrum is now used to identify the reference sample whose spectrum resembles it the most closely. The algorithm for complete spectrum recalibration described in Section 3.9.6.2 is used.

   The following calculation is carried out for every reference sample:

   A set of spectrum recalibration parameters consisting at least of pixel-specific profile offsets and intensity factors is determined. This parameter set allows it to convert the measured spectrum into a spectrum that is comparable to the reference spectrum with respect to intensity levels and line positions. The calculated parameters are applied to the spectrum of the unknown sample. Then, pixel by pixel, the difference between this resulting spectrum and that of the reference sample is determined. The amounts of the pixel deviations are summed.

   A sum of the intensity results for every reference sample is available as a result. The reference sample that is most similar to the unknown sample is the one for which the sum of the deviations is minimal.
4. There is a possibility that the reference sample found is not similar enough to the unknown sample. This can happen when an "exotic" material was measured, for which there is no equivalent in the reference sample library. For this reason, it only makes sense to continue the calculation when the quotient of the sum of the deviation amounts and the sum of all the pixel intensities does not exceed a specified upper limit.
5. If the reference sample found is similar enough to the unknown sample, the optimal excitation parameters stored with this reference sample are loaded and a measurement is conducted.

6. The dataset for the reference sample contains definitions for analyte lines and internal standards in the form of the pixel areas in which they are to be found in the spectrum. For these line pairs, the intensity ratios from the spectrum of the unknown sample are now determined.

7. For every relevant element, represented by a line pair, the following information is now available:
   – The intensity ratios $IR_U$ of the unknown sample
   – The intensity ratios $IR_R$ and concentration ratios $CR_R$ of the reference sample
   In addition, a calibration function f is available for every line pair. The function f can be a polynomial. The steps for the correction of line and interelement interferences described in Section 3.9.5 can usually be omitted, as the reference is very similar in composition to the sample to be analyzed. f is such that $f(IR_R) = CR_R$ applies. The intensity of the unknown sample is simply set into f, and the concentration of the unknown sample is obtained with $f(IR_U)$. Thus, the function f performs a polynomial interpolation.
   A simple example illustrates what is meant here: Assuming a linear 45°-curve behavior for the analyte line and assuming that the spectral background is negligible, then a half (double) intensity ratio of the unknown sample to the reference sample would lead to a half (double) concentration ratio with respect to this. Because a polynomial has been stored, the ever-present spectral background and the curvature of the calibration curve can be taken into account.

8. Finally, the concentration ratios are converted into concentrations using the 100% calculation described in Section 3.9.8.

It should be noted that it is possible to work with raw intensities and concentrations instead of intensity and concentration ratios. In this case, it is, of course, not necessary to store internal standards for the analyte lines and the function f has intensities as arguments and supplies concentrations as functional values. Here, the content of the base element is calculated in the eighth step by subtracting the sum of all analyte concentrations from 100%. However, the calculation using intensity and concentration ratios has the advantage the accuracy remains acceptable for all the analyte contents even with larger deviations from the best fitting reference sample stored. Thus, fewer reference samples are required for sufficient coverage of wider calibration ranges.

It is useful to issue the name of the reference sample together with its analysis if it contains the material grade that it represents. Alternatively, the instrument often provides the possibility of identifying a material from a database containing material specifications (see Section 7.7.4).

The spectrum recalibration algorithms are used to find the best fitting reference sample. This is done every time an unknown sample is measured. An offset and an additional sensitivity factor are calculated for every pixel. Thus, information about changes in the instrument hardware is available before the line intensity is calculated. It is possible to compensate for pixel drifts, due to, for example, fluctuationsin temperature or air pressure and changes in light transmittance resultingfrom soiling of the optical surfaces, before the line intensities are calculated.

A method that largely corresponds to the algorithm outlined above is described in the patent [19].

### 7.7.3 Checking the analysis for a match with a target material

In practice, it is often necessary to check whether a sample corresponds to an expected material. This task arises, for example, after the delivery of semi-finished products or when checking safety-relevant components in factories. Modern instrument software offers a mode of operation that supports this type of control.

First, the target material must be selected by entering the designation into the instrument's software (of course, these specs must be present in the material database of the instrument). In most cases, the permitted upper and lower concentration limits for every alloying element are immediately displayed on the screen, but at the latest, they are displayed after the first measurement. The tolerance windows are not necessarily fixed values, they frequently depend on the contents of other elements. For example, the titanium content for the steel grade 1.4571 must be at least five times as large as that of carbon but must not exceed a maximum limit [20].

After each measurement, the average of the contents is updated and compared with the limits. If an element violates the tolerances, the deviation is marked. This allows the user to see at a glance if the workpiece meets the specifications or not.

However, it should be noted that the measurement is subject to uncertainty. If the analysis lies less than the measurement uncertainty away from the upper or lower limit, it cannot be said that the workpiece is actually within the permissible range. Assessment of the result is then left to the experienced user.

### 7.7.4 Grade identification

Checking to see if there is a match to a target material is unproblematic apart from the unclarity of the results due to measurement uncertainty near the tolerance limits. It would, then, be assumed that it would be equally simple to assign an analysis to the fitting (and expected) grade specification by storing all the eligible grade specifications in a file and then comparing the analysis to see which one fits.

However, there are several issues that prevent a simple approach leading to this goal:
– The problem of incomplete specification of the element contents

- Assignment problems due to measurement uncertainties
- Problems with overlapping specifications for several qualities
- Assignment problems caused by elements that cannot be determined or that were not determined with the mobile spectrometer but that are relevant to the grade specifications

These problems are discussed in detail here and solutions, if any, are presented.

### The problem of incomplete specifications of element contents

Suppose the spectrometer delivers the following analysis (all concentrations in mass percentages):

| | | | | | |
|---|---|---|---|---|---|
| C 0,03 | Si 0,3 | Mn 1,2 | P 0,012 | S 0,015 | Cr 17,9 |
| Ni 9,8 | Mo 0,3 | Nb 0,05 | Ti 0,22 | Cu 0,3 | Al 0,02 |
| Fe Rest | | | | | |

Based on the analysis, a European user with previous experience recognizes that it is the grade 1.4541. An American would see in it an approximate equivalent to the American grade 321.

If the analysis above is compared with the entries in a database in which the permissible concentrations ranges are stored for all common grades, then grades that obviously do not fit would also be displayed. For example, the grade 1.4301, for which the tolerances shown in Table 7.4 apply, would be found.

The simple approach of stating that a material has been found when no tolerance limits have been violated leads to multiple suggestions in which materials such as 1.4301 are presented although they do not fit in reality. The user who expects the grade 1.4541, but is wrongly presented with 1.4301, would not be satisfied: The properties of grade 1.4301 are not as good in terms of weldability and acid resistance compared to 1.4541. 1.4301 has no place in the hit list.

The cause of the problem here is the fact that there is no information about the element titanium in the specifications for 1.4301.

Similar problems exist for other grades:
- The grade 1.4571 is no 1.4401, because it contains titanium.
- The grade 1.4580 is no 1.4401, because niobium has been added.
- The grade 1.4401 is no 1.4301, because molybdenum is present here to make the material acid resistant.
- The grade 1.4305 is no 1.4301, because it contains sulfur to improve the machinability.

The above-mentioned grades are not exotic, in fact, they belong to the most common chrome-nickel steels.

Table 7.4: Permissible concentration ranges for the grade 1.4301 [21].

| Element | from (%) | to (%) |
|---------|----------|--------|
| C | 0 | 0.07 |
| Si | 0 | 1 |
| Mn | 0 | 2 |
| P | 0 | 0.045 |
| S | 0 | 0.015 |
| N | 0 | 0.011 |
| Cr | 17 | 19.5 |
| Ni | 8 | 10.5 |

How is it possible to prevent unwanted materials from appearing in the hit list? The solution is quite simple and consists of two rules that must be followed when searching the specifications:

–   A cut-off concentration is determined for every element. For grade identification, concentrations below this limit are considered to be irrelevant contamination. However, if a limit is exceeded during the analysis of a sample, then the element has probably been intentionally added. A tolerance window for this element must, then, be included in the grade specifications. If this is not the case, then the grade does not match the analysis.
–   For the remaining, potentially matching grades, it is now only necessary to count how many elements meet the specifications. Only the grades with a maximum number of elements within the specified tolerances are included in the hit list.

Table 7.5 shows practice-oriented examples of limits for steels. For the elements that are not intended to be used in nonalloyed steels, the limits are identical to those named in DIN EN 10020 to distinguish between nonalloyed and alloyed steels [22]. However, the concentrations of some elements vary within nonalloyed steels. For this reason, it is reasonable to use lower limits for them than specified in DIN EN 10020. Please note that a tolerance window is always required for carbon. For some general structural steels (e.g., 1.0554 S355J0C St52-3 U), there are no specification entries for concentration tolerances. However, guide values are available for these materials that make it possible to estimate tolerances.

The limits in Table 7.5 are only meaningful for steels. Other limits must be applied to other material groups.

**Table 7.5:** Cut-off concentrations for ferrous materials above which an intended alloy is probable.

| Element | Limit | Element | Limit | Element | Limit | Element | Limit |
|---------|-------|---------|-------|---------|-------|---------|-------|
| C | 0.0 | Si | 0.6 | Mn | 0.6 | Cr | 0.3 |
| Ni | 0.3 | P | 0.015 | S | 0.015 | Mo | 0.08 |
| V | 0.05 | W | 0.3 | Nb | 0.05 | Al | 0.05 |
| Cu | 0.4 | Co | 0.3 | Pb | 0.1 | Ti | 0.05 |
| B | 0.0008 | Zr | 0.05 | Se | 0.1 | Te | 0.1 |
| Bi | 0.1 | La | 0.1 | Ce | 0.1 | Nd | 0.1 |
| Pr | 0.1 | N | 0.05 | Other | 0.1 | | |

Using the two above-described rules, the grade 1.4541 would be recognized while 1.4301 would no longer appear in the hit list, as the Ti concentration at 0.22% exceeds the limit of 0.05%.

The other above-mentioned conflicts no longer occur.

### Assignment problems due to measurement uncertainties

In the previous section, an analysis that could be clearly identified as the grade 1.4541 was performed. However, it is important to be aware that every measurement is subject to uncertainty. This measurement uncertainty depends of various factors.

The most important are:
- Stability, precision and detection sensitivity of the mobile spectrometer
- Representativeness and homogeneity of the measured sample
- Sample preparation
- Diligence and ability of the user

The measurement uncertainty can be estimated. Suppose there is an analysis that lies within the limits for the material 1.4301 (Table 7.4). The nickel concentration was found to be 8.1%.

If a measurement uncertainty of 0.2% for nickel is to be expected, then it is no longer possible to say with certainty whether the specifications according to Table 7.4 have been met or not. Of course, the reverse also applies: If 7.9% nickel is found, the true value of the sample could well be within the limits according to Table 7.4 although the analysis lies below the specifications. Unfortunately, this is aggravated by the fact that expensive alloying elements are used sparingly. Nickel belongs to these, so that situations like the one described here do occur in practice.

The problem can be circumvented by expanding the tolerance limits by the measurement uncertainty at the upper and lower limits. It makes sense to list the measurement uncertainties in one table per metal base. They must be determined separately for arc and spark methods, as the uncertainties are generally higher for arc mode.

As shown by the example with nickel, it is not a good idea to remove found grades with elements in a measurement uncertainty grey zone from the hit list or to list them there after the qualities without such tolerance violations. Marking the hit so that it can be seen that a given element is not clearly within the specifications would be helpful, for example, in the form of 1.4301 with the mark "Ni -" when the nickel concentration lies in the grey zone for the lower limit.

It is customary to specify the measurement uncertainties for confidence ranges of 68%, 95%, or 99%. It is possible to expand the tolerance limits by these measurement uncertainties.

### Problems with overlapping specifications for several grades

Table 7.4 shows the composition of the grade 1.4301. The quality 1.4307 is very similar, only the tolerances for C, Cr and Ni are slightly different. 1.4307 has maximal 0.03% carbon (for 1.4301 ≤ 0.07%), the Cr concentration must be at least 17.5% (instead of 17% for 1.4301) and a maximum of 10% instead of 10.5% nickel may be contained. The most important difference concerns carbon. A low carbon content is usually desirable, as this is connected to an improvement in the weldability and a higher corrosion resistance.

Most analyses of 1.4307 grades also fit the specifications for 1.4301. The hit list would then include both materials. If the database also includes the American grade 304 (corresponds to the European 1.4301) and 304L (similar to 1.4307), then these would also be displayed. The mobile spectrometer can do nothing but to suggest possible grades.

### Assignment problems caused by elements not measured, but important for grade identification

Not always can all the important elements be determined. In the arc mode, the important element carbon can only be determined when the ambient air is cleaned of carbon dioxide. Determination of sulfur in arc is not possible using the current technology. In the spark mode, sulfur can only be measured when a measuring probe is equipped with an optic for the vacuum UV range. This fact can lead to a hit list containing grades that are qualitatively very different and that are generally not interchangeable.

If, for example, the element sulfur cannot be determined, then an analysis of the grade 1.4305 (X8CrNiS18-9) usually also fits to 1.4301. The main difference in the composition is that the S content is between 0.15% and 0.35% instead of under 0.015% as is the case for 1.4301.

The two grades are intended for different applications: 1.4305 is a particularly easily machinable alloy that, compared to 1.4301, is well suited to processing with turning and milling machines. On the downside, 1.4305 is unsuitable for pressure tanks [21].

The fact remains that the correct match of a material identification to a spectrometer analysis is not a simple problem. If the search cannot be limited to a few clearly different materials, it is usually not possible to obtain exactly the designation under which the material originally came onto the market. However, it is possible to obtain a hit list with matching materials that will help the qualified user.

## 7.7.5 Sorting

Sorting was the first application for which mobile spectrometers were used beginning at the end of the 1970s. The aim was to ensure that all the workpieces were made of the same material before they were delivered or after goods receipt. In the beginning, the instruments were equipped with polychromators with exit slits and photomultiplier tubes for the most important spectral lines for the base metal to be tested. Initially only arc mode was used. The pre-burn time was typically one second, the measuring time 2 s. The output signal of the analyte photomultiplier tubes was integrated and divided be the integral derived from the spectral line of the base metal. Figure 7.12 shows such an early mobile spectrometer.

**Figure 7.12:** Early model of a mobile spectrometer for sorting, built in 1980, printed with the friendly permission of SPECTRO Analytical Instruments GmbH, Boschstr. 10, 47533 Kleve, Germany.

The test procedure using this type of instrument was as follows:

1. First, a section of a workpiece from the delivery was cut off and analyzed in the laboratory. If this reference piece was made of the expected material, the actual sorting could begin.
2. The reference piece was measured three times. Before each of the three measurements, the quotients of the analytical line and the internal standard were connected to a capacitor using a rotary switch.
3. Then the average of the three measurements was made by connecting the three capacitors assigned to each analyte with each other.
4. Now sequential measurement of the test pieces could begin. After each measurement, the quotient of analyte line and internal standard was formed for each analyte. Using simple analog electronics, it was then checked whether the quotient was in agreement with the average formed in point 3 within the allowed tolerances set with a switch. If at least one element exceeded the tolerances, a red lamp was switched on and further measurements were blocked until a reset button on the measuring probe was pressed.

While the instrument in Figure 7.12 was still equipped with analog electronics, the following generation of mobile spectrometers already had a microcomputer. The possibility for sorting was implemented there in the instrument software.

Even today, mobile spectrometers have a sorting mode based on the comparison of intensity ratios. Of course, it is no longer necessary to cut a piece off the test piece and analyze it in the laboratory. Instead, ideally in spark mode, the first test piece is analyzed and the analysis checked for agreement with the target specifications. Then the instrument is switched into arc mode and, in principle, the same procedure as for the first mobile spectrometers is followed. A sorting method is loaded that consists of a set of line pairs (analyte line with associated internal standard) and a set of permissible percentage deviations for each line pair.

The test proceeds as follows:

1. First, an average intensity ratio is recorded in arc mode for each line pair with the reference sample.
2. Then the test piece is measured with the same parameters.
3. After each measurement, it is checked whether the intensity ratios of all the lines being monitored lie within the specified intensity tolerances. If this is the case, then go to point 6.
4. If there is a deviation for at least one intensity ratio, then the measuring procedure is blocked until the reset button is pressed. Then the measurement is repeated, as the tolerance violation found may not be caused by a different material but may be due to statistical variation of the measured values. Incorrect placement of the measuring probe or measuring on a soiled or

corroded spot on the sample can also lead to deviations. If the measurement is now okay, the system continues with point 6. If not, the measurement is repeated again. If the tolerances are exceeded again, the counter for negatively tested workpieces is increased by one and the system continues with point 5, otherwise point 6. It may come as a surprise that a workpiece is considered correct after only one of three measurements are within the tolerance limits. However, it is very improbable that, in the case of an incorrect measurement on an incorrect material, coincidentally all the monitored channels are within the tolerance limits for the reference material.

5. This point in the procedure is reached if a material is considered incorrect. Modern spectrometers offer the possibility of conducting an analysis and then to perform a grade identification as described under Section 7.6.4. The incorrect sample must be acknowledged by the user. The test then continues with point 7.

6. The counter for positively tested workpieces is increased by one.

    If it has been ensured that the test pieces come from one heat, that is, that all have the same analysis, the intensity quotients of the test piece recognized to be correct can be used to improve the reference intensity ratios determined under point 1. This puts this data on a broader base because the average is then statistically sounder. In addition, particularly when testing larger lots, a longer amount of time may have passed since originally measuring the reference piece. However, the newly added measurements correspond to the current state of the instrument in terms of soiling of the inlet optics and the electrode degradation.

7. After testing a workpiece, the user can continue the measurement at point 2. The measurement can also be interrupted or completed. In the case of an interruption, the sorting method, reference intensities and counter readings are stored. The can be reloaded at a later time to continue testing.

    If the measurement is to be completed, a test protocol is created; it can be stored or printed out. The following information is commonly gathered in the protocol:

    – Date, time and duration of the testing
    – Name of the operator
    – Details about the tested lot (grade, dimensions, the name of the customer or supplier if applicable)
    – Name of the testing method used with measurement parameters
    – Number of test pieces classified as "good"
    – Number of test pieces classified as "bad" (as a rule, 0 should be here)

The question might arise as to why the entire test is not conducted in spark mode but switched into arc mode after identification of the reference piece. The reason for this is that testing in arc has lower requirements in terms of sample preparation and can be comparatively quickly performed.

In point 6, it was mentioned that it is advantageous to know if all the test pieces in a lot come from the same heat. For lots of the same heat, the measured values of the samples identified as correct can be used to update the expected intensity ratios. Information about heat fidelity is also advantageous when selecting the tolerances. If this is given, the tolerances can be selected to be relatively narrow. They are based on the uncertainty of measurement allowed by the mobile spectrometer on the test pieces to be measured.

If, on the other hand, the lot to be tested consists of a mixture of heats and if only the grade is known, then after analysis of the reference piece, it must be estimated in which range the intensity ratios for every channel may lie for the grade to check. This range must be expanded partially by the uncertainty of measurement. Thus, when testing heat mixtures, wider, usually asymmetric tolerances are obtained for the reference piece intensity ratios.

100% inspections are often conducted for structural components for which failure cannot be tolerated for reasons of safety or economy. If they always have similar forms, such as bars or tubes, for which only the diameter varies, then systems for automatic testing are installed directly into the production lines. Figure 7.13 illustrates the principle. In the initial position, the workpiece to be tested (1) in the line (2) is separated out, for example, by a prism support (3). A grinding machine (5) equipped with an abrasive mop (4) under the workpiece is started and raised. A width of a few centimeters is ground on the underside of the test piece. Then the grinder is lowered and switched off. Now the measuring probe (6), which is mounted together with the grinder on a movable carriage (8), is moved under the workpiece and raised and a measurement is performed. If a repeat measurement is required, the measuring probe is slightly lowered, moved a few millimeters along the workpiece axis and then placed on the test piece again. Then the workpiece is lifted onto the part of the line (7) on the other side of the testing equipment. If the piece is rejected, a barrier is opened and the workpiece falls into a reject position, otherwise it continues its way along the line.

## 7.8 Calibration, sample preparation and sampling

For laboratory spectrometers for melt management, the samples are poured into molds and then milled or ground. This procedure is standardized and always occurs in the same manner during routine operation. When using mobile spectrometers there is a larger variety of materials and shapes to be analyzed. This includes small

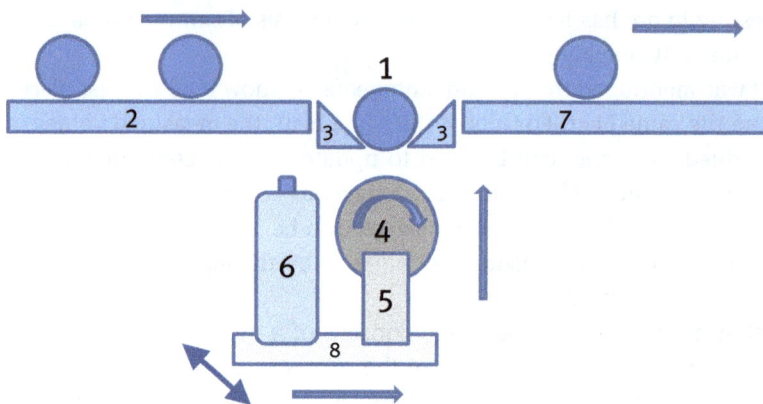

Figure 7.13: Mobile spectrometer integrated into a production line for sorting.

parts, wires, tubes, pipes, billets, flanges and sheets with different thicknesses down to foils and much more. Section 7.8.1 deals in general with noteworthy aspects of sample preparation and the selection of the measuring position. Section 7.8.2 covers typical scenarios for spark measurements with mobile spectrometers and discusses the analytical performance expected of spark mobile spectrometers. Section 7.8.3 deals in a similar fashion with the measurement tasks for which arc mode is used.

## 7.8.1 Sample preparation and selection of the measuring position

Before measurements using spark operation, the locations to be sparked must be ground to the bare metal. In most cases, quickly grinding with a belt or angle grinder is sufficient. When a belt grinder is used, the use of 60 grit abrasive belts has proven to be effective. Angle grinders must not be equipped with cutting discs but must be equipped with a rough grinding disc that complies with the local safety requirements. In Europe, such requirements are listed in the norm DIN EN 12413 [23]. During sample preparation, it is, of course, necessary to observe other safety precautions customary to metal processing, such as the wearing of safety glasses. The prepared section of the sample should be as flat as possible to enable gas-tight placement of the measuring probe while simultaneously ensuring that the gap between the sample surface and the counter electrode is correct.

In the following special cases, grinding can be dispensed with:

At the end of the production process for stainless steel, it is frequently possible to conduct a measurement without grinding. Even at later times, it is often possible to do without this, if the location to be measured has been cleaned and any grease removed through cleaning with isopropanol. In any case, the burn spot is to be critically observed. It should have a round melted center surrounded by a black condensate ring (Figure 3.23, lower center and right). Whitish zones, that indicate non-energetic discharges without material ablation (Figure 3.23, bottom left) should not occur. Additional cleaning with isopropanol or quickly sanding by hand is recommended.

Some alloys of nickel, cobalt and titanium base do not necessarily have to be ground either as long as the surface is clean. The same is true for some copper base alloys, such as cupro-nickel. However, it is always important to have a good burn spot.

Some distinctions must be made if the element carbon is to be measured:

In the case of steel products, a reduction of the carbon concentration (decarburization) can occur in areas that have been separated by flame cutting. In order to ensure measuring in regions where the carbon concentration is again representative, up to one millimeter of material must be ground down. However, this is difficult to do with an angle grinder. Therefore, in such cases, testing of the surfaces along the casting or the rolling direction is preferred. But here too, caution is required. The mill scale rolling skin must be removed by grinding. In the case of hot-rolled materials, there can be surface decarburization, so that it is not always sufficient to only grind down to the bare metal. A further problem can occur when semi-finished products are stored in stacks. Then, only the end faces cut with a burner and potentially decarburized are accessible for measurement. Frequently, the only option left is to separate the semi-finished products using a crane so that the measurement can be performed.

Another problem associated with carbon determination can arise when measuring hardened materials. For case hardening, the carbon concentration is deliberately increased on the surface to improve the hardness and wear resistance. If the surface concentration of carbon is of interest, the test piece must not be ground in an uncontrolled manner, as the carbon concentration decreases with the depth. The procedure of grinding by hand, described above, can be used here. If the carbon content in the core is to be determined, the entire hardened area must be removed. Its thickness can be between 0.1 mm and 4 mm. If in doubt, it must be ground and analyzed iteratively until the carbon concentration no longer falls.

Surfaces are also refined by nitriding as an alternative to case hardening. These layers increase not only the wear resistance but also the corrosion resistance. According to Merkel/Thomas [24], the nitrogen-enriched layer can have a thickness of up to 0.8 mm. Steels that are particularly suited to nitriding are standardized in DIN EN 10085 [25]. These are low-alloy steels that, in addition to C, Si, Mn, Cr and Mo, also occasionally contain Ni, V and especially aluminum in concentrations up to over 1%. However, other common steels, such as C15, C45, 16MnCr5, 42CrMo4 and S235, can also be nitrided. In addition to carburizing and nitriding, carbonitriding is a process in which C and N are simultaneously enriched. Boriding is less common. Here, a thin layer of $Fe_2B$ is generated on the surface of the workpiece. According to Merkel and Thomas [24], the boron-enriched layer can have a thickness of up to 0.25 mm.

The material surfaces may also have a different composition than the interior in the following cases:

1.  Galvanizing using Zn is the most commonly used coating method for the protection of thin sheets of carbon steels from corrosion. For hot batch galvanizing, layer thicknesses up to 0.15 mm can be expected; for strip galvanized materials thicknesses are between 5 and 40 µm.
2.  Galvanic and chemical nickel plating are also common methods for protecting metal surfaces. The layer thicknesses are usually below 50 µm. However, the nickel layer can be up to three millimeters for thick nickel plating.
3.  For hard chrome plating, a chromium layer of up to one millimeter thick is galvanically applied. Possible base materials are steels and cast iron as well as copper and aluminum alloys. Several layers of chromium and nickel on top of each other can also be found.
4.  The Sulf-Inuz process enriches the surfaces with sulfur with layer thicknesses of about 0.03 mm, thus improving the friction properties of the material. These surfaces are found on high speed steel and nodular cast iron parts.
5.  Occasionally the carbon concentration at the surface is deliberately reduced to achieve special magnetic properties or to improve the sealing properties.
6.  Aluminizing serves to improve the scale resistance. The Al-enriched layer can be up to several tenths of a millimeter thick. Nonalloyed case hardened steels with low carbon contents are suitable for aluminizing.
7.  Chrome plating increases the corrosion resistance of the steel. The chromium-enriched zone has a thickness between 5 and 200 µm.
8.  Occasionally, the surface layers of steels and nodular cast iron are enriched with silicon to improve corrosion resistance.
9.  Chromium aluminizing and chromium siliconizing combine the methods described in points 6 to 8.
10. When nitriding finished products, the areas that are not to be nitrided are covered with a thin layer of tin. This coating can lead to an increased tin value.
11. Eloxal, that is, electrolytically oxidized aluminum, must always be ground, as the oxidic surface does not conduct. The same applies to any other existing coating layers.

Apart from surface effects, the following materials cause problems:

–   The determination of lead, manganese and sulfur is problematic for free-cutting steels. It is possible to create special methods for these materials. The pre-burn times are increased to durations between 30 s and 1 min in these methods. This homogenizes the lead and manganese sulfide inclusions present in different sizes and lengths. The long sparking time, combined with the necessity of holding the measuring probe still on the test piece, makes this approach unattractive. With normal (short) measuring times, the Mn and S concentrations found are too high, but it is possible to recognize that it is free-cutting steel. This distinction is sufficient for many testing applications.

- Carbon is present as graphite in finished cast iron workpieces. This applies to the common cast iron types: grey cast iron (GJL), nodular cast iron (GJS), vermicular cast iron and black malleable cast iron. An exact carbon determination is not possible for these cast irons because the carbon is present as spheres (GJS) or lamellar structures (GJL) of different sizes. It is typical that the carbon concentration found is too low for GJS and too high for GJL. There are several graphite-free types of cast iron, labelled with the letter combination GJN, that are standardized in DIN EN 12513 [26]. The wear resistant chrome-hard casting materials are occasionally encountered in practice. These are characterized by a high chromium concentration. The carbon is bound as carbides in these alloys. Since no graphite precipitates are present, an exact carbon determination is possible provided the appropriate methods have been calibrated.

The condition of the surface must also meet minimum requirements for the examination of metals in *arc mode*. However, the arc is significantly less susceptible to slight impairments to the sample surface. Even thin oxide layers are penetrated by arc, but stronger milling skins should be removed by grinding.

## 7.8.2 Measurements in spark mode

Table 7.6 shows detection limits that are achievable with larger mobile spectrometers for several important elements in iron base with the spark mode. Determination is made according to DIN 32645 [27] and was established using the reproducibility of ultra pure iron. The values in Table 7.6 cannot be considered in isolation, as the dispersion of the calibration curve near the detection limit must also be taken into account for every element except the reproducibility of the background. Particularly for smaller mobile spectrometers, for example, handheld devices, it is more the dispersion of the calibration curves rather than the reproducibility of the background that determines the true achievable detectability of traces, as separation of the analyte lines from their spectral environment is less pronounced. The meaning of the detection limit according to DIN 32645 is reduced to the following: Suppose there are two workpieces with the same composition apart from the trace element S with a detection limit $LOD_S$. The trace element S is not present in the first of the two workpieces. Then the other workpiece for the determination of S must contain at least a concentration of $LOD_S$, to be able to say whether the concentration for S is higher

Table 7.6: Detection limits for larger mobile spectrometers, iron base in spark mode.

| Element | LOD (ppm) | Element | LOD (ppm) | Element | LOD (ppm) | Element | LOD (ppm) |
|---|---|---|---|---|---|---|---|
| C | 40 | Si | 10 | Mn | 30 | Cr | 10 |
| Ni | 20 | P | 10 | S | 10 | Mo | 10 |
| V | 10 | W | 250 | Nb | 20 | Al | 10 |
| Cu | 10 | Co | 10 | Pb | 10 | Ti | 10 |
| B | 3 | Sn | 40 | As | 30 | Ca | 3 |
| Ce | 20 | La | 3 | Zn | 20 | | |

than for the first workpiece. This statement is only valid if the analytical line used is undisturbed by overlapping of other lines and the reproducibility and background equivalent concentration of the workpiece corresponds to that of the pure metal with which the detection limit was determined. Additional information about detection limits can be found in Chapter 5.

The scatter of the calibration curves can be improved by not using a single set of calibration functions for all iron base grades, but by creating separate calibrations for individual material groups. Examples of such groups in iron base are: non-alloyed and low alloy steels, free-cutting steels, chromium steels, chrome-nickel steels, manganese steels, laminar and nodular cast iron. The real detection sensitivity approaches that determined according to DIN 32645 when such or further subdividing is used.

However, usually semi-finished or finished products are tested for compliance with concentration limits with mobile spectrometers. Especially for the larger mobile spectrometers, the detection sensitivity is in most cases sufficient for this. It is not as good as that achieved with laboratory spectrometers, but the application areas are different. Laboratory instruments are often used for melt management. Here, it is necessary to monitor for undesired concentrations of trace elements that can negatively influence subsequent processing steps. To increase the detection sensitivity, laboratory spectrometers use different sparking parameters adapted to the elements to be determined. This approach could also be chosen for mobile spectrometers, but it would lengthen the measuring time. The operator would have to hold the sparking probe motionless and gas-tight against the workpiece for a longer time. This increased effort would not be matched by any corresponding benefit.

At levels far above the detection limits, the repeatability is similarly good. The coefficients of variation of the measured concentration, in the middle concentration ranges, usually lie below 1%. For some elements, for example, Cr, the relative standard deviation of the concentration is even less than 0.2%.

Assuming the calibration has been carefully conducted, the accuracy is comparable to that which is achieved with laboratory spectrometers. However, this only applies to medium concentrations far above the detection limits.

Occasionally, it is necessary to conduct 100% inspections in spark mode. This is the case, for example, when elements such as P or S, which cannot be measured in arc mode, must be determined. The sparking times are often shortened to reduce the time normally required in spark mode from 8 s to values between 3 and 5 s. These measures worsen the detection limits and the repeatability for statistical reasons due to the smaller number of spark events. For this reason, a worsening of the repeatability by a factor of two must be expected if the measurement is carried out with only a quarter of the sparks. However, this can be tolerated for many testing applications. It is also necessary to note that some elements are not in a stationary spark state (state where the intensities per second stay stable) when the pre-spark time is reduced. This is especially true for the elements that are not dissolved in the metal matrix but are present as inclusions. $Al_2O_3$, MgO, CaO and $ZrO_2$ are examples of such inclusions. The elements in question tend to values that are too high when the measuring time is shortened, because the sparks prefer to hit on the borders between the inclusions and the metallic phase (see Chapter 6 for an explanation of this effect). The calibration must be adjusted if necessary.

Testing semi-finished products often requires a special approach:
1. It has already been mentioned in Section 7.2 that special spark stand attachments, with boron nitride or mica disks, and sample clamps are available for the measurement of small parts. Such an adapter is shown in Figure 7.5a. When using the adapter, it is necessary to ensure that the small parts do not overheat. The repeatability in spark under argon deteriorates dramatically at sample temperatures above 200 °C. This can be remedied by shortening the measuring time or reducing the spark frequency.
2. As described in Section 7.2, pipes are measured with special adapters in arc mode. For measurements of pipes in spark the attachments with boron nitride disks, mentioned in point 1, are used. However, the pipes frequently do not form a gas-tight closure with the spark stand opening, which can lead to problems for small tube diameters less than 10 mm. Then, leakage hinders material ablation and the measurement values cannot be used. On the bottom right, Figure 3.23 shows a usable burn spot, bottom left one that is unusable.
3. Wires are preferably measured on the end faces. In iron base, this is possible for diameters down to 0.8 mm. There are devices that enable the wire to be clamped and, at the same time, seal the environment around the wire from the outer atmosphere. Figure 7.5b and c show such devices for the measurement of wires. The analytical performance remains high even for thin wires. In this way, it is possible to analyze, for example small carbon concentrations, accurately enough to distinguish grades such as 1.4301 from 1.4307 or 1.4401 from

1.4404 based on the carbon concentrations. Once again, the prerequisite is that overheating of the wires is prevented, which can be done by reducing either the sparking times or the spark repetition frequency.

4. It is also possible to examine for the presence and composition of coatings. Of course, the pre-spark time must be chosen to be so short that the coating is still available with sufficient thickness during the measurement time. Conversely, a longer pre-spark time can be chosen if the material underneath the coating is to be analyzed. Of course, only electrically conductive coatings can be analyzed.

5. Even thin foils with a thickness of a few tenths of a millimeter can be measured if the excitation parameters have been selected to prevent overheating or even melting through. It is often useful to place a metal block on top of the foil. The metal dissipates the heat from the foil and ensures a flat surface.

### 7.8.3 Measurements in arc mode

Table 7.7 lists the detection limits for several elements in iron base. Again, they were determined with a medium-resolution mobile spectrometer according to DIN 32645, i.e., using the reproducibility of an ultra pure iron sample. Here too, the dispersion of the calibration curves used must be taken into account in order to be able to decide whether a separation is possible via an element (see Section 7.8.2). The parameters used to determine the data were not optimized to obtain the best possible detection limits. Rather they are a practice-oriented compromise that also take aspects such as measurement duration and minimizing of the electrode degradation into consideration.

**Table 7.7:** Detection limits for larger mobile spectrometers, iron base in arc mode.

| Element | LOD (ppm) | Element | LOD (ppm) | Element | LOD (ppm) | Element | LOD (ppm) |
|---------|-----------|---------|-----------|---------|-----------|---------|-----------|
| C       | 500       | Si      | 80        | Mn      | 80        | Cr      | 10        |
| Ni      | 60        | Mo      | 50        | Al      | 40        | Co      | 80        |
| Cu      | 30        | Nb      | 80        | Ti      | 50        | V       | 20        |
| W       | 180       | Pb      | 30        |         |           |         |           |

In addition to the scattering of the calibration curves, a further effect influences the detection of small concentrations in arc mode. If a sample with a high concentration of an alloying element is measured and then a sample without this element, a concentration is displayed for the element anyway during the subsequent measurement. Table 7.8 shows a series of measurements on a pure iron sample after a

Table 7.8: Example for memory effect with silver and copper electrodes in arc as well as in spark.

| | Si | Mn | Cr | Mo | Ni | Al | Co | Cu | Ti | V |
|---|---|---|---|---|---|---|---|---|---|---|
| Analysis 1.4571 (%) | 0.44 | 1.33 | 17.24 | 2.12 | 11.7 | 0.031 | 0.113 | 0.087 | 0.486 | 0.042 |
| **Measurement with Ag electrode** | Si | Mn | Cr | Mo | Ni | Al | Co | Cu | Ti | V |
| Pure iron before contamination (%) | <0.01 | 0.015 | <0.01 | 0.018 | 0.02 | <0.005 | <0.008 | <0.005 | 0.01 | 0.003 |
| Pure iron burn 1, after cont. (%) | 0.066 | 0.086 | 0.69 | 0.121 | 1.16 | 0.0057 | 0.038 | 0.0084 | 0.0151 | 0.01 |
| Pure iron burn 2, after cont. (%) | 0.06 | 0.059 | 0.62 | 0.086 | 0.8 | 0.008 | 0.036 | 0.0076 | 0.0161 | 0.0097 |
| Pure iron burn 3, after cont. (%) | 0.038 | 0.035 | 0.41 | 0.065 | 0.42 | 0.0059 | 0.0233 | 0.0055 | 0.0127 | 0.0077 |
| Pure iron burn 4, after cont. (%) | 0.0214 | 0.035 | 0.168 | 0.045 | 0.207 | <0.005 | 0.0154 | 0.0053 | 0.0121 | 0.0048 |
| Pure iron burn 5, after cont. (%) | 0.0288 | 0.049 | 0.276 | 0.052 | 0.34 | 0.0096 | 0.0202 | 0.0055 | 0.0204 | 0.0066 |
| Pure iron burn 6, after cont. (%) | 0.03 | 0.053 | 0.31 | 0.055 | 0.37 | 0.0059 | 0.023 | 0.0064 | 0.0137 | 0.0067 |
| Pure iron burn 7, after cont. (%) | 0.0102 | 0.0226 | 0.047 | 0.0247 | 0.066 | <0.005 | 0.0107 | 0.0050 | 0.0114 | 0.0046 |
| Pure iron burn 8, after cont. (%) | 0.0210 | 0.038 | 0.205 | 0.041 | 0.185 | 0.0057 | 0.0168 | 0.0062 | 0.0136 | 0.0054 |
| Pure iron burn 14, after cont. (%) | <0.01 | 0.064 | <0.01 | 0.0178 | 0.023 | <0.005 | <0.0008 | <0.005 | 0.01 | 0.0029 |
| **Measurement with Cu electrode** | Si | Mn | Cr | Mo | Ni | Al | Co | Pb | Ti | V |
| Pure iron before contamination (%) | 0.01 | 0.015 | <0.01 | 0.017 | 0.020 | <0.005 | 0.010 | 0.0039 | <0.008 | 0.006 |
| Pure iron burn 1, after cont. (%) | 0.011 | 0.071 | 0.227 | 0.056 | 0.168 | <0.005 | 0.012 | 0.0035 | <0.008 | 0.006 |
| Pure iron burn 2, after cont. (%) | <0.01 | 0.020 | 0.12 | 0.024 | 0.124 | <0.005 | 0.011 | 0.0038 | <0.008 | 0.006 |
| Pure iron burn 3, after cont. (%) | 0.015 | 0.021 | 0.193 | 0.047 | 0.102 | <0.005 | 0.013 | 0.0039 | <0.008 | 0.007 |
| Pure iron burn 4, after cont. (%) | 0.012 | 0.019 | 0.0261 | 0.018 | 0.055 | <0.005 | 0.010 | 0.0035 | <0.008 | 0.006 |
| Pure iron burn 5, after cont. (%) | <0.01 | 0.017 | 0.047 | 0.024 | 0.047 | <0.005 | 0.009 | 0.0035 | <0.008 | 0.0064 |
| Pure iron burn 6, after cont. (%) | 0.011 | 0.017 | 0.024 | 0.019 | 0.035 | <0.005 | 0.015 | 0.0037 | <0.008 | 0.0068 |
| Pure iron burn 7, after cont. (%) | <0.01 | 0.014 | 0.012 | 0.016 | 0.020 | <0.005 | <0.008 | 0.0039 | <0.008 | 0.0050 |
| **Measurement in spark mode** | C | Si | Mn | Cr | Mo | Ni | Al | Co | Cu | Ti |
| Pure iron before contamination (%) | <0.005 | <0.005 | <0.005 | <0.004 | 0.0053 | 0.0065 | 0.0079 | <0.003 | <0.005 | 0.0029 |
| Pure iron burn 1, after cont. (%) | <0.005 | <0.005 | <0.005 | <0.004 | 0.0132 | 0.0082 | 0.0137 | <0.003 | <0.005 | 0.0030 |
| Pure iron burn 2, after cont. (%) | <0.005 | <0.005 | <0.005 | <0.004 | 0.0055 | 0.0066 | 0.0083 | <0.003 | <0.005 | 0.0026 |

double sparking was conducted on a workpiece made of the alloy 1.4571 with about 17% Cr, 12% Ni and 2% Mo. The reason for this so-called "memory effect" is an alloying of the material on the tip of the electrode. Table 7.8 shows that this effect can still be observed after several measurements. When an Ag electrode is used, the contamination from the memory effect has completely subsided only after 14 measurements. A remedy is to re-grind or replace the electrode. From Table 7.8, it can also be seen that the memory effect is more pronounced when using silver electrodes than when using copper electrodes. In addition, longer lifetimes can be achieved with copper electrodes than with silver electrodes. Nevertheless, silver electrodes are more commonly used because only then can the element copper be determined. For the sake of completeness, in the last rows of Table 7.8, the same experiment was made with spark excitation. It can be seen that the memory effect plays no role here. In the spark mode, the electrode remains relatively cold so that no alloying can occur. As mentioned above, in spark operation, although high-melting tungsten electrodes are customarily used, the memory effect does not occur on the occasionally used silver electrodes either.

As already stated in Section 3.2.1, relative reproducibility of about 5–10% is achieved for the intensities in arc. The reproducibility can be improved to about 1–5% for intensity ratios through selection of suitable internal standard lines. At some distance from the background equivalent concentration, coefficients of variation of the concentrations on the same order of magnitude can be found. Thus, the reproducibility is significantly worse than in spark. In addition, the scattering of the calibration curves in arc is also worse. This has also already been mentioned and explained in Section 3.2.1. Several examples from practice are listed in Table 7.9 to give an impression of which deviations are to be expected. The first three examples are low-alloy steels. The deviations can often be tolerated in practice, whereby better accuracy would be desirable for carbon. It is also disadvantageous that the elements phosphorous and sulfur are not available in arc mode. The fourth example is an analysis of a chrome-nickel steel. A relative error of 7% for nickel would be considered disturbing here, as it accounts for almost 1% of the total concentration. The reasons for these deviations have been described in Section 3.2.1. The use of a fingerprint algorithm or the use of type recalibration can improve the accuracy but is associated with more effort. The fifth and last example shows the analysis of a high-speed steel.

It has already been mentioned in Section 7.7.5 that 100% inspections (sorting) of safety components made of steels are preferably conducted in arc mode because of the short testing duration and the low requirements on sample preparation. In addition, the arc mode is suitable for other test applications in iron base:

1.  Screening analyses and identification of low alloy materials.
    When a calibration has been carefully conducted, this method provides screening analyses for low alloyed steels that are useful for many applications. Table 7.9 gives an idea of which accuracies can be achieved. As has already been mentioned in Section 7.6, the element carbon can be determined with an

Table 7.9: Examples for the analysis of different steels in arc mode.

| Concentrations in % | C | Si | Mn | Cr | Mo | Ni | Al | Co | Cu | Ti | V | W |
|---|---|---|---|---|---|---|---|---|---|---|---|---|
| Expected analysis (34CrMo4) | 0.37 | 0.23 | 0.49 | 0.90 | 0.17 | 0.12 | 0.022 | <0.002 | 0.143 | <0.001 | 0.004 | 0.002 |
| Measured in arc mode | 0.42 | 0.21 | 0.53 | 0.81 | 0.21 | 0.04 | 0.027 | 0.015 | 0.155 | 0.01 | 0.005 | 0.015 |
| Expected analysis (24CrMoV55) | 0.26 | 0.27 | 0.50 | 1.59 | 0.25 | 0.17 | 0.018 | <0.002 | 0.084 | 0.002 | 0.146 | 0.056 |
| Measured in arc mode | 0.31 | 0.25 | 0.54 | 1.40 | 0.28 | 0.24 | 0.020 | 0.002 | 0.086 | 0.009 | 0.143 | 0.062 |
| Expected analysis (1.8550) | 0.34 | 0.27 | 0.52 | 1.69 | 0.20 | 0.98 | 1.21 | 0.02 | 0.05 | <0.005 | <0.01 | 0.04 |
| Measured in arc mode | 0.38 | 0.25 | 0.59 | 1.64 | 0.21 | 0.9 | 1.27 | 0.036 | 0.04 | 0.03 | 0.01 | 0.04 |
| Expected analysis (1.4571) | -.- | 0.44 | 1.33 | 17.24 | 2.12 | 11.7 | 0.031 | 0.113 | 0.087 | 0.486 | 0.14 | 0.042 |
| Measured in arc mode | -.- | 0.35 | 1.12 | 16.6 | 2.34 | 10.9 | 0.033 | 0.15 | 0.096 | 0.44 | 0.13 | <0.1 |
| Expected analysis (HSS) | 0.92 | 0.37 | 0.31 | 4.01 | 0.25 | 4.8 | 0.012 | 0.57 | 0.17 | -.- | 1.76 | 6.07 |
| Measured in arc mode | 0.96 | 0.33 | 0.31 | 3.98 | 0.29 | 4.7 | 0.012 | 0.51 | 0.14 | -.- | 1.65 | 6.79 |

accuracy of about 0.1% for concentrations to 0.5% when the atmosphere used is air cleaned of $CO_2$. The cyanide band head at 386 nm can be used for detection, as the commonly used carbon spectral line is in the vacuum UV region below 200 nm. If the air is not cleaned, the $CO_2$ contained in it leads to a significant increase in the background, which prevents the reliable determination of carbon. The possibility to determine carbon is an advantage that arc offers that competitive methods such as energy dispersive XRF do not.

2. Sorting of clearly different high-alloy grades.
   Absolute errors in percent ranges may occur for high concentrations of alloying elements, for example, for chromium steels and Cr/Ni steels. As a result, closely related qualities can frequently no longer be assigned using screening analyses. However, many important testing applications can be performed.

3. High boron-alloyed steels (B > 0.2%) can be distinguished from steel qualities without boron.

4. The elements vanadium, titanium and niobium must be controlled for micro-alloyed steels. This also can be done in arc mode. However, the monitoring of these elements is, especially for niobium, only possible with instruments with sufficient optical system performance.

5. For hot samples above 200 °C, the reproducibility in spark mode declines continuously with increasing temperature. Thus, if hot workpieces, for example, continuous cast products, need to be controlled in the production process, this is only possible in the arc mode for higher temperatures.

6. Steel scraps can be pre-sorted for recycling in arc mode. In Europe, the specifications for commercially available grades are laid down in the *Europäischen Stahlsortenliste* [28] (European List of Steel Grades). In addition to limiting values for admixtures, the list contains maximum concentrations of alloying elements. Thus, the sum of the elements Cu, Sn, Cr, Ni and Mo may not exceed 0.3% for new scraps of the classes E2, E6 and E8. By new scrap is meant, for example, punched out or cut remnants from metal processing. The good detection sensitivity of arc combined with a sufficient ability to be calibrated makes it possible to solve this testing application.

There are a number of other applications outside iron base for which arc excitation can be used:

1. Nickel base metals are frequently sorted in arc mode. Here it is often necessary to control for the presence or absence of certain elements, including elements such as yttrium, rhenium and ruthenium, as found in nickel-based superalloys.

2. Using arc, low alloyed copper grades can also be analyzed, sorted and distinguished from higher-alloyed copper grades.

3. Material ablation in refractory metals is low in spark mode. Because it is often only necessary to separate a few grades here, metals of the bases such as tungsten, tantalum and molybdenum are frequently sorted using arc excitation.

4. It is worthwhile to separate pure titanium from the more corrosion-resistant grade Ti 99.8-Pd, which contains 0.2% palladium. With a gram price of about 24 Euros per gram (as of April 2017), the palladium fraction of one kilogram of Ti 99.8-Pd has a value of about 48 Euro. Sorting can be performed with arc. Other titanium alloys can also be sorted using arc excitation, as a distinction between the most important alloys is usually be easily achieved with the main alloying elements V, Al, Sn, Zr, Mo, Cr, Nb, Fe and Si. Checking for the absence of tungsten, which can make some titanium base alloys unusable even with the smallest amounts, is an additional application in titanium base.

5. Also the few common Zr base alloys can be sorted using arc. These alloys were originally developed for reactor technology but were later also used in the chemical industry. The elements Sn, Fe, Cr, Ni and Nb are monitored, whereby only for tin are concentrations of more than 1% common.

In conclusion, it should be noted that many factors have to be considered in the construction and operation of mobile spectrometers. This includes a wide temperature range, the necessity for small dimensions and a low weight, the ability to measure quickly after being turned on and the possibility for mains-independent operation. The types of samples to be analyzed also vary widely in respect to shape and composition. All these factors make the mobile spectrometer technology quite complex despite the lower requirements for detection sensitivity and accuracy; confronting the designer and user with a wide spectrum of very different challenges.

## Bibliography

[1] Directive 2014/30/EU of the European Parliament and of the Council of 26. February 2014 on the harmonization of the laws of the member states relating to electromagnetic compatibility (new version). Official Journal EG Nr. L 96/79, March 29th, 2014.

[2] DIN EN 61010-1/A1:2015-04;VDE 0411-1/A1: 2015-04– Entwurf (Draft) Sicherheitsbestimmungen für elektrische Mess-, Steuer-, Regel- und Laborgeräte- Teil 1: Allgemeine Anforderungen (IEC 66/540/CD:2014). Berlin, Beuth Verlag, 2015.

[3] Saidel, Prokofjew, & Raiski. Spektraltabellen. Berlin, VEB Verlag Technik, 1955.

[4] Kammer, C. Aluminium-Taschenbuch, Band 1, Grundlagen und Werkstoffe, 16th edition. Düsseldorf, Aluminium-Verlag Marketing & Kommunikation GmbH, 2002.

[5] DIN EN 573-3:2013-12. Aluminium und Aluminiumlegierungen – Chemische Zusammensetzung und Form von Halbzeug, Teil 3, Chemische Zusammensetzung und Erzeugnisformen, German edition EN 573-3:2013. Berlin, Beuth Verlag, 2013.

[6] DIN EN 1780-1:2003-01. Aluminium und Aluminiumlegierungen – Bezeichnung von legiertem Aluminium in Masseln, Vorlegierungen und Gussstücken, Teil 1, Numerisches Bezeichnungssystem, German edition, EN 1780-1:2002. Berlin, Beuth Verlag, 2003.

[7] Freit M, Bohle W, Köppen H. Bestimmung von Phosphor und Schwefel – auch vor Ort möglich. Kontrolle 1995, 5.

[8]   Directive 2014/30/EU of the European Parliament and of the Council of February 26th 2014 on the harmonization of the laws of the member states relating to electromagnetic compatibility (new version).

[9]   Gesetz über die elektromagnetische Verträglichkeit von Betriebsmitteln „Elektromagnetische- Verträglichkeit-Gesetz" (Electromagnetic Compatibility Act "Electromagnetic Compatibility Law" ) of December 14th, 2016 (BGBl. I S. 2879).

[10]  DIN EN 55011:2017-03;VDE 0875-11: 2017-03. Industrielle, wissenschaftliche und medizinische Geräte – Funkstörungen – Grenzwerte und Messverfahren (CISPR 11:2015, modified), German edition EN 55011:2016. Berlin, Beuth Verlag, 2017.

[11]  DIN EN 61000-4-2:2009-12;VDE 0847-4-2: 2009-12: Elektromagnetische Verträglichkeit (EMV), Teil 4-2, Prüf- und Messverfahren Prüfung der Störfestigkeit gegen die Entladung statischer Elektrizität (IEC 61000-4-2:2008), German edition EN 61000-4-2:2009. Berlin, Beuth Verlag, 2009.

[12]  DIN EN 61000-4-3:2011-04;VDE 0847-4-3: 2011-04. Elektromagnetische Verträglichkeit (EMV), Teil 4-3, Prüf- und Messverfahren – Prüfung der Störfestigkeit gegen hochfrequente elektromagnetische Felder (IEC 61000-4-3:2006 + A1:2007 + A2:2010), German edition EN 61000-4-3:2006 + A1:2008 + A2:2010. Berlin, Beuth Verlag, 2010.

[13]  DIN EN 61000-4-4:2013-04;VDE 0847-4-4: 2013-04. Elektromagnetische Verträglichkeit (EMV), Teil 4-4, Prüf- und Messverfahren – Prüfung der Störfestigkeit gegen schnelle transiente elektrische Störgrößen/Burst (IEC 61000-4-4:2012), German edition EN 61000-4-4:2012. Berlin, Beuth Verlag, 2012.

[14]  DIN EN 61000-4-5 VDE 0847-4-5: 2015-03. Elektromagnetische Verträglichkeit (EMV), Teil 4–5, Prüf- und Messverfahren – Prüfung der Störfestigkeit gegen Stoßspannungen (IEC 61000-4- 5:2014), German edition EN 61000-4-5:2014. Berlin und Offenbach, VDE-Verlag, 2014

[15]  DIN EN 61000-4-6:2014-08;VDE 0847-4-6: 2014-08:. Elektromagnetische Verträglichkeit (EMV), Teil 4–6, Prüf- und Messverfahren – Störfestigkeit gegen leitungsgeführte Störgrößen, induziert durch hochfrequente Felder (IEC 61000-4-6:2013), German edition EN 61000-4-6:2014, Berlin, Beuth Verlag, 2014.

[16]  DIN EN 61000-4-1:2007-10;VDE 0847-4-1: 2007-10:. Elektromagnetische Verträglichkeit (EMV), Teil 4-1, Prüf- und Messverfahren – Übersicht über die Reihe IEC 61000-4 (IEC 61000-4-1:2006); German edition EN 61000-4-1:2007. Berlin, Beuth Verlag, 2007.

[17]  Directive 2011/65/EU of the European Parliament and of the Council of July 8th, 2011.

[18]  US patent US 7,227,636 B2. Apparatus and method for the spectroscopic determination of carbon. Date of patent Jun. 5th, 2007

[19]  German Patent DE102005002292 (B4)-2006-07-27. Verfahren zum Betrieb eines optischen Emissionsspektrometers. Patent granted Feb. 11th, 2016.

[20]  Specs of grade 1.4571. ThyssenKrupp Materials International, http://www.edelstahl-service-center.de/tl_files/ThyssenKrupp/PDF/Datenblaetter/1.4571.pdf. Downloaded from internet on March 25th, 2017.

[21]  Specs of grade 1.4301. ThyssenKrupp Materials International, http://www.edelstahl-service-center.de/tl_files/ThyssenKrupp/PDF/Datenblaetter/1.4301.pdf. Downloaded from the internet on March 25th, 2017.

[22]  DIN EN 10020:2000-07. Begriffsbestimmungen für die Einteilung der Stähle, German edition EN 10020:2000. Berlin, Beuth Verlag, 2014.

[23]  DIN EN 12413:2011-05. Sicherheitsanforderungen für Schleifkörper aus gebundenem Schleifmittel, German edition EN 12413:2007 + A1:2011. Berlin, Beuth Verlag, 2007.

[24]  Merkel M, Thomas K-H. Taschenbuch der Werkstoffe. München Wien, Fachbuchverlag Leipzig im Carl Hanser Verlag, 2003.

[25]   DIN EN 10085:2001-07. Nitrierstähle – Technische Lieferbedingungen, German edition EN
       10085:2001. Berlin, Beuth Verlag, 2001.
[26]   DIN EN 12513:2011-05. Verschleißbeständige Gusseisen, German edition EN 12513:2011.
       Berlin, Beuth Verlag, 2011.
[27]   DIN 32645:2008-11. Chemische Analytik – Nachweis-, Erfassungs- und Bestimmungsgrenze
       unter Wiederholbedingungen – Begriffe, Verfahren, Auswertung. Berlin, Beuth Verlag, 2008.
[28]   Europäische Stahlsortenliste. Bundesvereinigung Deutscher Stahlrecycling- und
       Entsorgungsunternehmen e.V., version of July 1st, 1995.

# 8 Statistics and quality assurance

Arc and spark spectrometers are literally used to clarify the quality of metallic materials because the term "quality" derives etymologically from the Latin "qualitas," which can be translated directly as quality or property [1]. An essential aspect of the properties of a metal is determined by its chemical composition.

The elemental contents decisively influence tensile strength, formability, ductility, corrosion resistance, wear resistance, low temperature resistance, cutting stability and other material properties.

Various applications for arc and spark spectrometers have already been presented in Chapters 3 and 7:
- In the context of quality control, samples are taken from the molten metal and analyzed, and depending on the result it may be necessary to intervene in the process, for example, by adding alloying elements, in order to achieve the desired target composition. Spark spectrometers are the measurement method of choice here because they are simultaneously fast and precise with good sensitivity.
- For incoming goods or before delivery, semi-finished products or workpieces are inspected for compliance with target analyses (see Sections 7.7.3 and 7.7.4). This can be done by random sampling. However, it is often necessary to conduct 100% inspections for safety relevant components (aerospace, medicine) (see Section 7.7.5).
- In the secondary raw material sector, the price of a scrap delivery and its possible use for certain metallurgical applications depends on the elemental composition.

It is not possible in practice to exactly determine the elemental composition of a sample. Every spark spectrometer analysis is associated with measurement uncertainty. The measurement uncertainty cannot be neglected in every case, as small deviations from the target element concentrations can have a major impact on the material properties. An example for this is given in Section 5.1.2: a mixed crystal strengthened GJS material, for which small excesses of the Si concentration can lead to a significant decrease in the elongation at break and tensile strength. Thus, the measurement uncertainty must be determined. An interval in which, with high probability, the "true" element concentration lies can be determined from the measured value and the measurement uncertainty.

For example, if a Si tolerance ranging from 4.05 to 4.35% is allowed, and 4.2% is measured with an uncertainty of 0.1%, the "true" value is most probably within the tolerated limits. If, however, for the same measurement value, the uncertainty is ±0.2%, then it may be outside of the tolerance limit.

The example shows that a measured value without specification of the measurement uncertainty is inconclusive.

https://doi.org/10.1515/9783110529692-008

Various indicators that enable characterization of a spectrometric testing method are introduced in Section 8.1. If these indicators are available, they can serve as a basis for estimation of the measurement uncertainty.

The working ranges for various common spark spectrometer methods were presented in Chapter 5 in the form of concentration ranges. The measurement uncertainty must be separately determined for each element here. If the elements are measured over a wide concentration range, which is often the case in spark spectrometry, estimation of the measurement uncertainty must be carried out separately for different segments of the total concentration range, that is, traces, medium and high concentrations.

If the working range and the associated measurement uncertainty are known, then it can be proven that a testing method is suitable for a specific application. This proof, referred to as validation, is discussed in Section 8.4. Section 8.5 shows how validated analytical methods fit into modern quality management systems.

## 8.1 Statistical indicators and their determination

This section explains the definitions and meaning of important indicators and describes how they are determined. Further descriptions can be found in the relevant technical literature [2–5] as well as the *Handbuch für das Eisenhüttenlaboratorium* (Handbook for the Ironworks Laboratory) [6]. As mentioned at the outset, the statistical indicators of a method are required in particular for validation as a testing method (see Section 8.2).

### 8.1.1 Working range

A complete description of a test method includes the specification of the working range. The lowest and the highest concentration, for which the accuracy and precision of the testing method is sufficient, is given for every element. Verification of the linearity can also be used to determine the working range [6, 7]. If the calibration function is, for example, too steep, it is no longer usable. The limit of quantification (LOQ) can also serve as the lower limit of the working range. Categorically, analyses may only be conducted in the statistically secured working range.

### 8.1.2 Detection limit of a method

The detection limit of an analyte within a method is understood to be the smallest measured value $M_{LOD}$ that can be distinguished with sufficient statistical certainty from the scattering of the blank values. In the context of spark spectrometry, the blank value is the value measured for a sample that does not contain the analyte but is otherwise within the usable range of the method.

$M_{LOD}$ must be $k$ times the standard deviation of the blank value [6]. The selection of $k$ and the distribution function for the blank value determines the probability with which the analyte is actually contained in the sample when $M_{LOD}$ is exceeded when the sample is analyzed. As a rule, $k = 3$ is chosen.

Conversion of the measured value into concentration or mass is carried out using the calibration function as has been described in Section 3.9 of this book.

In practice, the detection limit of an analyte is estimated by analyzing a sample that does not contain the analyte or only at a negligibly low concentration 6 to 10 times and then calculating the standard deviation s from the concentrations obtained according to eq. (8.1).

$$s = \sqrt{\frac{1}{n-1} \sum_{i=1}^{n} (x_i - \bar{x})^2} \tag{8.1}$$

Here $n$ is the number of individual measurements, $x_i$ the value of the $i$th individual measurement and $\bar{x}$ the mathematical average of all the individual measurements.

The measurements must take place under repeatability conditions. This means that they must be carried out in immediate succession on the same instrument, by the same operator, with the same measurement parameters [6]. In addition, the system must be in a stable state, that is, the argon supply should not have been recently interrupted and the instrument should not be in a warm-up phase. Of course, the measurement series must not be interrupted by a recalibration. For measurements under repeatability conditions, it is ensured that the state of the instrument does not change during a short measurement series of 6 to 10 consecutive measurements. Thus, the measurement series contains only random errors.

If $k = 3$ is chosen, then three times the standard deviation is, based on DIN 32 645 [8], an estimation of the detection limit:

$$DL = 3 \cdot s \tag{8.2}$$

The relationships between the estimates of averages and standard deviations determined from short measuring series and the true averages or standard deviations are discussed in Section 8.1.5.1.

In spark spectrometry, this type of determination of limits of detection is only conclusive for pure materials, for example, in connection with the analysis of electrolytic copper. Here, impurities in the ppm range can influence, for example, the conductivity (see Section 5.3).

A detection sensitivity is implied for alloyed materials that cannot really be achieved. It makes more sense to estimate the detection limit based on the dispersion in the lower range of the calibration function, as has already been explained at the beginning of Chapter 5.

Numerically, this can be done by calculating the standard deviation of the residuals, which is then also multiplied by a factor $k = 3$. However, it is often justifiable to work with a smaller factor, for example, $k = 2$, as the measurement uncertainty of the certified values for the reference materials used in the calibration must be considered. In this context, residuals are understood to be the deviations between the certified values and the calibration function (see Figure 3.86, the residuals correspond to the lengths of the deviation squares drawn there).

### 8.1.3 Limit of quantification

The limit of quantification is the smallest analyte concentration that can be determined with a specified precision. The determination limit is ten times the standard deviation of a blank sample or a reference material with an elemental concentration near or below the limit of the working range (based on DIN 32645 [8]).

$$BG = 10 \cdot s \tag{8.3}$$

Occasionally, the determination limit is given as nine times $s$. An older version of DIN 32645 also uses this definition. Just as in Section 8.1.2, the assertion that calculation of the determination limit using the standard deviation is only conclusive for nearly pure materials is also valid here.

In the case of alloyed materials, a more realistic determination limit is again obtained when this is derived from the dispersion of the calibration function. For this purpose, the detection limit should first be calculated from the dispersion in the lower range of the curve as described in Section 8.1.2. The limit of quantification is higher. It can be estimated by adding the limit of quantification calculated with eq. (8.3) to the residual dispersion in the lower range of the calibration curve.

### 8.1.4 Accuracy

The difference of an individual measurement value from the true value is determined by systematic and random errors (see Figure 8.1).

The accuracy is then high when both bias and precision are high, that is, when both systematic and random errors are low [6, 7].

### 8.1.5 Precision

A measurement is precise when the fraction of the random error in the total error is small. When determining the precision, all the steps of the testing method must be considered. This includes, in addition to the measurement error of the spark spectrometer,

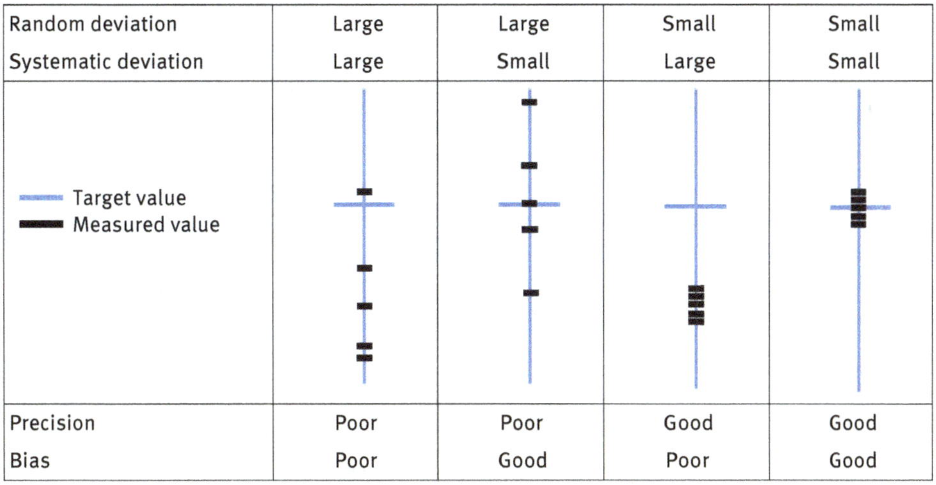

| Random deviation | Large | Large | Small | Small |
|---|---|---|---|---|
| Systematic deviation | Large | Small | Large | Small |

| | | | | |
|---|---|---|---|---|
| — Target value | | | | |
| ▬ Measured value | | | | |

| Precision | Poor | Poor | Good | Good |
|---|---|---|---|---|
| Bias | Poor | Good | Poor | Good |

Figure 8.1: Illustration of precision and accuracy.

sample taking and sample preparation. To assess the precision, repeatability and comparability (reproducibility) are determined using a constant set of samples, preferably (certified) reference materials [6, 7]. The precision is best in the middle of the usable range.

### 8.1.5.1 Repeatability and repetition limit r

The repeatability is an indicator for the agreement between the individual results measured in a short time interval under repeatability conditions (the explanation of repeatability conditions can be found in Section 8.1.2). To determine the repeatability, a sample is measured and analyzed 6 to 10 times under repeatability conditions.

The repeatability $W$ corresponds to the variation coefficient determined from the measurement series. The variation coefficient is often referred to as the relative standard deviation.

$$W = \frac{s_r}{\bar{x}} \cdot 100\% \qquad (8.4)$$

where
$W$ is Repeatability;
$s_r$ the standard deviation under repeatability conditions;
$\bar{x}$ the mathematical average under repeatability conditions.

It can be assumed and verified with measurements that the random errors are normally distributed (at least approximately) for measurements made with spark spectrometers under repeatability conditions. The standard deviation $s_r$ is an estimate of the standard deviation $\sigma$ of this normal distribution whose average is $\bar{x}_N$.

It can be shown that a measured value lies with a 95.45% probability in the interval $[\bar{x}_N - 2\sigma, \bar{x}_N + 2\sigma]$, that is, no further than $2\sigma$ from the average. $\bar{x}$ and $s_r$, determined with a measurement series of 6 to 10 measurements, serve as estimates for the average or standard deviation of the normal distribution and can replace $\bar{x}_N$ and $\sigma$ in the interval given above. For a correct estimate, the percentage deviation between the individual measurements and the average value falls below the limit of 2 * W in approximately 95% of all cases.

$\bar{x}$ is usually close to $\bar{x}_N$. If only a few individual measurements are carried out, $s_r$ can differ significantly from the standard deviation of the normal distribution $\sigma$. The experienced user of the spark spectrometer is aware of this fact. The $s_r$ for consecutive measurement series under repeatability conditions vary much more than the averages. This fact should always be kept in mind when working with estimates of standard deviations determined from just a few individual measurements. For example, if a detection limit is determined using eq. (8.2) in Section 8.1.2 and if a specified target value is slightly exceeded, then the "true" detection limit must not necessarily exceed the target value. Conversely, this means that being slightly below the target value can also be random.

Often the question arises as to whether two alloys can be distinguished using an arc/spark spectrometer (see the description of sorting in Section 7.7.5). This question arises, for example, when a lot of semi-finished products is to be investigated, which because of a mix-up of materials consists of a mixture of two similar grades that can only be distinguished from each other using one element E.

Separation is only possible when the tolerances for E for both grades differ by at least a concentration difference $d$. Example: If the tolerances overlap for all elements except carbon and if the carbon values for the first material are between 0.16% and 0.2% and for the second between 0.25–0.29%, then $d = 0.05\,\%$.

In such cases, knowledge of a repetition limit $r$ helps. With a very low probability, the normal distribution allows even measured values that are far from the average. For this reason, in our example, even with very high precision, a wrong assignment is possible. This is why a probability is assigned to $r$. Normally, a probability of 95% is chosen for the repetition limit and $r_{95}$ is the designation used. It is intuitively clear that the repetition limit must be larger than twice the standard deviation, as with significant probability, the sample with the lower true value will be measured too high and at the same time, the sample with the higher true value will be measured too low.

The repetition limit is calculated as follows:

$$r_{95} = 2.8 \cdot s_r \qquad (8.5)$$

The derivation of the factor 2.8 shall be omitted here.

The above-mentioned materials that can only be distinguished using the carbon concentration can, then, with at least 95% probability, be distinguished when the difference $d$ between them is larger than $r_{95}$.

### 8.1.5.2 Reproducibility and reproducibility limit R

Reproducibility is an indicator for the agreement between individual measurements made at different times under reproducible conditions.

Reproducible conditions differ from repeatability conditions in the following ways [6]:
- The measurements are conducted on different measuring systems of the same or similar types. Thus, spark spectrometers can be from different manufacturers as long as they are calibrated for the same application.
- Measurement and sample preparation can be performed by different people at different locations at different times.

The analysis must always be conducted with the same method, in our case with spark spectrometry. Accordingly, an analysis using inductively coupled plasma optical emission spectrometry (ICP-OES) cannot be used to determine the reproducibility.

Since the use of different spark spectrometers is permitted under reproducible conditions, the analyses are generated using different calibrations based on the same pool of commercially available reference samples. This is an important difference to the repeatability conditions. In addition, the measured values are determined under reproducible conditions, that is, based on different recalibrations.

The reproducibility is given as a variation coefficient $V$.

$$V = \frac{s_R}{\bar{x}} \cdot 100\% \qquad (8.6)$$

where
$V$ is Reproducibility;
$s_R$ the Standard deviation under reproducible conditions;
$\bar{x}$ the Average under reproducible conditions.

In the original sense, the determination of the reproducibility requires the involvement of several laboratories.

As a modification of the original definition, multiple examinations can be carried out on one sample under changing conditions (laboratory precision). Hereby, a sample is examined several times on different days under different conditions (different sample preparers, operators, environmental conditions, etc.).

The following approaches can be applied to determine the laboratory precision:
- The average and the standard deviation are determined from 20 values outlier-cleaned using the Grubbs test. If average value control charts are kept, the data can be taken from them.
- Alternatively, a sample can be analyzed at least ten times over a period of at least three different days.

However, it must be remembered that errors that result from differences in determining the calibration function, such as through selection of reference materials, do not enter into the laboratory precision, because normally always the same spectrometer is used.

The Grubbs outlier test is easy to perform:
- First, the average and the standard deviation are determined for the measurement series.
- Then the measured value that is furthest from the average is determined.
- The difference of this value from the average is divided by the standard deviation.
- If the quotient $q$ determined in this way exceeds a given limit, then the measured value is an outlier.

The limits depend on the number of measured values and the probability required for outlier determination. They can be found in relevant tables and on the internet. According to [9], for a measurement series of 20 measurements if $q > 3.0008$, it is with 99% probability an outlier.

Analogous to the repetition limit $r$, a reproducibility limit $R$ can be defined, to which in turn a probability is assigned. A probability of 95% is usually also chosen for the reproducibility limit:

$$R_{95} = 2.8 \cdot s_R \tag{8.7}$$

If the reproducibility limit is determined using the reproducibility, it indicates which deviations are to be expected for analytical results conducted in different laboratories. If, for example, a sample is analyzed once in the supplier's laboratory and subsequently in the customer's laboratory, deviations up to $R_{95}$ can be tolerated.

If the reproducibility limit is determined with the laboratory precision, deviations up to $R_{95}$ may occur when repeating an analysis at a later date.

## 8.1.6 Bias

The bias of a testing method depends on the fraction of the systematic error in the test result. As an indicator, the bias is a measure for the deviation between the "true" value and the test result measured, which must be determined by a large number of measurements [10]. The data in the certificate for a reference material, for example, which is considered as correct by definition, serves as a reference. In practice, the "true" value cannot be identified for metallic samples.

The bias can be determined with the following methods:
- Analogous to determination of the reproducibility (see Section 8.1.5.2) at least ten analytical results must be measured for a sample under reproducible conditions on at least three different days.

- A fixed number, for example, 20, of outlier-cleaned values from an average control chart are used.
- A certified reference material is measured six to ten times under repeatability conditions (see Section 8.1.5.1).

The value for the bias is calculated as follows:

$$bias = \frac{\bar{x}}{x_R} \cdot 100\% - 100\% \qquad (8.8)$$

where $\bar{x}$ is the average of the values measured for the sample; $x_R$ the average from an average control chart or the target value for a reference material.

It should be noted that the value of the bias is extremely dependent on the choice of reference sample. Reference samples often lie systematically above or below the calibration curve. The reasons for this have been explained in Section 3.2.2.3. Modern instrument software makes it possible to view the position of the reference materials used with respect to the calibration curve. Thus, it is possible to estimate in which dispersion band around the calibration function the reference materials lie and how far the reference material used to calculate the bias deviates from the calibration curve. The bias, as determined with eq. (8.8), can be used to determine the expansion of this band. Such an expansion is required because the calibration was (approximately) carried out under repeatability conditions.

### 8.1.7 Recovery rate

The recovery rate or recovery is closely related to the bias described in Section 8.1.6. It is the ratio of the average determined under repeatability conditions to the "true" value of the quantity in the analytical sample [7]. Ideally, the recovery rate is 100%.

The recovery rate enables evaluation of the entire testing method. It can be used to verify the method's bias, selectivity and robustness. The smaller the difference between the recovery rate and 100%, the smaller the probability that systematic errors occur when using the testing method.

The recovery rate $RR$ can, in a way similar as for bias, be determined by measuring a reference material R:

$$RR = \frac{\bar{x}}{x_R} \cdot 100\% \qquad (8.9)$$

Here, $\bar{x}$ denotes the average of the test results and $x_R$ the expected value for the sample.

It should be noted that the results for the reference materials used to calibrate the spark spectrometer fluctuate to different degrees around the calibration function. Thus, the determination of $RR$ depends on the selection of the reference material. Furthermore, *RR* is not constant over the entire calibration range. *RR* is usually higher for medium concentrations than near the detection limit and, in the case of curved calibration functions, for high concentrations. Therefore, it may make sense to establish a function for which the recovery is dependent on the concentration.

### 8.1.8 Selectivity/specificity

Selectivity and specificity are related terms that are often used synonymously.

The "ability of a method to detect different indicators to be determined in parallel without mutual interference" is referred to as the selectivity [7].

The specificity describes "the capacity of a testing procedure to determine different test values to be determined in parallel without mutual interference and thus to clearly identify them" (corresponding translation of the definition contained in [6]).

When applied to spark emission spectrometry, this means than an analyte can be determined without having significant interferences caused by overlaps from other analyte lines. While line interferences are unavoidable for many important analyte lines, they must be kept within limits and their effects minimized using correction algorithms (see Section 3.9.5).

However, there is no clear mathematical definition for either of these terms. Rather, an indirect assessment must be made by observing the bias, reproducibility or recovery rate.

### 8.1.9 Robustness

The robustness is a "measure of the ability of a testing method to deliver a test result that is not or only slightly falsified by small changes to the test parameters or environmental conditions" (corresponding translation of the definition contained in [6]). In this sense, the robustness describes the independence of a testing method and its results from influencing variables, such as operator, environmental conditions or location.

The collective term "robustness" captures the aspects of method robustness and process stability.

While the method robustness captures the susceptibility to failure due to changing of the method parameters, the process stability examines the changes to the results based on a longer series of analyses.

When determining the reproducibility, measurements were recorded under reproducible conditions. These allow changes to the parameters, for example, analysis and sample preparation can be performed by different operators on different similarly built instruments under variable environmental conditions.

After each variation, it is possible to check the influence on the results using a reference material.

In many cases, particularly when the analytical system is operated exclusively under controlled laboratory conditions, assessment of the robustness can be dispensed with.

## 8.1.10 Linearity

The "capability of a method to deliver results that are proportional to the concentration of the analyte within a given concentration range" is referred to as the linearity (corresponding translation of the definition contained in [7]). Therefore, there must be a clear mathematical relationship between the concentration of the analyte and the measured signal that can be described with an evaluation function f that is at least monotone. This means formalized: From $x > x'$ it follows that $f(x) > f(x')$, where $x$ and $x'$ are intensities and $f(x)$ and $f(x')$ are the associated concentrations calculated from the calibration function.

The term "linearity" wrongly suggests that only linear regression models are allowed. Other mathematical models are normally required for spark emission spectrometry, for example, a higher order regression function with additional corrections to compensate interelement effects to generate an effective, precise evaluation function.

As a rule, the calibration can be used to determine the linearity for spark emission spectrometry. Here, usually more than 20 samples with known concentrations are used. They are either certified or tied to SI units through reference methods.

Evaluation of the linearity can either be done visually based on the correlation coefficients or using a residuals analysis.

Only larger deviations can be detected with a visual evaluation. The correlation coefficient should be as close to one as possible.

Residuals analysis is a valuable method for deciding which model best describes the relationship between measured signal and concentration. Such functions are usually implemented in modern spark spectrometer software and have already been described in Section 3.9.5. For all models, the squares of the differences between the points determined (signal, concentration) and the calculated function are added. The least squares method is the preferred model for simple cases (given the presence of a linear function and the absence of additive and multiplicative corrections). For models with a larger number of degrees of freedom (higher polynomial degree, several interelement and line interferences), the function determined must be examined for plausibility. If the model is a second-degree polynomial with two additive and two multiplicative interfering elements and if only seven samples have been measured, a calibration function is determined that yields zero as the sum of the deviation squares. However, the calculated interferences and polynomial coefficients do not necessarily reflect reality.

### 8.1.11 Measurement uncertainty

Laboratories accredited according to DIN EN ISO/IEC 17025 [11] must have procedures for estimating the measurement uncertainty for measurements with spark spectrometers and they have to apply them. As explained in the previous sections, a spark spectrometer can never deliver the actual "true" value $x$. Rather, the analytical result is influenced by both systematic and random errors. If the measurement uncertainty $u$ is properly determined, the true value lies (with a given, high probability) within an interval $[x - u, x + u]$. If this interval is narrow, it proves the competence of the testing laboratory for the method in question and enables the client to judge the test results. Example: A client has ordered a material of the grade 1.4401 and received it together with a certificate indicating a molybdenum value of 2.12% with a measurement uncertainty of ±0.1%. According to standards, Mo concentrations between 2.0 and 2.5% are permitted for this material. The client can be quite sure that the Mo concentration lies within the permissible limits. If the measurement uncertainty is higher, for example, by 0.2%, this decision could not be made with high certainty. The factors that influence the measurement uncertainty have already been discussed in the previous sections. Every one of these factors must be considered in order to quantify and minimize its contribution to the measurement uncertainty.

There are two different approaches that can be used to determine the total measurement uncertainty.

#### The bottom-up method

This method, described in the GUM (Guide to the Expression of Uncertainty in Measurement) [12] requires complete understanding of and a mathematical formulation of the measurement uncertainty for all work steps of the testing procedure, in our case for each of the analytical methods for spark spectrometry. All relevant actions must be covered – from sampling and sample preparation to every aspect of the instrument technology. Subsequently, the partial errors, the so-called measurement uncertainty components, are combined into a total measurement uncertainty. The error propagation law is used to combine them.

The bottom-up method is unsuitable for calculation of the total measurement uncertainty for analyses with spark spectrometers. The interrelations, especially the interaction between material ablation, atomization, and ionization are very complex (see Sections 3.2.1 and 3.2.2) and have still not been fully understood even today. It is very hard to generate realistic mathematical models.

#### The top-down method

The top-down method uses a different approach. Certified reference samples are measured and the procedures described in Sections 8.1.2 to 8.1.10, especially the bias (8.1.6) and the recovery rate (8.1.7), are used to estimate the measurement uncertainty.

All error components, including the contributions from sampling and sample preparation, enter into the measurement uncertainty determined in this way. For this reason, the top-down method offers good results for reasonable effort.

The top-down method is described in detail in the Nordtest Report TR537 *Handbook for Calculation of Uncertainty in Environmental Laboratories* [13], further information dealing with this subject, some with examples, can be found in various publications [14–18].

## 8.2 Reference materials

Reference materials, abbreviated to RM, and certified reference materials (CRM) have already been mentioned in various parts of this book. The ISO Guide 30 refers to a reference material as a "material, sufficiently homogeneous and stable with respect to one or more specified properties, which has been established to be fit for its intended use in a measurement process" [19]. Other standards also go along with this definition [20–22].

Reference samples for purposes in spark spectrometry can be categorized as follows:
- Calibration samples are used to determine the calibration functions. Here, the best possible bias of the concentrations stated in the certificate is of central importance. Whenever possible, certified reference materials are used for the calibration. Certified reference materials are issued by government institutions or commercial providers. Examples for government institutions are: the *Bundesanstalt für Materialprüfung* (BAM) (Federal Institute for Materials Testing) in Germany or the *National Institute of Standards and Technology* (NIST) in the United States. Extensive round robin testing is required before certified reference materials can be issued, which is reflected in the high prices. To assess whether a certified reference material is usable for a specific purpose, it is advisable to survey the measurement uncertainty stated in the certificate. In addition, it is a measure of quality when a large number of renowned laboratories have participated in the round robin tests. Requirements for certificate contents are specified in ISO-Guide 31 [22].
- Even for certified reference materials, the true concentration values are not known with absolute certainty. However, it can be assumed that the concentrations stated for these materials approximate the true values. In this context, they are referred to as correct values. This makes reference materials an indispensable tool for investigating the accuracy of testing methods using spark spectrometers.
- Recalibration samples are used to return the current state of the instrument to the state at the time of the calibration (see Section 3.9.6). Here, the main focus

is placed on the homogeneity of the sample. Inhomogeneity can lead to variations in the recalibration measurement average values even if the state of the instrument has not changed. The accuracy of all the analyses subsequent to a faulty recalibration is then compromised. As a rule, recalibration samples are much less expensive than certified reference materials. If the homogeneity of recalibration samples has been demonstrated, they can be used to determine the precision of the testing method using spark spectrometers.

– Control samples serve to check the methods for spark spectrometer methods at regular intervals. Such measurements are carried out periodically, the measurement results flow into quality control charts. Quality control charts are described in the next section. Certified reference materials are less likely to be chosen for such control purposes, as these are expensive and they are quickly consumed through constant sample preparation and sparking. This results in the need to produce internal reference materials for instrument controls. Such samples are referred to as *in-house standards, secondary standards* or *secondary normals*. However, one hurdle is the fact that it is generally necessary to trace the analysis of secondary standards to SI units. Testing laboratories that operate in accordance with DIN EN ISO 17025 [23] are required to demonstrate traceability to SI units before using secondary standards.

To demonstrate this, a procedure that enables samples to be synthetically produced is used. One such method is inductively coupled plasma optical emission spectroscopy (ICP-OES). For this method, primary substances are dissolved and mixed to obtain the concentrations desired. Primary substances are very pure materials, the analysis of which amounts to the chemical formula.

Traceability to SI units proceeds as follows:
1. The material intended to be a secondary standard is brought into solution.
2. Calibration solutions, that correspond to the secondary standard in terms of the matrix, are created from primary substances. The analytes of interest must be present in graduated levels above and below the concentration of the secondary standard.
3. The ICP-OES instrument is calibrated using the solutions.
4. For the plausibility test, certified reference materials that have compositions as close as possible to those of the secondary standard are brought into solution and used to check the calibration of the ICP-OES instrument.
5. If, in the context of the measurement uncertainty, the analysis of the CRM does not result in the expected values, then a mistake has been made. Steps 2 to 3 must be repeated.
6. If the analytical results for the CRM lie within the scope of the expected measurement uncertainty, then the secondary standard can be measured on the ICP-OES using the method established in Step 3.

7.  The measured values as well as the procedure for their determination must be documented in a test report. The secondary standard can now be used in the context of control measurements for spark emission spectrometry.

Of course, proof must also be provided that the secondary standard is sufficiently homogeneous. This test can be performed on the spark spectrometer itself. In general, no statistically significant dependence of the analytical results on the location of the burn spot should be discernible. However, a small amount of inhomogeneity is usually tolerated at the center of the sample for round materials. Here, it is customary to position the burn spots on a circle that is about 20 mm smaller than the diameter of the sample, thus omitting the center of the sample.

## 8.3 Quality control charts

Quality control charts (QCC) were first used for production monitoring by the American, Walter A. Shewart, in the twenties of the last century. The examination of production fluctuations using statistical means had already been considered in the 19th century [10, 24]. One type of quality control chart, the average value chart, which is occasionally also called the $\bar{x}$ control chart ($\bar{x}$-CC), is particularly suitable for the monitoring and optimizing of measuring methods for spark spectrometers. For this purpose, in average value charts, for example, the averages from analytical control samples are plotted against time. Strictly speaking, an entire set of $\bar{x}$-CC is kept for each analytical control sample – every element being monitored has a chart. The analytical control samples are often reference materials (RM) or certified reference materials (CRM). However, in principle, secondary standards can also be used if they are homogenous enough.

To be able to work with $\bar{x}$-CC, so-called warning and intervention limits must be specified before the control phase (see Figure 8.2). This can be done by calculating the averages and standard deviations for element concentrations from 20 to 30 averages.

Warning and intervention limits are set as follows:
– The upper warning limits (UWL) are obtained by adding twice the standard deviation to the average determined in the preliminary phase. The lower warning limit (LWL) results from subtraction of twice the standard deviation from the average.
– The upper intervention limit (UCL) and the lower intervention limit (LCL) are calculated analogously. Here three times the standard deviation is added or subtracted.

It is problematic to use a target value, for example, a certified value from a reference material, instead of the average determined in the preliminary phase. As has already been explained in Section 8.1, many reference materials lie systematically above or

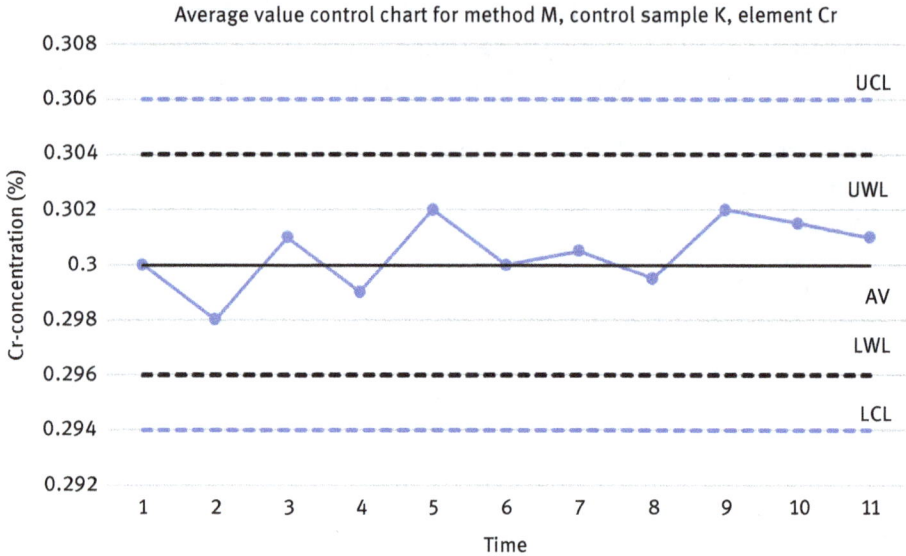

Figure 8.2: Example of an average value control chart.

below the calibration function. A properly determined calibration is, however, a compromise by which the curve is calculated so that the deviations over the collective of all the reference materials is as small as possible. Thus, it is better to use the average value determined in the preliminary phase.

In the utilization phase, it should be promptly checked after every new measurement of an analytical control sample whether there is a so-called out-of-control situation. Out-of-control situations are discussed, for example, in [25], page 36 et seq.

An out-of-control situation is usually assumed when:
-   The value for an element is above the upper intervention limit or below the lower intervention limit.
-   The warning limits are frequently exceeded, for example, when two from three measured values are above the UWL or below the LWL.
-   A specified number of subsequent measured values are above or below the average value (usually, a limit of seven or eight is chosen here).
-   When a larger number of measured values are observed and they are found to be predominantly (e.g., 10 of 11 or 18 of 20) above or below the average value.
-   The average values rise or fall n times in succession, that is, there is a trend. A value of seven or eight is usually chosen as a value for $n$.

Figure 8.2 shows an example of an average value control chart.

The utilization of software for statistical process control simplifies timely evaluation of average value control charts. Immediate intervention is required if out-of-control situations are detected. In the simplest case, this could be the recalibration of the spark emission spectrometer. Other possible actions are routine maintenance procedures, such as cleaning of the spark stand or the inlet to the optics. If the possible remedial measures do not lead to success, the spectrometer must be shut down and blocked for further use. It is advisable to define a catalog of possible actions in advance to ensure a quick and efficient response when necessary.

The average value control chart described above is, in connection with spark spectrometers, the most commonly used type of control chart. In addition, there are other forms, however, they find little application in spark analysis.

Detailed information on the structure and use of quality control charts can be found in the relevant specialist literature, for example, in Timischl [10] and Faes [25].

## 8.4 Validation

If the laboratory conducts a validation, it establishes that the testing method is suitable for fulfilling a defined testing task in accordance with the client's testing requirements. The client may be another department in the company or an external customer. Thus, it is not possible to conduct a validation if the requirements of the client are not known.

The validation proceeds according to the following scheme:
1. First, in communication with the client, it must be clarified what the testing method must do to solve the analytical problem. A validation plan is created.
2. Then, the testing method that appears to be the most appropriate is selected.
3. If the indicators are known (see Section 8.1) for the selected testing method, continue the process with point 5.
4. If the indicators for the testing method are unknown, they must first be determined. The data recorded for this must be traceably documented and archived.
5. The indicators are now checked to be sure that they are such that the requirements determined in point 1 can be met.
6. If this check is negative, return to step 2 and try to find an alternative method.
7. If the indications allow fulfillment of the requirements, a validation report must be drawn up to demonstrate compliance with the indicators necessary for the validation plan. The testing method is now validated.
8. Finally, approval must be granted by an authorized person.

The requirements for a validation are summarized in document 71 SD 4 019 from the *Deutschen Akkreditierungsstelle* (DAkkS) [26] (German Accreditation Body). This can be viewed and downloaded from the DAkkS Homepage.

A practical example explains the validation process. The numbers in parentheses indicate the associated validation steps in the description above:

A company contacts the laboratory because they suspect a mix-up in a stack of pipes and requests clarification as to whether the wrong workpieces are actually in the lot. In dialogue with the company, it becomes clear that the target grade is 1.7225 (42CrMo4) and the wrong material can only be the grade 1.7218 (25CrMo4), because pipes of the same dimensions are only produced with these materials. Sorting must be performed using the element carbon, as the tolerances for all other elements overlap. According to the standard, the maximum carbon concentration for the material 1.7218 is 0.29%, the minimum carbon concentration of the grade 1.7225 is 0.38%. The difference between the two grades is, thus, 0.09%. This results in the requirement to determine the carbon concentration with a measurement uncertainty of better than 0.04%. This must also be possible on the curved pipe surface, which can only be prepared with a hand grinder. These conditions are included in the validation plan (Point 1 of the validation process).

The testing methods that come into question are measurements with a handheld XRF instrument and a mobile spectrometer in arc or spark mode. One of these three methods must be chosen. In the situation assumed in this example, the handheld XRF spectrometer is eliminated, because the element carbon cannot be determined with it.

As a next step, the possibility of sorting with an arc spectrometer is investigated, as less effort is required for the sample preparation than for spark measurements (Point 2).

The indicators for measurements on low-alloy steel in arc on ground pipe surfaces are known in the laboratory. Their check shows that the bias achievable with arc is not sufficient. Even sorting with the sorting method described in Section 6.7.5, for which the repeatability limit is the relevant indicator, does not appear to be feasible (Point 5).

Therefore, they consider spark. The indicators for low-alloy steel are known only for flat samples and not for curved surfaces like those for pipes. That is why the influences of this curved surface on the indicator must be determined. Measurements are carried out on sample pipes for which the analysis is known. The measured data are stored. It turns out that the bias is not affected compared to the flat, optimally prepared samples. The precision is only marginally worse (Point 4). To compensate for the slightly worse precision, a double measurement is performed on every pipe. Then, it is possible to determine the C concentration with a measurement uncertainty of 0.02%.

A short validation report in the form of a table, which also specifies where the validation data can be found, is created. A testing procedure is also prepared, specifying the modalities of the test including sample preparation and the required qualification of the tester. Herewith is the validation complete (Point 6).

The laboratory supervisor approves the validation with date and signature. Testing can begin (Point 7).

This example describes a fairly simple case. As a rule, validation involves a more complex analytical task, such as the analysis of low-alloy steel, high-alloy steel or cast iron for the monitoring of melt processes. In these cases, the validation is much more complicated, but, in principle, follows the same scheme.

The DAkkS document on validation and verification of test methods [26] mentioned above explains that there are several levels of validation. The validation requirements of standardized procedures are much lower than for a basic validation of a non-standardized testing method.

The DAkkS document 71 SD 1 005, which defines the minimum requirements for the accreditation of spark optical emission spectrometry [27], specifies that testing methods for spark emission spectrometers must always be treated as in-house methods and require primary validation before being used. As a rule, traces, medium and high concentrations are determined in parallel with spark emission spectrometer. For a primary validation, this means that the indicators discussed in Section 8.1 for working range, detection and determination limit, precision and accuracy must be determined. In addition, the selectivity should be investigated. It is also advisable to check the robustness of the method.

A so-called routine validation (tertiary validation) should be conducted during normal operations. On one hand, this is done by measuring control samples and recording the results in the quality control charts known from Section 8.3. On the other hand, the laboratory should regularly participate in round robin testing and proficiency tests. The type and frequency of routine validation measures should be defined in a plan.

## 8.5 Quality management systems

Currently (2017), in Europe, DIN EN ISO 17025:2005 [11] is the central standard regulating all quality assurance activities in analytical laboratories. In the introductory article to DIN EN ISO/IEC 17025:2005-08 [28] the standard is demarcated from other quality standards, for example, DIN ISO 9001. When drawing up DIN EN ISO 17025 it was ensured that it would be possible to work simultaneously with a quality management system in accordance with DIN EN ISO 17025:2005 [11] and to the then valid version of DIN ISO 9001:2000 [29]. Both standards are based on similar principles, but DIN EN ISO/IEC 17025:2005-08 provides a framework that helps in properly conducting measuring and testing applications.

However, in the meantime, DIN EN 9001:2015 [30], in which additional requirements are defined, applies. Compatibility to this standard is important because it defines rules for company-wide quality management meaning that, in addition to the laboratory, departments such as production, sales, service and research and development are included. Many customers expect certification according to DIN EN ISO 9001. DIN EN ISO/IEC 17025 is, in turn, the basis for accreditation. Therefore, conformity to this standard is often essential. A revision, which is currently (2017) in the drafting stage, is also planned for DIN EN ISO/IEC 17025 [31].

The IATF 16949 [32] is another generic standard that is often used in parallel to DIN EN ISO/IEC 17025. It is similar to DIN EN 9001:2015, but has an internationally oriented approach, which is advantageous for companies because they do not then have to work in conformity with several different national quality standards at the same time. This standard can be understood to be an extension of DIN EN 9001:2015. An explanation of these extensions can be found in P. Strompen [33].

The regulations in DIN EN ISO/IEC 17025 are of both an organizational and a technical nature.

Key organizational requirements:
– The quality policy must be defined and laid down in writing.
– The organizational structure and all relevant processes used in the laboratory must also be recorded in writing. The processes to be described include not only testing, calibration and sample preparation tasks, but also the handling of training and further education and managing of customer orders.
– All employees of the laboratory, insofar as they perform analytical tasks, must be familiar with quality documentation and must work in accordance with the processes defined there.
– There must be a document control system that ensures that valid versions of the documents are used. Here the term "document" covers a wide range and includes, for example, work instructions, meeting protocols, descriptions of methods and procedures, technical drawings, software, databases of analyses, etc.
– The way in which customer complaints are dealt with must also be regulated.
– The management system should be regularly evaluated according to a specified plan, whereby the results of internal audits, customer feedback and complaints, results of round robin and proficiency tests should be included.

Key technical requirements:
– Persons performing measuring, testing and calibration tasks must have the necessary qualifications for this work.
– Adequate rooms must be available.
– Measuring and testing methods must be validated and their measurement uncertainty estimated before being used.
– The laboratory must have all necessary equipment required for proper performance of the testing. This includes, in addition to measuring instruments, equipment for sampling and sample preparation. The equipment must be regularly inspected and maintained.
– Calibrations and measurements must be traceable to SI units. If this is not possible, a calibration can be conducted with certified reference materials.

The quality management manual specifies how the above-listed requirements are to be implemented in the laboratory and how to act in specific cases. This is a company internal set of rules for the laboratory. Figure 8.3 shows the structure of such a quality management manual. The establishment of a quality management handbook is complicated, details can be found in the relevant literature [29–31, 34, 35].

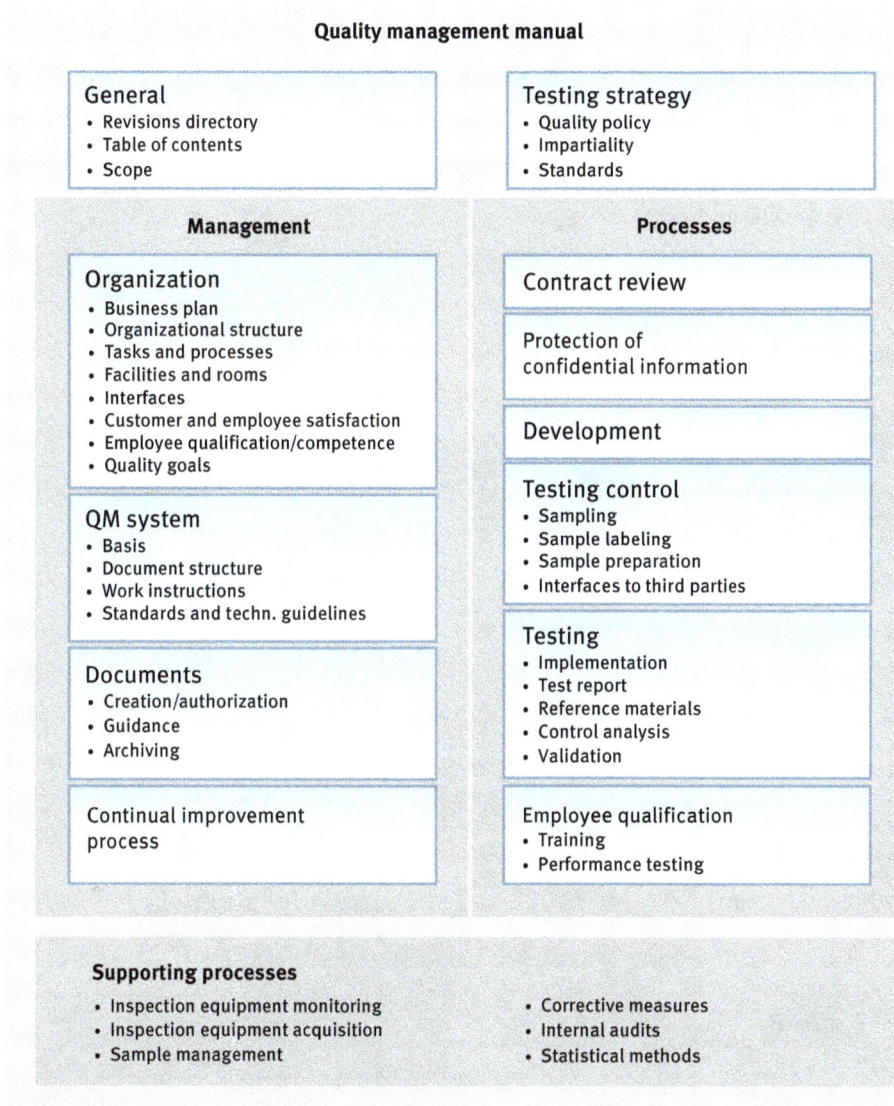

**Quality management manual**

**General**
- Revisions directory
- Table of contents
- Scope

**Testing strategy**
- Quality policy
- Impartiality
- Standards

**Management**

**Processes**

**Organization**
- Business plan
- Organizational structure
- Tasks and processes
- Facilities and rooms
- Interfaces
- Customer and employee satisfaction
- Employee qualification/competence
- Quality goals

**Contract review**

**Protection of confidential information**

**Development**

**QM system**
- Basis
- Document structure
- Work instructions
- Standards and techn. guidelines

**Testing control**
- Sampling
- Sample labeling
- Sample preparation
- Interfaces to third parties

**Testing**
- Implementation
- Test report
- Reference materials
- Control analysis
- Validation

**Documents**
- Creation/authorization
- Guidance
- Archiving

**Continual improvement process**

**Employee qualification**
- Training
- Performance testing

**Supporting processes**
- Inspection equipment monitoring
- Inspection equipment acquisition
- Sample management

- Corrective measures
- Internal audits
- Statistical methods

**Figure 8.3:** Structure of a quality management manual.

Testing instructions must be established for all testing methods used in the laboratory. Examples for such testing methods for arc and spark spectrometers could be:

- The analysis of low-alloy steels with a stationary laboratory spark spectrometer
- Material control with mobile spark spectrometers
- Sorting with a handheld arc instrument

Work instructions must include numerous formal (revision number, testing purpose, scope, etc.) and technical aspects (instruments and tools/aids, sample preparation, analyses, etc.). Figure 8.4 shows a serviceable structure.

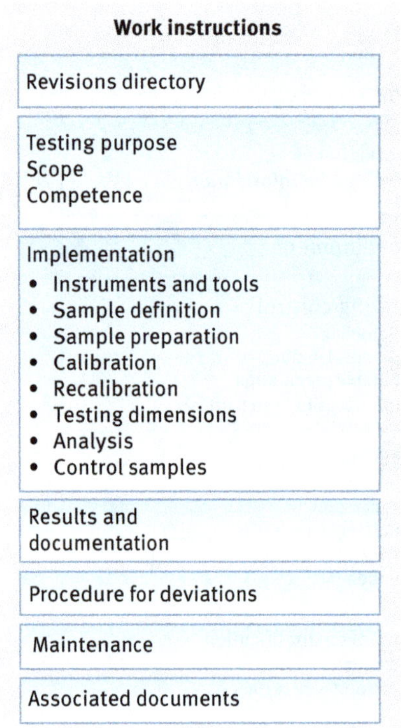

**Work instructions**

Revisions directory

Testing purpose
Scope
Competence

Implementation
- Instruments and tools
- Sample definition
- Sample preparation
- Calibration
- Recalibration
- Testing dimensions
- Analysis
- Control samples

Results and
documentation

Procedure for deviations

Maintenance

Associated documents

Figure 8.4: A set of working instructions.

## 8.6 Accreditation

The word accreditation is derived from the Latin "*accredere*," which can be translated as "believe" or "(give) credit" [1]. It is always individual methods that are accredited, blanket accreditation of an entire laboratory is not possible. Thus, accreditation of a spectral laboratory in regard to spark spectrometry is more or less a certification that the laboratory has the expertise to conduct the method of spark spectral analysis. The accreditation refers to precisely defined material groups. "Low alloy steels" or "cast irons" are examples of such material groups.

Measuring and testing conducted by laboratories often serve to prove that products comply with legal requirements. This applies, for example, to medicinal products or for the monitoring of limiting values in the environmental sector. Thus, based on EU directive no. 765/2008 [36], all EU member states have set up national institutions responsible for accreditation. In Germany, the federal government established the *Deutsche Akkreditierungsstelle* (DAkkS, German Accreditation Body), in other countries of the EU there are similar accreditation bodies. The DAkkS performs tasks of public authority although it is a private sector organization. Further information can be found on the organization's homepage [37].

DAkkS also carries out accreditations in the field of spark optical emission spectrometry. Spark emission spectrometers are not usually used to verify compliance with legal requirements. Therefore, accreditation is generally not absolutely necessary. But, the acceptance of its test results by its customers and suppliers increases if the laboratory is accredited.

The standard DIN EN ISO 17025 [11], which has already been discussed in Section 8.5, is the basis for accreditation of spark optical emission spectrometry. The document 71 SD 1 005 is a guide from the DAkkS [27] that specifies the minimum requirements to be met in order for a laboratory using a spark spectrometric method to operate in compliance with DIN EN ISO 17025 thus enabling accreditation.

This compact guide addresses the following points:
- Scope of accreditation
- Process description
- Contents of the work instructions
- Qualification requirements for the personnel
- Requirements for the rooms
- Requirements for the certified reference materials
- Requirements for the base calibration
- Participation in proficiency tests
- Requirements for the form of testing reports

The section "Process Description" stipulates that spark spectrometric testing methods are always to be treated as in-house procedures. The consequence of this classification on the validation has already been discussed in Section 8.4.

How the metrological traceability must be demonstrated is specified in the *Merkblatt zur messtechnischen Rückführung im Rahmen von Akkreditierungsverfahren* (Technical Note for the Metrological Traceability in the Accreditation Process) [38]. This document is available on the DAkkS Homepage. It also deals with the requirements for calibration and traceability of measuring and testing equipment in the context of DIN EN ISO 9000 [39] or DIN EN ISO 9001 [30]. This is of interest because spark emission spectrometers are frequently used as measuring equipment in DIN EN ISO 9000 or IATF 16949 compliant quality management systems.

The DAkkS document *Einbeziehung von Eignungsprüfungen in die Akkreditierung* (Use of Proficiency Testing in Accreditation) [40] describes what is meant by proficiency testing, which is absolutely necessary for accreditation, and which rules must be followed. This document is also available from the DAkkS Homepage.

## Bibliography

[1]   Hau R, Ender A. Pons Schülerwörterbuch, 2. Auflage, 2003, 2nd reprint. Stuttgart, Ernst Klett Sprachen, 2005.

[2]   Funk W, Dammann V, Donnevert G. Qualitätssicherung in der Analytischen Chemie. Weinheim, Wiley-VCH, 2005.

[3]   Doerffel K. Statistik in der analytischen Chemie. Leipzig, VEB Deutscher Verlag für Grundstoffindustrie, 1966.

[4]   Günzler H. Akkreditierung und Qualitätssicherung in der Analytischen Chemie. Berlin, Springer-Verlag, 1994.

[5]   Deutsches Einheitsverfahren DEV AO-4 Leitfaden zur Abschätzung der Messunsicherheit aus Validierungsdaten. Berlin, Beuth-Verlag, 2006.

[6]   Handbuch für das Eisenhüttenlaboratorium, Vol. Band 6, Begriffe und Definitionen, Regelwerke, Referenzmaterialien, Wissenschaftliche Tabellen. Düsseldorf, Verlag Stahleisen, 2013.

[7]   Kromidas S. Handbuch Validierung in der Analytik, Wirtschaftlichkeit, Praktische Fallbeispiele, Alternativen. Weinheim, Wiley-VCH, 2011.

[8]   DIN 32645:2008-11. Chemische Analytik – Nachweis-, Erfassungs- und Bestimmungsgrenze unter Wiederholbedingungen – Begriffe, Verfahren, Auswertung. Berlin, Beuth-Verlag, 2008.

[9]   Lohninger H. Ausreißertest nach Grubbs. www.statistics4u.info/fundstat_germ/ee_grubbs_outliertest.html, reviewed September 9th, 2017.

[10]  Timischl W. Qualitätssicherung – Statistische Methoden. München, Carl Hanser Verlag, 2002.

[11]  ISO/IEC 17025:2005-05. General requirements for the competence of testing and calibration laboratories. Berlin, Beuth-Verlag, 2005.

[12]  GUM Evaluation of measurement data – Guide to the expression of uncertainty in measurement, JCGM 100:2008. www.bipm.org, reviewed September 10th, 2017.

[13]  Nordtest Report TR 537 Handbook for calculation of uncertainty in environmental laboratories, 2004-02.

[14]  DIN ISO 11352:2011-03EN DE. Wasserbeschaffenheit – Bestimmung der Messunsicherheit basierend auf Validierungsdaten. Berlin, Beuth-Verlag, 2005.

[15]  Deutsches Einheitsverfahren DEV AO-4 Leitfaden zur Abschätzung der Messunsicherheit aus Validierungsdaten (2006). Berlin, Beuth-Verlag, 2006.

[16]  DIN V ENV 13005:1999-06, Guide to the expression of uncertainty in measurement; German version ENV 13005:1999. Berlin, Beuth-Verlag, 1999.

[17]  EURACHEM CITAC GUIDE Ermittlung der Messunsicherheit bei analytischen Messungen, 2004.

[18]  EUROLAB Leitfaden zur Ermittlung von Messunsicherheiten bei quantitativen Prüfergebnissen (TB_2_2006). http://www.austrolab.at/fileadmin/user_upload/Leitfaeden/TB_2_2006.pdf, reviewed September 10th, 2017.

[19]  International Organization for Standardization. ISO Guide 30:2015 – Referenzmaterialien – Reference materials – Selected terms and definitions. Berlin, Beuth Verlag, 2015.

[20]   International Organization for Standardization. ISO Guide 33:2015 Reference materials –
       Good practice in using reference materials. Geneva, Switzerland, 2015.
[21]   International Organization for Standardization. ISO Guide 35:2006 Reference materials –
       General and statistical principles for certification. Geneva, Switzerland, 2006.
[22]   International Organization for Standardization. ISO Guide 31:2015-11,Reference
       materials – Contents of certificates, labels and accompanying documentation. Berlin,
       Beuth Verlag, 2015.
[23]   DIN EN ISO/IEC 17025:2005. Allgemeine Anforderungen an die Kompetenz von Prüf- und
       Kalibrierlaboratorien. Berlin, Beuth-Verlag, 2005.
[24]   Best M, Neuhauser D. Walter A Shewhart, 1924, and the Hawthorne factory. BMJ Quality &
       Safety. 2006, 15, 142–143.
[25]   Faes G. SPC Statistische Qualitätskontrolle. Norderstedt, BoD Books on Demand GmbH,
       2009.
[26]   Validierung und Verifizierung von Prüfverfahren nach den Anforderungen der DIN ISO/IEC
       17025 für Prüflaboratorien auf dem Gebiet der chemischen und chemisch-physikalischen
       Analytik im Bereich der Abteilung 4 (Gesundheitlicher Verbraucherschutz / Agrarsektor /
       Chemie / Umwelt), Document 71 SD 4 019, Revision 1.1, January 14th, 2015. www.dakks.de/
       sites/default/files/71_sd_4_019_validierung_20150114_v1.1_0.pdf,
       reviewed September 10,2017.
[27]   Mindestanforderungen zur Akkreditierung der optischen Funkenemissionsspektrometrie
       (OES), Dokument 71 SD 1 005, Revision 1.1, February 24th 2014. Berlin, Deutsche
       Akkreditierungsstelle GmbH. www.dakks.de/sites/default/files/dokumente/71_sd_1_005_
       funkenemissionsspektrometrie_20140224_v1.1.pdf, reviewed September 10th, 2017.
[28]   Inductory text to the DIN EN ISO/IEC 17025:2005-08. Berlin, Beuth Verlag, www.beuth.de/de/
       norm/din-en-iso-iec-17025/77196483, reviewed September 10th, 2017.
[29]   DIN EN ISO 9001:2000-12 Qualitätsmanagementsysteme – Anforderungen (ISO 9001: 2000-09).
       Berlin, Beuth Verlag, 2000.
[30]   DIN EN ISO 9001:2015-11 Qualitätsmanagementsysteme – Anforderungen (ISO 9001:2015).
       Berlin, Beuth Verlag, 2015.
[31]   DIN EN ISO/IEC 17025: 2017-02– Entwurf Allgemeine Anforderungen an die Kompetenz von
       Prüf- und Kalibrierlaboratorien (ISO/IEC DIS 17025:2016). Berlin, Beuth Verlag, 2017.
[32]   International Automotive Taskforce. IATF 16949:2016. Anforderungen an
       Qualitätsmanagementsysteme für die Serien- und Ersatzteilproduktion in der
       Automobilindustrie, 1st edition, October 2016. Available via the webshop of the German
       Association of the Automotive Industry (Verband der Automobilindustrie, VDA), www.web
       shop.vda.de/QMC/de/IATF-169492016.
[33]   Strompen P. Interpretation der Anforderungen der IATF 16949:2016. Köln, TÜV media Group,
       2017.
[34]   DIN EN 45001:1997-06:General requirements for the competence of testing and calibration
       laboratories; Trilingual version prEN 45001:1997 – document withdrawn. Berlin, Beuth-Verlag,
       1997.
[35]   International Organization for Standardization. ISO Guide 25:1990, General requirements for
       the competence of calibration and testing laboratories. Geneva, 1990.
[36]   Verordnung (EG) Nr. 765/2008 des Europäischen Parlaments und des Rates vom 09.07.2008
       über die Vorschriften für die Akkreditierung und Marktüberwachung im Zusammenhang mit
       der Vermarktung von Produkten und zur Aufhebung der Verordnung (EWG) Nr. 339/93 des
       Rates (Regulation (EC) No 765/2008 of the European Parliament and of the Council
       of July 9th, 2008 on the rules for accreditation and market surveillance relating to the
       marketing of products and repealing Council Regulation (EEC) No 339/93).

[37] Auf einen Blick: Allgemeine Informationen zur DAkkS. Berlin, Deutsche Akkreditierungsstelle GmbH. From the internet, www.dakks.de/content/auf-einen-blick-allgemeine-informationen-zur-dakks, reviewed September 20th, 2017.

[38] Merkblatt zur messtechnischen Rückführung im Rahmen von Akkreditierungsverfahren, Dokument 71 SD 0 005, Revision 0. Berlin, Deutsche Akkreditierungsstelle GmbH. www.dakks.de/userfiles/71%20SD%200%20005_merkblatt_r%C3%BCckf%C3%BChrung.pdf, reviewed September 20th, 2017.

[39] DIN EN ISO 9000:2015-11:Quality management systems – Fundamentals and vocabulary (ISO 9000:2015); German and English version EN ISO 9000:2015. Berlin, Beuth Verlag, 2015.

[40] Einbeziehung von Eignungsprüfungen in die Akkreditierung, Dokument 71 SD 0 010, Revision 1.2, April 14th 2016. Berlin, Deutsche Akkreditierungsstelle GmbH. www.dakks.de/sites/default/files/dokumente/71_sd_0_010_eignungspruefungen_20160414_v1.2.pdf, reviewed September 20th, 2017.

[41] Griepink B, Quevauviller P. Referenzmaterialien und chemische Standards: Bedeutung für QS; Herstellung, Qualitätsanforderung, Lieferanten in: Die Akkreditierung chemischer Laboratorien, editor H. Günzler. Berlin, Springer Verlag, 1994.

# Index

https://doi.org/10.1515/9783110529692-009

www.ingramcontent.com/pod-product-compliance
Lightning Source LLC
Chambersburg PA
CBHW080910220326
41598CB00034B/5538